ORBITING THE MOONS OF PLUTO

Complex Solutions to the Einstein, Maxwell, Schrödinger and Dirac Equations

K&E Series on Knots and Everything — Vol. 45

ORBITING THE MOONS OF PLUTO

Complex Solutions to the Einstein, Maxwell, Schrödinger and Dirac Equations

Elizabeth A Rauscher
Tecnic Research Laboratory, USA

Richard L Amoroso
Noetic Advanced Studies Institute, USA

World Scientific

NEW JERSEY • LONDON • SINGAPORE • BEIJING • SHANGHAI • HONG KONG • TAIPEI • CHENNAI

Published by

World Scientific Publishing Co. Pte. Ltd.
5 Toh Tuck Link, Singapore 596224
USA office: 27 Warren Street, Suite 401-402, Hackensack, NJ 07601
UK office: 57 Shelton Street, Covent Garden, London WC2H 9HE

Library of Congress Cataloging-in-Publication Data
Rauscher, Elizabeth A.
Orbiting the moons of Pluto : complex solutions to the Einstein, Maxwell, Schrödinger, and Dirac equations / by Elizabeth A. Rauscher & Richard L. Amoroso.
p. cm. -- (Series on knots and everything ; v. 45)
Includes bibliographical references and index.
ISBN-13: 978-981-4324-24-3 (hardcover : alk. paper)
ISBN-10: 981-4324-24-8 (hardcover : alk. paper)
1. Generalized spaces. 2. Gravitational fields. 3. Electromagnetic theory. 4. Quantum field theory. I. Amoroso, Richard L. II. Title.
QC173.55.R34 2011
530.1--dc22

2010050427

British Library Cataloguing-in-Publication Data
A catalogue record for this book is available from the British Library.

Printed in Singapore by World Scientific Printers.

To my son Brent A. Rauscher -
for his many contributions and love in my life…

The Gift of Life

In the face of life, there is a glimmer of truth
and one must hunt along a darkened, shadowed path
feeling one's way there;

One can go "sloop shouldered",
tapping one's cane on the rocky path

Or one can run in Zen joy
and skip flying over rocks – feeling in tune
with all reality as one goes,

So that even in the dark
The path illumination is found

Life is worth nothing without joy
And joy comes from wading deep in
The rivers that cross one's path
And thrill to the forces of nature

Although many elements of life
Are unclear to us at many times,

A beam of light breaks through
and there is a path of clear vision
to let us know that this reality exists

Love to all things of the Universe !

E.A. Rauscher

Preface

Complex spacetime got its start with the discoveries of Cartan, Pauli and Dirac relating to the spinor and its utility in many areas of physics. String Theory, which has its origins in the early Kaluza-Klein Theory of the 5th dimension, has been in trouble in many sectors in recent decades because there is as yet no experimental confirmation and the current state of the theory allows for a nearly infinite number of possible vacuum solutions. This volume stretches the exploration of complex spacetime through all of the fundamental equations of Physics considered of paramount importance – The Einstein, Maxwell, Schrödinger and Dirac equations. We hope we have done our part to help tighten the noose on "the" search for fundamental theory. This kind of exploration of the nature of complex spacetime has for the most part been neglected by M-Theorists and physicists in general and is taken to its penultimate form here. This volume is not a survey or review of existing or well-known theoretical models but presents unique ground breaking new work on the structure and implications of complex higher dimensional spacetime by the authors building on over four decades of original work.

This volume provides unifying insights for graduate students and seasoned professionals alike in the broad spectrum of fields touched by complex spacetime and should thus be a boon to extending the horizons of their studies and professional work. Currently a lot of effort is being performed in General and Special Relativity, Quantum Theory, Electromagnetism Astrophysics and Cosmology but this volume introduces a unique perspective providing putative solutions to numerous unanswered questions in these and related fields of theoretical physics taking great strides toward finding the ultimate fundamental theory.

The search for truth and knowledge and its dissemination for the betterment of all existences is the basis of my endeavors. As Socrates so importantly stated "the unexamined life is not worth living" and I would add, the Universe should also not be left unexamined. According to John A. Wheeler formally of Princeton University who facetiously said the unexamined or unobserved universe may not exist! In this volume we present a theoretical foundation that not only allows a domain for the existence of but also allows a role for the action of consciousness in the physical world in both the micro and macro cosmos. Such paradoxes as the wave particle paradox and Schrödinger cat paradox and other so termed paradox are examined in detail and resolutions determined. Nature does not admit of paradox, and hence paradoxes are resolved by a more comprehensive working of nature.

The foundation of the knowledge and comprehension of the fundamental workings of natural world appears to proceed from physics with its methodology of systematic exploratory and predictive power expressed in the logical and reproducible 'language' of mathematics.

Two great theoretical edifices of physics were developed in the last century, the quantum theory and the theory of relativity and a geometrization of gravity. From the quantum theory arose two fundamental findings, the Heisenberg uncertainty principle and quantum entanglement and the quantum nonlocality. Much scientific debate and philosophical discussion has transpired over the uncertainty principle but the full richness and vast implications of the principle of nonlocality has not been fully realized. Earlier in the late 1800s Maxwell unified electric and magnetic phenomena.

Einstein's dream and those who followed him was to determine a scheme that will unify the four force fields (FFF) of the strong and weak short range nuclear forces, and the long range electromagnetic and gravitational forces. Current attempts at a Grand Unification Theory (GUT), which is an attempt to unify the strong, weak and electromagnetic (electroweak forces) with the theory of Everything (TOE) which also attempts to include gravity. The current theories include additional "rolled up" small Planck size dimensions in string theory and superstring theory. Any unified theory must take nonlocality into account.

We have developed a complexification of Minkowski space consistent with major principles of physics, Lorentz invariance, analyticity and unitarity. In order to accommodate quantum nonlocality and nonlocality in general, we complexify M_4 space as a complex C_4 8D space having certain fundamental symmetry principles. Minkowski space is one of the backbones of special and general relativity. We formulate and solve Maxwell's equations, the Dirac and Schrödinger equations using our hyperdimensional 8D and 12D space demonstrating a unification approach that incorporates nonlocality at a deep and fundamental level which is well described in the context of extended extra, $n > 4$ dimensions as well as the group theoretical mapping to the Kaluza-Klein 5D theory, the spinor calculus and twister algebra which relates to the M theories and superstring multi vacuum theories. Uniqueness of the complex Minkowski space is demonstrated in our new theoretical framework in "the moons of Pluto: Complex Solution to Maxwell, Schrödinger, Einstein and Dirac Equations". This work comprises over four decades of original research and presents a new unique ground breaking theoretical development in the foundations of physics. Some background of current standard theories are presented to demonstrate the framework of the meaningful interpretations of our work in the context of modern physics as well as the manner in which it relates and advances the ongoing process to better kin our universe. Complexifying Maxwell's, Schrödinger's and the Dirac equation, geometrizes these equations. Therefore from the complexified Minkowski space these formulations yield a more fundamental relation between the quantum theory, QED and QCD and gravity.

In Chapter 15 we let our hair down a bit and enjoy some speculation on FTL warp drive travel. To some it will appear too far afield; but we remind the community that superluminal properties of Lorentz transformations have been around for a long time. This has only been strengthened when we complexified equations here. Also it should be noted that Cramer's transactional interpretation of quantum theory, while unfortunately ignored for the most part, although not overtly stated by Cramer requires superluminal connectedness in a transaction.

I acknowledge the valuable communication with some of the many colleagues with whom I have discussed some of the fundamental issues of the foundations of physics, philosophical interpretations and the role for action of the observer and the nature of consciousness in the physical world. A special appreciation goes to my Ph.D. faculty advisor Glenn T. Seaborg at the University of California, Berkeley for his thoughtfulness, and broad comprehension of nuclear and particle physics. He was UC provost, head of the Atomic Energy Commission, now called DOE, and held more honorary degrees on the planet than anyone else, and he was head of the Lawrence Berkeley National Laboratory (LBNL) Nuclear Science Division, yet he was an attentive and careful listener to all who spoke with him. I also acknowledge attentive and rigorous research lessons I learned from my undergraduate and graduate master's degree advisor and collaborations John O. Rasmussen formally of the Nuclear Science Department Theory Group of LBNL; the kind and thoughtful advice and discussions with John Archibald Wheeler, Princeton and the Universal of Texas at Austin. His encouragement was of great value as was that of Eugene P. Wigner, formally Princeton University.

David Bohm and John Hasted formally of the University of London Physics Department, Burbeck College not only delved into the depths of physics, the issue of nonlocal interactions in a rigorous and heartfelt manner because can we truly separate our passion for knowledge and the new interactions we develop. Also of Berkeley, LBNL is the careful thinking and many discussions with Geoffrey Chew, Chairman of the Department of Theoretical Physics and a mentor of mine when I was on the LBNL staff for many years. I enjoyed the many discussions of the history of physics and unique stories of discoveries with Edwin McMillan, former head of LBNL and the discussions about the future of science with Andrew Sessler, also former head of LBNL when he was our director. Also, the meeting and discussion of the experimental verifications of Bell's inequality and the fundamental nature of quantum nonlocality is greatly appreciated with John

Clauser of UC Berkeley and Alan Aspect of Orsay, France and also examining their excellent experimental set up. I especially acknowledge the wonderful productive collaboration, the search for truth and knowledge, shared warmth, friendship and love of my late husband of many years, William Van Bise. Also the productive collaboration with William Tiller, formerly of Stanford University is acknowledged as well as James B. Beal formerly of NASA.

In the exciting era of consciousness research and its context from the work of physics has generated many colleagues and co-researchers including Russell Targ and Saul Paul Sirag. Targ and I and others conducted extensive research related to attributes, properties and nonlocality of conscious thought and intentionality at SRI International and UC Berkeley. There was much discussion for development of theoretical concepts and about experimental data, in particular the basic nature of the science of nonlocality. I founded the Fundamental "Fysiks" theory group at LBNL with forty physicist participants to study the foundations of physics, nonlocality, remote interactions and consciousness and the role of the observer. There are many colleagues who are too numerous to mention in this preface, but I wish to thank them all for their struggles and triumphs over the deep understanding of the most fundamental problems in current physics. As the work here and the correspondence it makes to M-Theory, it becomes more and more reasonable to assume that the role of complex spacetime and solutions to the prominent equations of physics will play an increasingly important role.

Elizabeth A. Rauscher
Tecnic Research Laboratories
Apache Junction, AZ USA
December 2010

Contents

Chapter 1

Introduction – Orbiting the Moons of Pluto

He never stopped exploring new "strange" ideas, even when they were likely to destabilize an interpretation he had himself put in orbit with his crucial discoveries...[1].

We begin our 'Orbit of the Moons of Pluto' with a brief introduction to the intended processes for complexifying Maxwell's equations and the nature of multidimensional Minkowski space, $\hat{M}_4 \subset \mathbb{C}_4$ where $\hat{M}_4$ is the energy dependent subspace of the complex 8-space, $\mathbb{C}_4$. The complex solutions to Maxwell's equations are given in Hertzian as well as non-Hertzian form which in ensuing chapters we will show lead to the relativistic and nonrelativistic forms.

1.1 Introduction

Although there has been a great deal of technological development since the inception of Maxwell's unification of electrostatic and magnetic phenomena, much of the field of fundamental theoretical electro-magnetism has remained unexplored. Here we will present some new approaches to the examination of non-linear electromagnetic phenomena as well as reexamining certain areas of exploration, such as non-Hertzian waves, that have been explored in the past but have subsequently been abandoned [2,3]. Some recent areas of theoretical examination appear to shed new light on possible electromagnetic excitations, such as longitudinal "acoustic-like" waves or 'acoustiton', a charged acoustic wave with soliton-like properties [4], and non-linear excitations that are coherent and collective in nature and may be identifiable as having magnetic monopole-like properties. These acousticons may relate to natural phenomena like gaseous plasmas, ionosphere resonance and non-Hertzian and Hertzian modes in Biology (Chap. 5). Specific formulations of Maxwell's equations are developed for their structure and solution in multi-dimensional geometries with a dimensional space of $n > 4$, where we have the usual space-time geometry of $n = 4$ dimensions [4-11].

Specifically, we will examine formalisms of electromagnetic phenomena in a complex Minkowski space. This geometry is presented in detail in [4-15]. The specific Minkowski space that we examine in most detail, presently, is a complex 8-space of the four real dimensions of space-time expanded over the domain to include four imaginary space-time components. 4-space is seen as a slice through the 8D Minkowski space, rather than as a projective geometric subspace. Each 4-space dimension is transformed to a complex dimension of real plus imaginary components of space and time [5-15]. Certain properties of the expanded domain of 8-space have intriguing properties, such as remote connectedness [3,5,8] and apparent superluminal signals in the lower dimension, projected as sliced space [8,14]. The higher dimensional cover space may have luminal and subluminal propagation velocities only [12,13]. See Chap. 9.

The use of the complex formalism is not new to physics. Design specifications for LRC circuits by the use of Kirkhoff's laws is often performed by utilizing complex equations in which, of course, the final results involve real currents, resistances, etc. Also, the conditions on analytic continuation in the complex space are utilized in elementary particle physics to formulate generalized causality conditions on stable and unstable (virtual) particle properties. In fact we use the analytic properties and Lorentz invariance conditions (as related to the set of unimodular transformations of Maxwell's equations) are central to describing the predicted phenomena by the current formalism in this volume [3,13-15].

Some of the motivation in examining the properties of Maxwell's equations in the expanded space of the complex Minkowski space involves the earlier success of the use of this and other similar geometries in which we have:

- Formulation of Maxwell's equations in complex 8-space,
- Formulated the Schrödinger equation [10,13],
- Solved the Dirac equation,
- Developed a relationship of electromagnetic and gravitational phenomena [11],
- Examination of supercoherence phenomena in the field theoretic second quantized formalism in plasma physics (MHD), electromagnetic [2,3] and superconductivity (BCS) theory [16,17].

Central to these approaches are:

- Collective, coherent phenomena,
- Non-linearity in various media,

- Non-equilibrium phenomena,
- Dynamic process oriented phenomena.

See [20] for a list of implications and applications for these physical properties in multidimensional geometries.

For example, dynamic processes such as the dispersion and recoher-ence which occurs due to nonlinearities in the plasma media lead to the so-called instabilities in plasmas which are explained in terms of the coupling of collective 'acoustiton' [2,3] or plasmon modes to electron excitation modes which form a coherent dynamically stable state [12]. A similar approach in which the nonlinearities are formulated in terms of the imaginary components of space and time in the complex Minkowski space lead to a term in Maxwell's equations which appears to be a kink soliton monopole. Implications of such structures are also discussed here and involve the remote connectedness and superluminal aspects of the multidimensional space.

We will present several approaches to the formulation and solution to Maxwell's equations in the complex Minkowski geometry. The possible implications of these approaches will also be presented. This includes a number of approaches, formulations and speculations. These approaches are amenable to experimental testing and specific equipment design. Also explanations of certain physical phenomena appear to unify certain concepts in physics and these unifying approaches can lead to possible technologies.

We first examine the complexification of the electromagnetic fields $\underline{E}$ and $\underline{B}$, or $F^{\mu\nu}$, and the implications for complex metric space. Several interpretations of the magnetic monopole are made. Second, we examine the complexification of $F^{\mu\nu}$ and A^{μ} directly in terms of the complex derivations in Maxwell's classical and relativistic equations. We examine a 'kink' soliton magnetic monopole form in complex Minkowski space and the generating Lagrangian for the complex form of the Dirac equation. Third, we examine the transformations of the vector and scalar potential $A^{\mu} = (\underline{A}, \phi)$ in complex Minkowski spaces. Mixing of $\underline{A}$ and ϕ is examined by considering harmonic exponential forms for $\underline{A}$ and ϕ.

We also discuss the relationship of the complex 8D Minkowski space and its associated twistor algebra mapable with the spinor calculus of the Kaluza-Klein 5D geometry with associated Lorentz invariant metric of relativity (gravity) with the set of unimodular transformations of Maxwell's equations (electromagnetic field) [6,18].

Implications and applications are discussed in terms of non-Hertzian waves; coherent non-dispersive, non-linear phenomena; possible magnetic monopoles; issues of gauge invariance and renormalization in quantum electrodynamics and the relationship of $F^{\mu\nu}$ and A^{μ}.

One of the key issues in considering possible longitudinal modes of excitation of the electromagnetic field is calculating energy fluxes and other electromagnetic parameters in terms of the vector and scalar potential 4-vector A_{μ} as (A,ϕ), rather than $\underline{E}$ or $\underline{B}$ or the 4-vector form, $F_{\mu\nu}$, (the electromagnetic field operator) where indices μ,ν run 1 to 4 and $F_{\mu\nu}$ is an antisymmetric second rank tensor. We use the signature (+++-) throughout this volume.

In conventional 'classical' electromagnetic theory, the formulation of electromagnetic phenomena can be made either in terms of $F_{\mu\nu}$ or A_{μ}. It has been suggested that the properties of A_{μ} may act in domains in which formalisms in terms of $F_{\mu\nu}$ do not apply. The possible physical effects manifest from A_{μ} outside the domain of $F_{\mu\nu}$ have implications for gauge invariance conditions. The usual gauge condition is given as $A_{\mu} \rightarrow A_{\mu} + \partial\Lambda / \partial x^{\mu}$ for Λ is an arbitrary function which must satisfy $\Box\Lambda = 0$. The requirement of gauge invariance is that all physically observable field quantities must remain invariant under the above transformation of A_{μ}. See Chap. 8.

To understand the issue of gauge invariance and the possible action of A_{μ} outside of $F^{\mu\nu}$ let us briefly examine the relativistic form of Maxwell's equations. The gauge invariant Lagrangian, L for the electromagnetic field is given by

$$L = -\frac{1}{4} F_{\mu\nu}(x) F^{\mu\nu}(x) \tag{1.1}$$

where $F^{\mu\nu}$ is the electromagnetic 4 x 4 field tensor which is related to the potentials by

$$F^{\mu\nu}(x) = \frac{\partial A^{\mu}(x)}{\partial x_{\nu}} - \frac{\partial A^{\nu}(x)}{\partial x_{\mu}} \tag{1.2}$$

where $F^{0k} = E^{k}$ is the electric field vector and $F^{k\ell} = \varepsilon_{k\ell j} H^{j}$ is the magnetic vector, and $L = -1/2(\underline{E}^2 - \underline{H}^2)$.

The variation of the Lagrangian, L with respect to the potentials then yields Maxwell's equations

$$\frac{\partial F^{\mu\nu}}{\partial x^{\nu}} = 0 \tag{1.3}$$

which, in terms of the potentials, are given as

$$\Box A^{\mu}(x) - \partial^{\mu}\chi(x) = 0 \tag{1.4}$$

for $\chi(x) = \partial^{\nu} A_{\nu}(x)$ and in classical theory the Lorentz gauge $\chi = 0$. Then Maxwell's equations are equivalent to $\Box A^{\mu}(x) = 0$ with $\chi = 0$ and $\Box$ the usual 4-vector D'Alembertian operator.

Since the potentials A_{μ} are not uniquely defined and since we have the usual assumption that only their 4D $\nabla \times F^{\mu\nu}(x)$ is observable, i.e. only the electric and magnetic fields are observable, then we see that gauge invariance guarantees any observation would not find the action of A^{μ} outside the observation of E or H (or B). In considering the possibility of longitudinal modes of the electromagnetic field and the possible action of a scalar potential outside the observation of E and H, we must address the relaxation of the gauge condition and the implication of this for Lorentz invariance [19]. See Chap. 8.

1.2 Multidimensional Minkowski Space

We have developed a multidimensional Minkowski space of $n > 4$ which has properties of remote connectedness or non-locality and superluminal signals in the projected or slice through 8D forming 4-space, M_4 from the complex cover space. We have examined both an 8D and 12D geometry [5,11,15]. The complexified 8D cover space involves the transformations $\underline{x}' \to \underline{x}_{\text{Re}} + i\underline{x}_{\text{Im}}$ and $t' \to t_{\text{Re}} + it_{\text{Im}}$ where x represents x, y, z. The 12D space involves the use of time symmetry of t_x, t_y, t_z which is complexified.

We have solved Maxwell's equations in the complex 8-space which complexifies the electric and magnetic fields. There are possible field configurations in which there is an action of the A^{μ} (or $\underline{A}, \phi$) fields where $\underline{E}$ and $\underline{B}$ are not observed. Modification of the gauge invariance condition is examined.

In this model we formulate advanced and retarded potentials (Chap. 6) as well as space-like and time-like causality conditions are used to formulate Bell's theorem, Young's double slit experiment and other supercoherence, non-local phenomena. See Chap. 4. Connection of the formalism is also made to the Higgs field magnetic monopole as a 'kink' soliton which relates to the remote connectedness properties of the complex Minkowski monopole. There are a number of classes of experiments for monopoles. We consider certain classes of experiments for determining the monopole's existence which is significant n determining certain properties relating $\underline{E}$ and $\underline{B}$ to $(\underline{A}, \phi)$. The magnetic monopole demands certain symmetry conditions of Maxwell's equations given by their solution in the complex geometry [5,12].

We also find that there are longitudinal as well as transverse components of the $\underline{E}$ and $\underline{B}$ fields. Recall that in about 1888, Hertz suggested that radio and light rays were part of the same spectrum and that electromagnetic waves are composed of transverse components of E and B to the direction of propagation of the wave. Faraday's earlier work laid the foundation for Maxwell's theoretical electric and magnetic unification [20]. We also find longitudinal components of $\underline{E}$ and $\underline{B}$ which comprise an acoustic-like mode which relates to a soliton mode. Early publications (1859 to 1873) of Maxwell appear to not preclude non-Hertzian waves [21]. We examine these wave forms in conjunction with Tesla's non-Hertzian wave hypothesis and detected extremely low frequency (ELF) waves [17,22-26]. See Chap. 5. Energy and communication systems are examined in conjunction with this model. The properties of the non-Hertzian waves, including their remote connectedness, appears promising for developing communication systems, including underwater and through-the-earth systems.

The existence of non-Hertzian waves implies a relaxation of gauge invariance in much the same manner that the Lorentz group for translations is extended to the inhomogeneous Lorentz group which includes rotation. Some of the issues to consider in regard to non-Hertzian or transverse wave phenomena are:

- The existence of some forms of monopole-like phenomena, where $\nabla \cdot \underline{B} \neq 0$ for a generalized complex vector $\underline{B}$,
- A small but finite rest mass for the photon,
- The apparent existence of some form of potential interaction of $(\underline{A}, \phi)$ outside the domain of interaction of $\underline{E}$ and $\underline{B}$, such as in the Aharonov-Bohm experiment,
- Theoretical and experimental design to detect non-Hertzian longitudinal-like waves.
- Possible model of biological informational processing

Specific existing current experimental information will be presented, including forms of the Aharonov-Bohm experiment, magnetic monopole data, finite rest mass of the photon experiment, and non-Hertzian (soliton) wave "detectors" which match in form. For example, neuronal functioning may detect such waves. We consider the soliton neuron information propagation model where this system acts as a non-linear, far-from-thermodynamic equilibrium, non-Hertzian wave detector [25].

References

[1] Petroni, N.C & Vigier, J-P (1983) (discussing Dirac) Dirac's aether in relativistic quantum mechanics, Foundations of Physics, 13:2; 253-285.
[2] Rauscher, E.A. (1868) J. Plasma Phys. 2, 517.
[3] Haramein, N. & Rauscher, E.A. (2005) Collective coherent oscillation plasma modes in surrounding media of black holes and vacuum structure – Quantum processes with considerations of spacetime torque and Coreolis forces, in R.L. Amoroso, B. Lehnert, and J-P Vigier (eds.) Beyond the Standard Model, Oakland: The Noetic Press.
[4] Rauscher, E.A. (1981) Conceptual changes in reality models from new discoveries in physics, Tecnic Research Laboratories PSRL-1076, presented to the M/B Conference, Palo Alto, October 10-11, 1981, Proceedings of the 1st Intl Symposium on Non-Conventional Energy Technology, October 23-24, 1981, Univ. of Toronto.
[5] Ramon, C. & Rauscher, E.A. (1980) Found. Physics 10, 661.
[6] Rauscher, E.A. (1980) Some physical models potentially applicable to remote information access, in A. Puharich (ed.) The Iceland Papers, Amherst: Essentia Research Associates; and (1997) Ottawa: P.A.C.E.
[7] Rauscher, E.A. & Targ, R. (2006) Investigations of a complex spacetime metric to describe anticipatory systems, in D.P. Sheehan (ed.) Frontiers of Time: Retrocausation, Experiment and Theory, AIP Conference Proceedings.
[8] Hansen, R.O. & Newman, E.T. (1975) Gen. Rel. and Gravitation 6, 216.
[9] Penrose, P. & Newman, E.J. (1978) Proc. Royal Soc., A364, 445.
[10] Rauscher, E.A. & Amoroso, R.L. (2009) Relativistic physics in complex Minkowski space, nonlocality, ether models and quantum physics, in M.C. Duffy, & J. Levy (eds.) Ether, Spacetime and Cosmology, Montreal: Aperion.
[11] Amoroso, R. L. & Rauscher, E.A. (2009) The Holographic Anthropic Multiverse: Formulating the Ultimate Geometry of Reality, Singapore: World Scientific.
[12] Rauscher, E.A. (1983) Electromagnetic Phenomena in Complex Geometries and Nonlinear Phenomena, Non-Hertzian Waves and Magnetic Monopoles, Tesla Book Co., Millbrae, CA; and (2008) Oakland: The Noetic Press.
[13] Rauscher, E.A. & Amoroso, R. L. (2005) The Schrödinger equation in complex Minkowski, nonlocal and anticipatory systems, in R. L. Amoroso, I. Dienes & C. Varges (eds.) Proceedings of the 1st Unified Theories, Symposium, Budapest, Hungary, Oakland: The Noetic Press.
[14] Rauscher, E.A. & Amoroso, R. L. (2006) The physical implications of multidimensional geometries and measurement, in D.M. Dubois (ed.) IJCAS, Liege Belgium, 19: 161-177.
[15] Ramon, C. & Rauscher, E.A. (1982) Remote connection in complex geometries and itsd application to the human mind, in N.A. Salaonen (ed.) pp. 1423-1442, New

York: FCF Press.
[16] Rauscher, E.A. (1982) Application of soliton physics to plasma – MHD and superconductivity BCS: Theoretical implications for primary energy, TRL Report, PSRL-3107, Los Angeles, Primary Energy Tech.
[17] Rauscher, E.A. (1985) Prospects of nonlinear coherent modes in MHD plasmas and possible resolution of the plasma confinement problem, pp. 108-118, in E.A. Rauscher & T. Grotz (eds.) IEEE 1st International Tesla Society Meeting, Colorado Springs.
[18] Appelquist, T., Gregory, A, & Freund, P.G.O. (1987) Modern Kaluza-Klein Theories, Reading: Addison-Wesley.
[19] Rauscher, E.A. (1972) Proposal on complexification of the spacetime metric and a fundamental description of nonlocality, TRL Report 3761A, The Radio Physics Laboratory, Palo Alto: SRI International.
[20] Faraday, M. (1855) Experimental Researches in Electricity, Vols. 1-3, R. Taylor & W. Francis (eds.) London: University of London.
[21] Maxwell, J.C. (1873) A Treatise on Electricity and Magnetism, Vols. I & II, Oxford: Clarendon Press; and (1865 A dynamic theory of the electromagnetic field, Royal Society Transactions, Vol. CLV, p.459.
[22] Tesla, N. (1983) Dr. Nikola Tesla Complete Patents, Vols. I & II, Millbrae: Tesla Book Co.; and (1982) Dr. Nikola Tesla Selected Patent Wrappers, Vols. I- IV, J.T. Ratzlaff (ed.) Millbrae: Tesla Book Co.
[23] Rauscher, E.A. & Van Bise, W.L. (1988) Fundamental excitatory modes of the Earth and Earth ionosphere resonant cavity, in S.R. Elswick (ed.) Proceedings of the 1988 International Tesla Symposium, pp. 34-69, Denver.
[24] Rauscher, E.A. & Van Bise, W.L. (2002) Harnessing the Earth ionosphere energy cavity energy for wireless transmission, in T. Valone (ed.) Harnessing the Wheelwork of Nature: Tesla Science and Energy, pp. 233-268, Chicago: Adventures Unlimited Press.
[25] Rauscher, E.A. & Van Bise, W.L. (1988) Non-superconducting apparatus for detecting magnetic and electromagnetic fields, US Patent, 4-724-390.
[26] Rauscher, E.A. & Van Bise, W.L. (1999) The relationship of extremely low frequency electromagnetic and magnetic fields associated with seismic and volcanic natural activity and artificial ionospheric disturbances, in E.M. Hayakawa (ed.) Atmospheric and Ionospheric Electromagnetic Phenomena, London: Terra Publishing Co.

Chapter 2

Structure, Properties and Implications of Complex Minkowski Spaces

We consider the properties and implications of three $n > 4$ multidimensional geometries. These are Descartes geometry [1], the properties and implications of which are enumerated in [2-6]. Both macroscopic and microscopic implications of these geometries are presented. We also develop several forms of complex Minkowski space in terms of a generalized metric containing terms derived from real and imaginary coordinates. The metric of the space is real and therefore physical [7-17]. This geometry is found to be one-to-one with Kaluza-Klein geometry [18-20] in which there has been much recent interest in developing M-Theory, in particular in the apparent relationship between the gravitational and electromagnetic fields often called Quantum Gravity. We have discussed the properties and implications of complex geometries in a number of works. The basic structure of the geometries is based on the construction of complexified dimensions, consisting of orthogonal real and imaginary parts. We examine the implication of a complex 8-space geometry in which we introduce imaginary components for each real spatial dimension, $X = (x,y,z)$ and temporal dimension, t.

2.1 Some Predictions of Complex Geometries

The complexification of Minkowski space, M_4 yields an 8D geometry, $M_4 \subset \mathbb{C}_4$. The 8D space is the least number of dimensions to accommodate nonlocality and anticipatory incursion in complex symmetry. This 8-space is also Lorentz invariant. Additional dimensional spaces (XD) are also considered, such as 12D spaces that also yield an approach to a unification of macro and micro processes [7,11,18].

We have solved the Schrödinger equation and Dirac equation in this complex 8D Minkowski space [16,17] and also formulated a field theoretical model that has implications for both MHD and BCS type phenomena [19].

For additional symmetry considerations, we have also introduced a 12D space in which we consider a 3-component time which is complexified [11,18]. Let us briefly list some of the implications of $n > 4$ geometries.

- Remote connectedness properties exist between physical events and processes in spacetime.
- Anticipatory or precognitive-like processes are allowed in temporal processes.
- Superluminal 'signals' appear to exist in 4-space.
- Tachyonic 'particles' are predicted.
- Coherent nondispersive phenomena exist, such as plasma oscillations of individual particle states like 'ball lightning' or solar activity, possibly 'sun spots' [20,21].
- A mechanism for physical effects such as conductivity and dielectric properties of plasmas in 4-space based on vacuum polarization properties in complex space [19,20].
- A model for unification of electromagnetic and gravitational phenomena [12,18] through the one-to-one mapping of the spinor calculus and twister algebra of the complex space [22].
- A mechanism of formulating the so-called 'collapse of the wave function' in terms of the geometric structure of space and interpretation of the 'observer effect' [23]. See Chap. 14.
- Possible interpretation of nonlinear effects in multidimensional geometries as an interpretation of the mechanism of the collapse of the wave function to a particular state. One possible interpretation of such a model is that 'consciousness' generates 'geometric reality' (or constraints on reality) which relate to a particular 'potentia' (Heisenberg's term or de Broglie pilot wave) being actualized as a specific event, which is a possible mechanism of the physical effects or a manifestation of thought [14,22]. Such an interpretation is key to a model which may include *Psi Phenomena*, since it now demonstrates a mechanism in which intention (goal) can be physically actualized [14].
- Application may be made to antenna theory [21].
- Formulation of certain processes in biological tissue is examined [24-27].

We list some examples of remote connectedness:

- Bell's theorem [28],
- Young's double slit experiment,
- Aharanov-Bohm experiment [29],
- Various supercoherence phenomena, such as plasma coherent states and superfluidity [19,20],

- Remote perception [30].

Some examples of coherent 'non-dispersive' phenomena in which dispersion is overcome by recoherence are:

- Soliton-like phenomena of plasma-phonon-electron interactions [20].
- Solving the Dirac equation in complex 8-space and the Fermi-Dirac vacuum state model
- Complexifying $F_{\mu\nu}$ in 8-space, Non-Hertzian and Hertzian waves [12].
- Ball lightning is a coherent electrostatic soliton-like phenomena. These phenomena are modeled after Prigogine [31] dissipative structures and catastrophe phenomena [32].
- Vortices in helium II represent soliton-like structures [22].

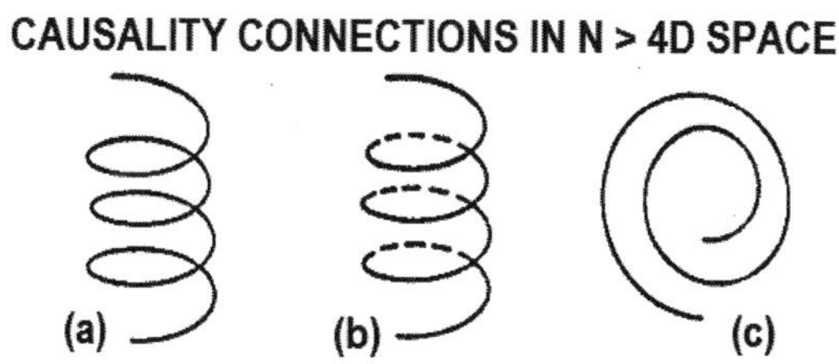

Figure 2.1. Causality in HD space. Representing a lightcone schematic. (a) Represents closed time-like loops in 4-space where the vertical dimension is time, t and the horizontal dimension is $X = x,y,z$. (b) Represents a non-multiply connected world line in $n > 4$D where no CTL or multi-connected 'nows' exist. (c) Represents the view of Fig. 2.1b where the future time dimension comes out of the paper, X is in the plane of the paper and no multi-connected world lines exist and only single valued 'nows' exist.

Certain spacetime relationships that involve Closed Time-like Loops (CTL) paradoxes can be resolved utilizing formulations in terms of multidimensional geometries (Fig. 2.1). The issues involved are presented and extensively discussed in [3]. In the so-called twin paradox, only future time travel is possible in non-inertial frames because time dilation only occurs in the rapidly accelerating frame (Fig. 2.2).

For time machines of the relativistic Twin Paradox, time moves into the future. At each point along a world line in spacetime there are a number of potential states in which one is actualized with preferential probability (or equal as in the Schrödinger cat paradox). Time machines that move into the past from the future represent CTLs (Fig. 2.3). Figure 2.4 represents various spacetime connections in a subset of 3D spacetime of the 4D X, ict, (+++-)

Minkowski spacetime. Figure 2.5 represents various world line connections on a Minkowski light cone including an unconnected past and future, a single valued world line and a multivalued CTL world line. See Fig. 2.5.

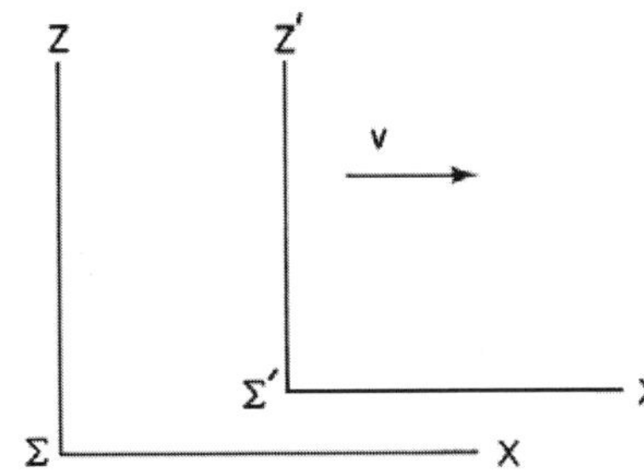

Figure 2.2. The relative velocities of two frames of reference, Σ, taken as the rest frame and, Σ' as the moving frame, having a relative velocity, v. The x,z plane is represented as the abscissa and ordinate respectively in 2-space. Not represented is the y coordinate extending out of the plane in this representation of two relative 3-space Euclidian coordinates (x,y,z) for Σ and $(x',y'z')$ for Σ'.

In order to describe processes involving apparent future-to-past lightcone connections, one has paradoxes involving CTL, multi-valued 'nows' and 'accelerated times' which involve the paradox of moving more slowly than a rest frame! These paradoxes can not be resolved in the usual Minkowski 4-space metric. In $n > 4$D spaces we have the possibility of the resolution of these paradoxes and the possibility of a more definitive formalism and description of some of the previously listed phenomena (Fig. 2.3).

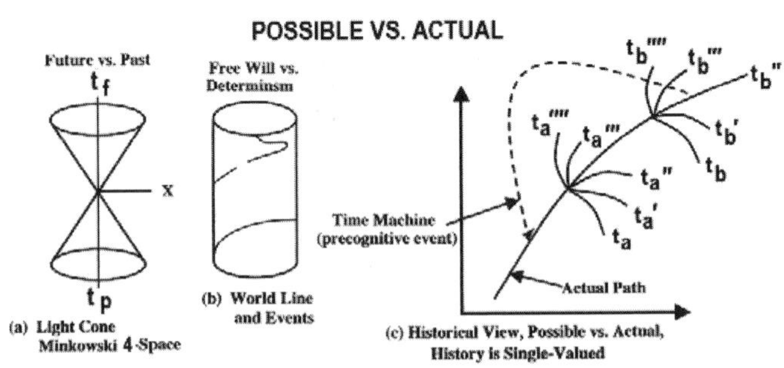

Figure 2.3. Possible versus actual. Several types of world lines are depicted. Figure 2.3a depicts a worldline with a single-valued "now", but Figs. 2.3b and 2.3c depict a multi-valued present. There is a dual world: constancy and change, absolute versus relativistic and Mach's Principle, and certainty versus uncertainty in terms of Einstein and Bohr $(\Delta x, \Delta p \geq \hbar)$.

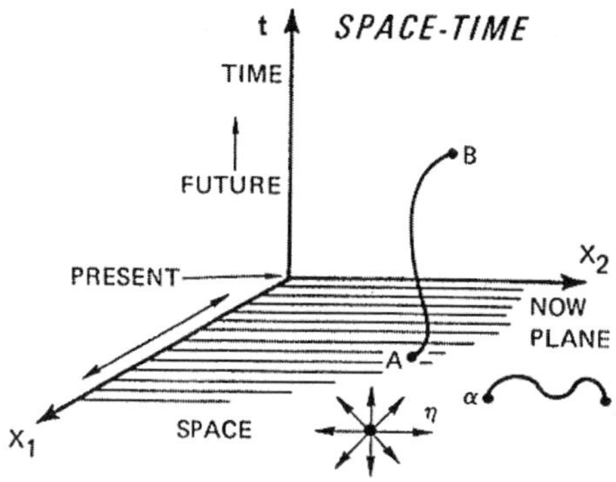

Figure 2.4. Our location is measured in space and time. In this figure we represent two dimensions of *x,y,z* as x_1 and x_2 and of time, *t* on the ordinate. Causal connections in real space are such that A can cause B by passage of time; at one time, α and β are correlated as effects at a distance. η represents instantaneous connections.

We have examined several forms of complex geometries. The complexification of Minkowski 4-space, M^4, gives rise to an 8D complex Minkowski space, $\hat{M}^4 \subset \mathbb{C}^4$ in which we take each of the 8D as an independent orthogonal dimension and that the real and imaginary components can be considered as two independent 4-space lightcones, $(X_{\text{Re}}, t_{\text{Re}})$ and $(X_{\text{Im}}, t_{\text{Im}})$; 4-space is then a slice through 8-space, rather than a subset or subspace formed by a projected geometry distorting the projection, causing variation in the defined variable length and vector orientation whereas orthogonal slices maintain uniformity (Fig. 2.4) [33,34].

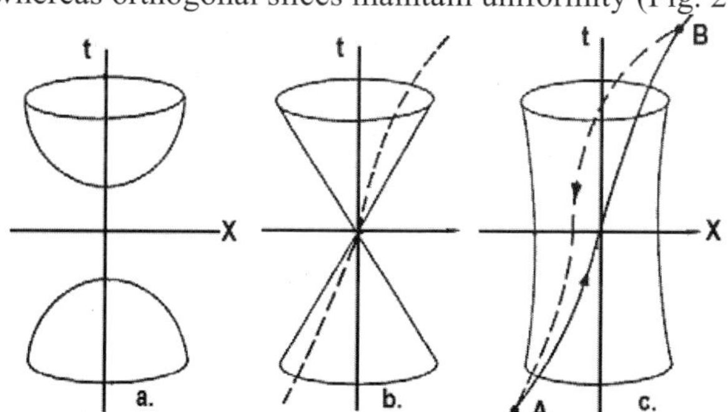

Figure 2.5. One can consider three classes of Minkowski diagrams with three types of causal connections of events along world lines. In Fig. 2.5a no connection exists between past and future. In Fig. 2.5b only one connection between past and future exists for a single valued now for the usual lightcone world line connection; and in Fig. 2.5c more than one connection of past and future exists as CTL. For example one path to the future and another from the future hook into its point "B" past to point A.

We have also examined other forms of the complex geometric model [11,12]. For symmetry considerations, we consider an extension of the temporal variable as the pseudoscalar $\underset{\sim}{t} = t_x\hat{x} + t_y\hat{y} + t_z\hat{z}$. Then we complexify each of these three temporal dimensions as $t_x = t_{x\mathrm{Re}} + it_{x\mathrm{Im}}$, $t_y = t_{y\mathrm{Re}} + it_{y\mathrm{Im}}$, $t_z = t_{z\mathrm{Re}} + it_{z\mathrm{Im}}$. We have handled the complexification of spatial dimensions similarly in [7-18]. We now have a complex form of Minkowski space which is a 12D space with the 12D listed respectively as: $x_{\mathrm{Re}}, y_{\mathrm{Re}}, z_{\mathrm{Re}}, x_{\mathrm{Im}}, y_{\mathrm{Im}}, z_{\mathrm{Im}}$, $t_{x\mathrm{Re}}, t_{y\mathrm{Re}}, t_{z\mathrm{Re}}, t_{x\mathrm{Im}}, t_{y\mathrm{Im}}, t_{z\mathrm{Im}}$.

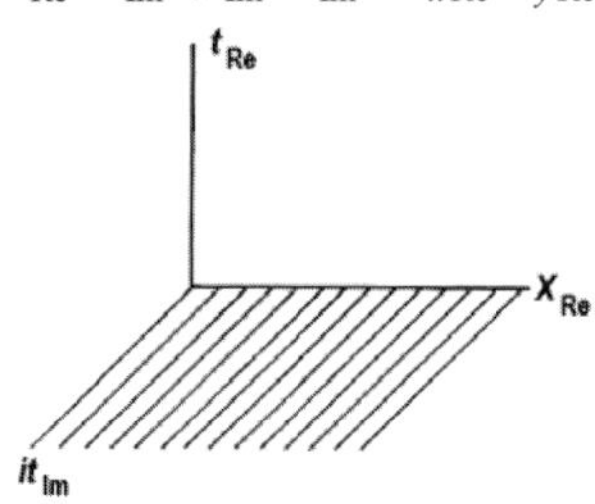

Figure. 2.6. The figure provides a relatively simple representation of real and imaginary components for a x_{Re}, t_{Re}, t_{Im} space. We extend this picture to a real and imaginary vertical time plane. t_{Re}, t_{Im} and a real and imaginary space plane of x_{Re}, x_{Im} for a 5D complex plane. See Fig. 2.7.

In considering certain classes of transformations we examine the manner in which moving and rest frames, $(\underline{A}, \phi)$ form and transform. We consider the fact that both these geometries appear to demand remote connectedness and superluminal signal propagation in the subset real 4-space, M_4. In Fig. 2.6 we represent various of events in real space and time. In Fig. 2.3a is represented the usual Minkowski spacetime metric, which is like a 4D Pythagorean theorem for a right triangle where $h^2 = a^2 + b^2$, the sum of the square of the hypotenuse is equal to the sum of the other two sides.

2.2 Multidimensional Geometric Models and Macroscopic Remote Connectedness

It appears that a resolution of the problem of closed time-like loops (CTL) lies in developing a model in terms of a space of higher dimensionality, HD. What appears to be a closed loop in 4D spacetime may in fact *not* have an intersection in an HD space [8,11,18]. See Fig. 2.3. Normal macroscopic causality demands that no point in the forward lightcone is connected to another point outside the forward lightcone; that is, all signals are time-like [8,21]. Real events involve simultaneity which is defined by signals that do not exceed the velocity of light, $v \leq c$ where v is the velocity of propagation and c is the velocity of light. Causality conditions for superluminal signals in constructing a Lorentz invariant quantum field theory are given in [7,8,12]. Tipler examines the problem of CTLs in general relativity for a rapidly rotating gravitational field [35]. The relationship of causality and locality conditions is discussed in [8].

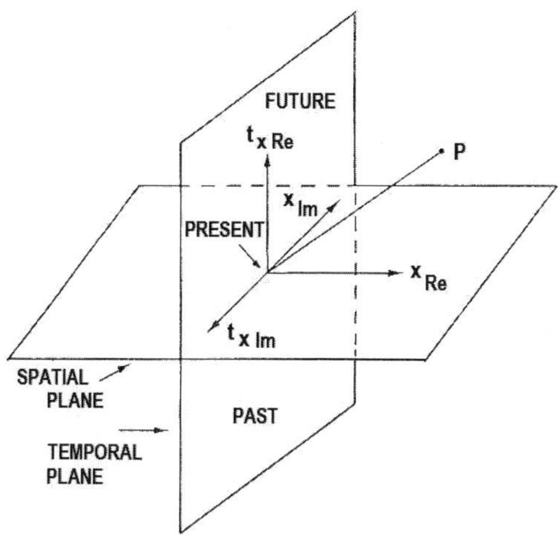

Figure 2.7. Spacetime of two intersecting complex planes in 8D described by the x components of space and time. Here $x = x_{Re} + ix_{Im}$ and time, t is $t = t_{xRe} + it_{xIm}$. Event P in this complex plane will be represented as: P($x + itx$). This figure can be extended to 3D of space and time where P will be represented as: P($x + itx, y + ity, z + itz$).

- First, the case in which there is no connection of past and future is represented, i.e., there is no causal connection.
- Second, the usual Minkowski diagram for a single valued present. In quantum mechanical terms, the collapse of the wave function describing the system under consideration allows only one world line.
- Third, the present or 'now' condition is not single valued. The event wave function no longer collapses to a point, localized region of spacetime, and more than one world line can represent the present.

In fact, for point-like events, one could conceivably have an infinite number of world lines passing through the present. Everett, Graham and Wheeler have examined the quantum mechanical implications of a multi-valued universe theory [36]. More information about a future event may then be traced back to the present via another world line and that actual time sequencing experienced is associated with the first world line or possibly a third world line. See Fig. 2.3.

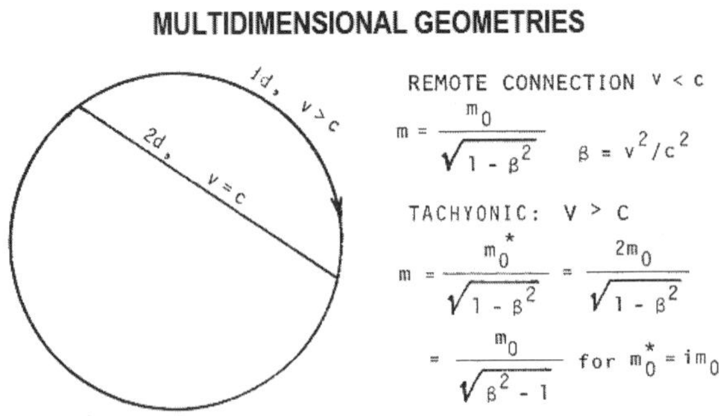

Figure 2.8. Represents a 1D circumference and a chord across the circle in a 2D space. In a lower dimensional space the velocity of propagation appears superluminal, $v > c$ and luminal in the hyperdimensional space, $v \leq c$. In analogy the velocity of propagation for Bell's Theorem in 4-space is $v > c$ or instantaneous, but in complex 8-space signaling can be luminal, $v = c$.

Of course one of the major problems of a theory containing multivalue solutions is the difficulty in defining a reasonable and useful causal relationship. The 4-space description gives us CTL which yield difficulties in describing prior and post event occurrences [27], Fig. 2.1. Intuitively, considering HD geometric models appears to reconcile the problem of CTL. For example, a helical world line in a 3-space would be single valued but would appear to contain multiple intersections if viewed at a 45° angle to the vertical helical axis as represented in a 2D space. This representation would contain multiple intersections even with a large pitch of the angle to the perpendicular to axis radius and hence act like a CTL [29]. See Fig. 2.3.

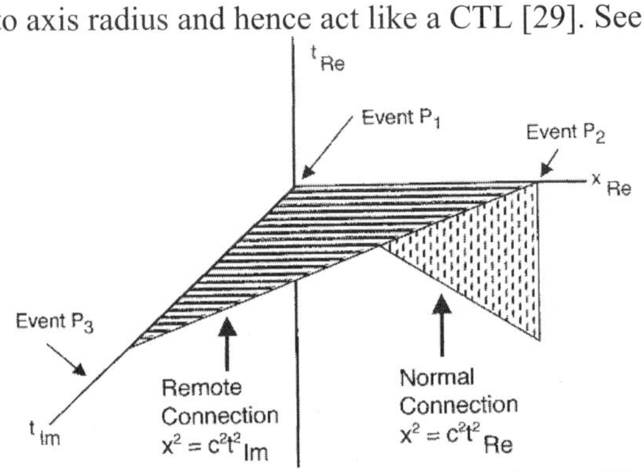

Figure 2.9. Complex time model of remote connectedness. We have the usual physical spatial separation of events on the x axis in the x_{Re}, t_{Re} plane which appears separated by a zero separation by 'moving" to the x_{Im}, t_{Im} plane. The separation between event P_1 and P_2 appears contiguous or simultaneously nonlocally correlated from the perspective of P_3. In an $n > 4$ space or an 8D space, nonlocal events can be correlated in such a manner as to not require standard signal propagation.

A number of HD geometries have been examined, in terms of reconciling complex anticipation and precognition-like signaling and causality as well as their possible relationship to superluminal signals [8,37-39]. In particular we have examined some 5 and 6D geometries where the additional dimensions, XD are space-like and time-like. In [8], instead of hypothesizing a model which involves energy transmission and associated problems of energy conservation, we chose to develop a model in which remote information is accessed in 4-space as though it was not remote in a HD geometry.

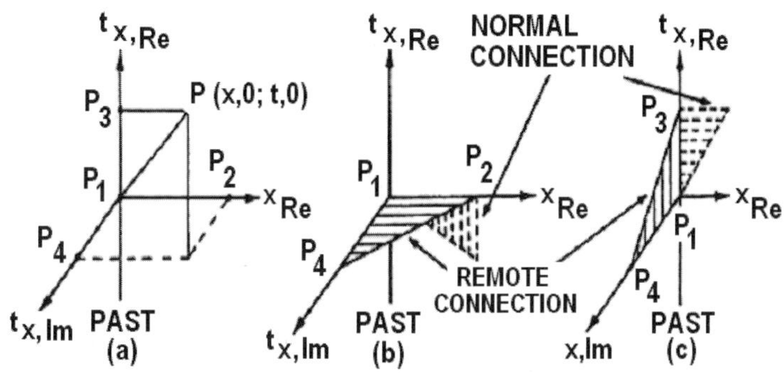

Figure 2.10. Four events in a complex plane. P_1 is at the origin. Event P is marked by non-zero spatial and temporal separation from the origin. P_1 and P_2 are separated in space but synchronous in time. P_1 and P_3 are separated in time, but there is no spatial separation. Event P_4 is located on the imaginary time axis; (b) Remote and normal connections of events P_1 and P_2 as viewed by an observer at P_4 such that space-like separation, $x(P_2) - x(P_1)$, between the events P_1 and P_2 is zero; (c) Remote and normal connections for zero time-like separation between the events P_3 and P_1 as viewed by an observer at P_4, such that, $t(P_3) - t(P_1) = 0$. (Figure received in a dream by EAR)

Relativity theory formally describes the relationship of macroscopic events in spacetime and, in particular, their causal connection is well specified. HD geometries appear to reconcile anticipation or precognition and causality and define a formalism in which the spatial and temporal separation of events in 4-space appear to be in juxtaposition in the HD geometry. This model can well accommodate information and perhaps energy transmission conditions as we will discuss in more detail in this volume. See Fig. 2.9 which represents a subset of the geometry we use in the present approach. There appears to be a reasonable relationship between these complex spaces and real 4, 5 and 6D spaces. The generalized causal relations in the complex space are consistent with the usual causality conditions, and exclude the CTL paradox. Multidimensional models appear to reconcile Maxwell's equations with the structure of general relativity in the weak gravitational field limit having some quantum mechanical features such as quantum nonlocality.

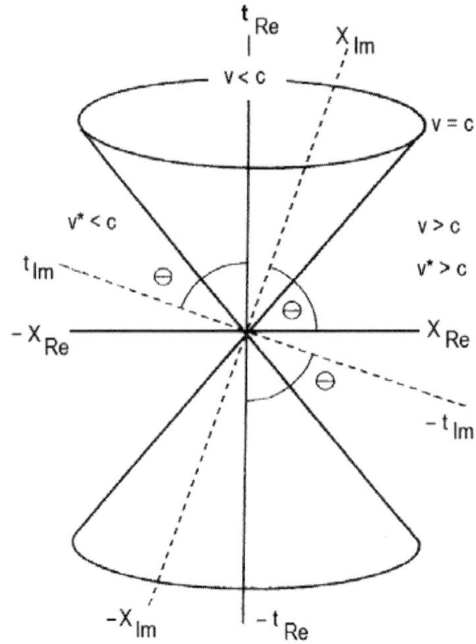

Figure 2.11. In the complex space multidimensional model, we introduce, in addition to the usual orthogonal 4-space, four imaginary components, three spatial and one temporal. This is necessary in order to model remote connectedness and to retain the causality and symmetry conditions in physics. We consider the eight orthogonal dimensions to be constituents of two intersecting lightcones, one real (x_{Re}, t_{Re}) and the other (y_{Im}, t_{Im}) coordinates.

We introduce a complex 8D matrix in which the real components comprise the usual 4-space of three real space components and a real time component and four imaginary components composed of three imaginary space components and one imaginary time component. See Fig. 2.10. Hansen and Newman [33,34] and Rauscher [7-19] developed the properties of a complex Minkowski space and explored the properties of this geometry in detail. The formalism involves defining a complex space $Z^{\mu} = X^{\mu}_{\text{Re}} + iX^{\mu}_{\text{Im}}$ where the metric of the space is obtained for the line element $ds^2 = g_{\mu\nu} dZ^{\mu} dZ^{*\nu}$ where indices μ and ν run 1 to 4.

In defining conditions of causality for $ds^2 = 0$ for the metrical form we have the usual 4-space Minkowski metric with signature (+++-)

$$ds^2 = g_{\mu\nu} dx^{\mu} dx^{\nu} \tag{2.1a}$$

using units $c = 1$ and $dx_1 = dx,\ dx_2 = dy,\ dx_3 = dz$ and $dx_4 = cdt$ where the indices μ and ν run 1 to 4; where also

$$g_{\mu\nu} = \begin{bmatrix} 1 & 0 & 0 & 0 \\ 0 & 1 & 0 & 0 \\ 0 & 0 & 1 & 0 \\ 0 & 0 & 0 & -1 \end{bmatrix} \tag{2.1b}$$

which is a 16-element matrix where the trace, $tr = 2$.

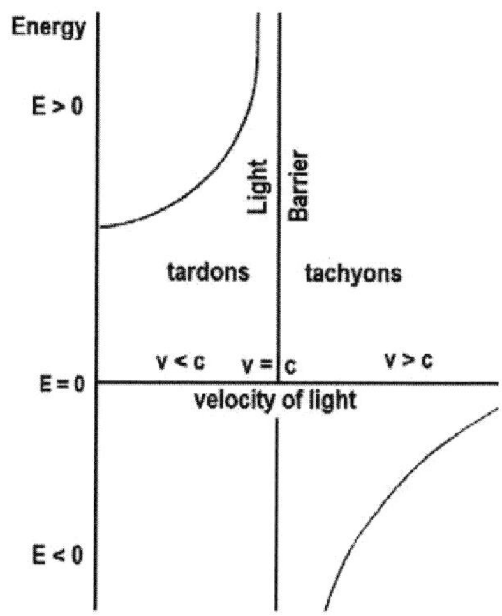

Figure 2.12. Tachyon and tardon signals are represented in the plot of energy versus velocity, as $v \to c$, $E \to \infty$. Perhaps tunneling through the velocity barrier from $v < c$ to $v > c$ can occur so that $E \neq \infty$.

In complex 8D space, we have for our differential line element with coordinates labeled $dZ^{\mu} = dX^{\mu}_{\text{Re}} + idX^{\mu}_{\text{Im}}$ (in which dZ is complex and dX_{Re} and dX_{Im} are themselves real), with a complex matrix where $\eta_{\iota\nu}$ is analogous to $g_{\mu\nu}$ such that

$$ds^2 = \eta_{\mu\nu} dZ^{\mu} dZ^{*\mu} \tag{2.2}$$

so that, for example, $dZ^{\mu} dZ^{*\mu} = (dX^{\mu}_{\text{Re}})^2 + (dX^{\mu}_{\text{Im}})^2$ where $\eta_{\mu\nu}$ is a 64-element matrix. We can write in general for real and imaginary space and time components:

$$\begin{aligned} ds^2 &= \left(dx^2_{\text{Re}} + dx^2_{\text{Im}}\right) + \left(dy^2_{\text{Re}} + dy^2_{\text{Im}}\right) \\ &\quad + \left(dz^2_{\text{Re}} + dz^2_{\text{Im}}\right) - c^2\left(dt^2_{\text{Re}} + dt^2_{\text{Im}}\right) \end{aligned} \tag{2.3}$$

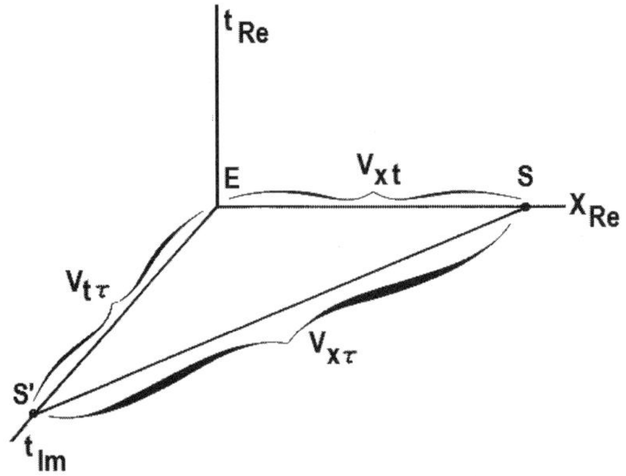

Figure 2.13. Real time separation between event E and event S on the real x axis can be made to appear contiguous by accessing the t_{Im} axis in 5D space as a subset of 8D space. The apparent velocity, v_{Re} is denoted as $v_x t = v_{x\mathrm{Re}} t_{\mathrm{Re}}$. Access to t_{Im} through a velocity $v_{t\mathrm{Re}} t_{\mathrm{Im}}$ along the t_{Im} axis run via this signal propagation can make $v_{x\mathrm{Re}} t_{\mathrm{Re}}$ appear instantaneous as $v_{x\mathrm{Re}} t_{\mathrm{Re}}$ goes to infinity. This figure corresponds to the remote connection points P_1 and P_2 via access to P_3 in Fig. 2.9 and also 2.10b.

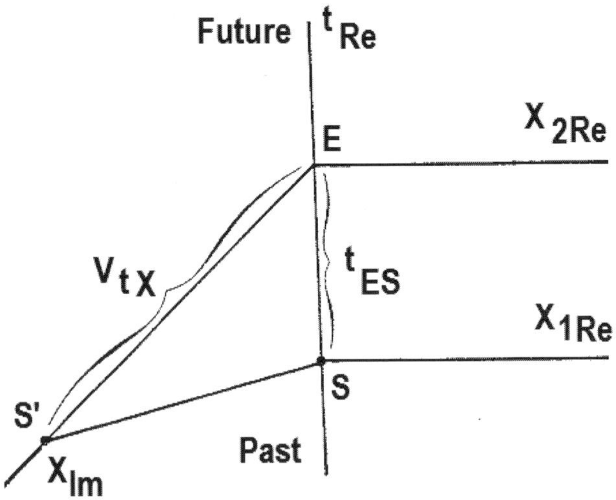

Figure 2.14. The separation of two events S and E along the real time axis, t_{ES}. The anticipatory time separation does not violate CTL, if we have access to the imaginary space coordinates, $X_{\mathrm{Im}} = x_{\mathrm{Im}}, y_{\mathrm{Im}}, z_{\mathrm{Im}}$. The velocity of propagation on $x_{\mathrm{Im}}, t_{\mathrm{Re}}$ space is $v_{t\mathrm{Re}}, x_{\mathrm{Im}}$. Comparison with Fig. 2.10c the event P_1 corresponds to S, and P_3 to E, and S′ to E at a velocity of $vt_{\mathrm{Re}}, x_{\mathrm{Im}}$ from S′ to P_4. Then E is an anticipatory event from the S frame of reference. At the vantage point of E at the future time can appear present and past events S can be anticipated when having access to S′ is possible. The space comprises a 7D geometry.

In [7] we represent the three real spatial components, dx_{Re}, dy_{Re}, dz_{Re}, as dX and the three imaginary spatial components, dx_{Im}, dy_{Im}, dz_{Im} as dX_{Im} and similarly for the real time component $dt_{\text{Re}} \equiv dt$ and $dt_{\text{Im}} = d\tau$. We then introduce complex spacetime-like coordinates as a space-like part $x_{\text{Im}} = \chi$ and a time-like part $t_{\text{Im}} = \tau$ as imaginary parts of X and t [8].

Now we have the invariant line elements as

$$s^2 = |x'|^2 - c|t'|^2 = |x'|^2 - |t'|^2 \tag{2.4}$$

again where we choose units where $c^2 = c = 1$ and

$$x' = X_{\text{Re}} + iX_{\text{Im}} \tag{2.5}$$

and

$$t' = t_{\text{Re}} + it_{\text{Im}} \tag{2.6}$$

as our complex dimensional component [7,8]. We use

$$x'^2 = |x'|^2 = X_{\text{Re}}^2 + X_{\text{Im}}^2 \tag{2.7}$$

and

$$t'^2 = |t'|^2 = t_{\text{Re}}^2 + t_{\text{Im}}^2 . \tag{2.8}$$

Recalling that the square of a complex number is given as the modulus

$$|x'| = x'x'^* = (X_{\text{Re}} + iX_{\text{Im}})(X_{\text{Re}} - iX_{\text{Im}}) \tag{2.9}$$

for X_{Re} and X_{Im} real. The fundamental key to this set of calculations is that the modulus of the product of complex numbers is real. Therefore, we have the 8-space line element

$$\begin{aligned} s^2 &= x_{\text{Re}}^2 - c^2 t_{\text{Re}}^2 + x_{\text{Im}}^2 - c^2 t_{\text{Im}}^2 \\ &= x_{\text{Re}}^2 - t_{\text{Re}}^2 + x_{\text{Im}}^2 - t_{\text{Im}}^2 \end{aligned} \tag{2.10}$$

Causality is defined by remaining on the right cone, in real spacetime, as

$$s^2 = x_{\text{Re}}^2 - c^2 t_{\text{Re}}^2 = x_{\text{Re}}^2 - t_{\text{Re}}^2 \tag{2.11}$$

using the condition c = 1. Then generalized causality in complex spacetime is defined by

$$s^2 = x_{\text{Re}}^2 - t_{\text{Re}}^2 + x_{\text{Im}}^2 - t_{\text{Im}}^2 \tag{2.12}$$

in the $x_{\mathrm{Re}}, t_{\mathrm{Re}}, x_{\mathrm{Im}}, t_{\mathrm{Im}}$ generalized light cone 8D space. See Fig. 2.11.

Let us calculate the interval separation between two events or occurrences Z_1 and Z_2 with real separation $\Delta x_{\mathrm{Re}} = x_{2\mathrm{Re}} - x_{1\mathrm{Re}}$ and imaginary separation $\Delta x_{\mathrm{Im}} = x_{2\mathrm{Im}} - x_{1\mathrm{Im}}$. Then the distance along the line element is $\Delta s^2 = \Delta(x_{\mathrm{Re}}^2 + x_{\mathrm{Im}}^2 - t_{\mathrm{Re}}^2 - t_{\mathrm{Im}}^2)$ and it must be true that the line interval is a real separation. Then

$$\begin{aligned}\Delta s^2 &= (x_{2,\mathrm{Re}} - x_{1,\mathrm{Re}})^2 + (x_{2,\mathrm{Im}} - x_{1,\mathrm{Im}})^2 \\ &\quad -(t_{2,\mathrm{Re}} - t_{1,\mathrm{Re}})^2 - (t_{2,\mathrm{Im}} - t_{1,\mathrm{Im}})^2\end{aligned} \tag{2.13}$$

or

$$\Delta s^2 = (x_{2,\mathrm{Re}} - x_{1,\mathrm{Re}})^2 + (x_{2,\mathrm{Im}} - x_{1,\mathrm{Im}})^2$$

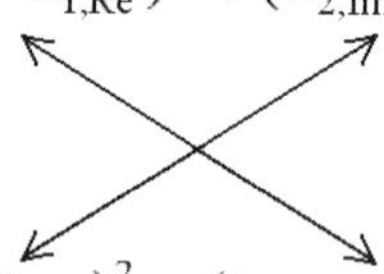

(2.14)

$$-(t_{2,\mathrm{Re}} - t_{1,\mathrm{Re}})^2 - (t_{2,\mathrm{Im}} - t_{1,\mathrm{Im}})^2$$

Because of the relative signs of the real and imaginary space and time components and in order to achieve the causality connectedness condition between the two events, or Δs^2, we must "mix" space and time. That is, we use the imaginary time component to effect a zero space separation. We identify $(x_{1,\mathrm{Re}}, t_{1,\mathrm{Re}})$ with one spacetime event causally correlated with another spacetime event, $(x_{2,\mathrm{Re}}, t_{2,\mathrm{Re}})$ [8]. See Fig. 2.9. By introducing the imaginary time component, one can achieve a condition in which the apparent separation in the real physical plane defined by x_{Re}, t_{Re} is zero, given access to the imaginary time, t_{im}, or the x_{Re}, t_{im} plane yielding spatial nonlocality.

The lightcone metric representation may imply superluminal signal propagation between an event A transmitter and even in the four real subset space by the event B (receiver) or two simultaneously remotely connected events. Separation will not appear superluminal in the 8-space representation. The causality conditions, which do not contain closed time-like loops, are for the complex 8-space geometry, where 4-space is a cut through the 8-space [8]. Newton examines causality conditions in 4-space with superluminal signals [40] and the problem of closed time-like loops posed by Feinberg's classic "Tachyon" paper [41,42]. These problems appear to be resolved by considering spaces of higher (> 4D) dimensions and are consistent with subliminal and superluminal signals. See Fig. 2.12.

In Chap. 4 we will discuss the relationship between subliminal, time-like, and superluminal, space-like, interpretation of the remote connectedness phenomena, such as the nonlocality test of Bell's theorem.

2.3 The Lorentz Condition in Complex 8-Space Geometry and Tachyonic Signaling

In order to examine as the consequences of the relativity hypothesis that time is the fourth dimension of space, and that we have a particular form of transformation called the Lorentz transformation, we must define velocity in the complex space. That is, the Lorentz transformation and its consequences, the Lorentz contradiction and mass dilation, etc., are a consequence of time as the fourth dimension of space and are observed in three spaces [43]. These attributes of 4-space in 3-space are expressed in terms of velocity, as in the form $\gamma = \left(1-\beta^2\right)^{-1/2}$ for $\beta \equiv v_{\mathrm{Re}} / c$ where *c* is always taken as real.

If complex 8-space can be projected into 4-space, what are the consequences? We can also consider a 4D slice through the complex 8D space. Each approach has its advantages and disadvantages. In projective geometries information about the space is distorted or lost. What is the comparison of a subset geometry formed from a projected geometry or a subspace formed as a slice through an XD geometry? What does a generalized Lorentz transformation "look like"? We will define complex derivatives and therefore we can define velocity in a complex plane [8].

Consider the generalized Lorentz transformation in the system of x_{Re} and t_{Im} for the real time remote connectedness case in the $x_{\mathrm{Re}}, t_{\mathrm{Im}}$ plane. We define our substitutions from 4-to 8-space before us,

$$\begin{aligned} x &\rightarrow x' = x_{\mathrm{Re}} + i x_{im} \\ t &\rightarrow t' = t_{\mathrm{Re}} + i t_{im} \end{aligned} \tag{2.15}$$

and we represented the case for no imaginary component of x_{Re} or $x_{\mathrm{Im}} = 0$ where the $x_{\mathrm{Re}}, t_{\mathrm{Re}}$ plane comprises the ordinary 4-space plane.

Let us recall that the usual Lorentz transformation conditions defined in four real space. Consider two frames of reference, Σ, at rest and Σ' moving at relative uniform velocity *v*. We call *v* the velocity of the origin of Σ' moving relative to Σ. A light signal along the *x* direction is transmitted by $x = ct$ or $x - ct = 0$ and also in Σ' as $x' = ct'$ or $x' - ct' = 0$, since the velocity of light in vacuo is constant in any frame of reference in 4-space. See Fig. 2.2. For the usual 4D Lorentz transformation, we have as shown in Eq. (2.6) and (2.8), $x = x_{\mathrm{Re}}, t = t_{\mathrm{Re}}$ and $v_{\mathrm{Re}} = x_{\mathrm{Re}} / t_{\mathrm{Re}}$.

$$x' = \frac{x - vt}{\sqrt{1 - v^2 / c^2}} = \gamma(x - vt)$$
$$y' = y$$
$$z' = z \tag{2.16}$$
$$t' = \frac{t - (v/c^2)x}{\sqrt{1 - v^2/c^2}} = \gamma\left(t - \left(\frac{v}{c^2}x\right)\right)$$

for $\gamma = (1 - \beta^2)^{-1/2}$ and $\beta = v/c$. Here x and t stand for x_{Re} and t_{Re} and v is the real velocity.

We consider the x_{Re}, t_{Im} plane and write the expression for the Lorentz conditions for this plane. Since again t_{Im} like t_{Re} is orthogonal to x_{Im} and t'_{Im} is orthogonal to x'_{Im} we can write

$$x' = \frac{x - ivt_{Im}}{\sqrt{1 - v^2 / c^2}} = \gamma_v(x - vt_{Im})$$
$$y' = y$$
$$z' = z \tag{2.17}$$
$$t' = \frac{t - (v/c^2)x}{\sqrt{1 - v^2/c^2}} = \gamma_v\left(t - \left(\frac{v}{c^2}x\right)\right)$$

where γ_v represents the definition of γ in terms of the velocity v; also $\beta_{v\,Im} \equiv v_{Im}/c$ where c is always taken as real [7] where v can be real or imaginary.

In Eq. 2.17 for simplicity we let x', x, t' and t denote x'_{Re}, x_{Re}, t'_{Re} and t_{Re} and we denote script v as v_{Im}. For velocity, v is $v_{Re} = x_{Re}/t_{Re}$ and $v = v_{Im} = ix_{Im}/it_{Im}$ where the i drops out so that $v = v_{Im} = x_{Im}/t_{Im}$ is a real-valued function. In all cases the velocity of light c is c. We use this alternative notation here for simplicity in the complex Lorentz transformation. The symmetry properties of the topology of the complex 8-space gives us the properties that allow Lorentz conditions in 4D, 8D and ultimately 12D space. The example we consider here is a subspace of the 8-space of x_{Re}, t_{Re}, x_{Im} and t_{Im}. In some cases we let $x_{Im} = 0$ and just consider temporal remote connectedness and anticipation; but likewise we can formulate remote, nonlocal connectedness solutions for $x_{\mathrm{Im}} \neq 0$ and $t_{\mathrm{Im}} = 0$ or $t_{\mathrm{Im}} \neq 0$. The anticipatory case for $x_{\mathrm{Im}} = 0$ is a 5D space as the space for $x_{\mathrm{Im}} \neq 0$ *and* $t_{\mathrm{Im}} = 0$ is a 7D space and for $t_{\mathrm{Im}} \neq 0$ as well as the other real and imaginary spacetime dimensions, we have our complex 8D space. See Fig. 2.11.

It is important to define the complex derivative so that we can define velocity, v_{Im}. In the $x_{\mathrm{Re}}t_{\mathrm{Im}}$ plane then, we define a velocity of $v_{\mathrm{Im}} = dx/dit_{\mathrm{Im}}$. In Sec. 2.4 we detail the velocity expression for v_{Im} and define the derivative of a complex function in detail [38]. For $v_{\mathrm{Im}} = dx / idt_{\mathrm{Im}} = -idx / dt_{\mathrm{Im}} = -iv_{\mathrm{Re}}$ for v_{Re} as a real quantity, we substitute into our $x_{\mathrm{Re}}, t_{\mathrm{Im}}$ plane Lorentz transformation conditions as

$$
\begin{aligned}
x' &= \frac{x_{\mathrm{Re}} - v_{\mathrm{Re}} t_{\mathrm{Im}}}{\sqrt{1 + v_{\mathrm{Re}}^2 / c^2}} \\
y' &= y \\
z' &= z \\
t'_{\mathrm{Im}} &= \frac{t_{\mathrm{Re}} - v_{\mathrm{Re}} x_{\mathrm{Re}}}{\sqrt{1 + v_{\mathrm{Re}}^2 / c^2}} .
\end{aligned}
\tag{2.18}
$$

These conditions will be valid for any velocity, $v_{\mathrm{Re}} = -v$.

Let us examine the way this form of the Lorentz transformation relates to the properties of mass dilation. We will compare this case to the ordinary mass dilation formula and the tachyonic mass formula of Feinberg [41] which nicely results from the complex 8-space. See Fig. 2.7. In the ordinary $x_{\mathrm{Re}}\, t_{\mathrm{Re}}$ plane then, we have the usual Einstein mass relationship of

$$m = \frac{m_0}{\sqrt{1 - v_{\mathrm{Re}}^2 / c^2}} \quad \text{for } v_{\mathrm{Re}} \leq c \tag{2.19}$$

and we can compare this to the tachyonic mass relationship in the x_{Im}, t_{Im} plane

$$m = \frac{m_0^*}{\sqrt{1 - v_{\mathrm{Re}}^2 / c^2}} = \frac{im_0}{\sqrt{1 - v_{\mathrm{Re}}^2 / c^2}} = \frac{m_0}{\sqrt{v_{\mathrm{Re}}^2 / c^2 - 1}} \tag{2.20}$$

for v_{Re} now $v_{\mathrm{Re}} \geq c$ and where m^* or m_{Im} stands for $m^* = im$ and we define m as m_{Re},

$$m = \frac{m_0}{\sqrt{1+v^2/c^2}} \quad . \tag{2.21}$$

For m real (m_{Re}), we examine two cases on v as $v < c$ or $v > c$, so we let v be any value from $-\infty < v < \infty$, where the velocity, v, is taken as real, or v_{Re}.

Consider the case of v as imaginary (or v_{Im}) and examine the consequences of this assumption. Also we examine the consequences for both v and m imaginary and compare to the above cases. If we choose v imaginary or $v* = iv$ (which we can term v_{Im}) the $v^{*2}/c^2 = -v^2/c^2$ and $\sqrt{1+v^{*2}/c^2}$ becomes $\sqrt{1-v^{*2}/c^2}$ or

$$m = \frac{m_0}{\sqrt{1-v_{Re}^2/c^2}} \quad . \tag{2.22}$$

We get the form of this normal Lorentz transformation if v is imaginary $(v^* = v_{Im})$.

If both v and m are imaginary, as $v* = iv$ and $m* = im$, then we have

$$m = \frac{m_0^*}{\sqrt{1+v^{*2}/c^2}} = \frac{im_0}{\sqrt{1-v^2/c^2}} = \frac{m_0}{\sqrt{v^2/c^2-1}} \tag{2.23}$$

or the tachyonic condition.

If' we go "off" into x_{Re} t_{Re} t_{im} planes, then we have to define a velocity "cutting across" these planes, and it is much more complicated to define the complex derivative for the velocities. For subliminal relative systems Σ and Σ' we can use vector addition such as $W = v_{Re} + iv_{Im}$ for $v_{Re} < c$, $v_{Re} < x$, $v_{Im} < c$ and $W < c$. In general there will be four complex velocities. The relationship of these four velocities is given by the Cauchy-Riemann relations in the next section. These two are equivalent. The actual magnitude of v may be expressed as $v = [vv^*]^{\frac{1}{2}}\hat{v}$ (where $\hat{v}$ is the unit vector velocity) which can be formed using either of the Cauchy-Riemann equations. It is important that a detailed analysis not predict any extraneous consequences of the theory. Any possibly new phenomenon that is hypothesized should be formulated in such a manner as to be easily experimentally testable.

Feinberg suggests several experiments to test for the existence of tachyons [8,41,42]. He describes the following experiment. Consider in the laboratory, atom A, at time, t_0 is in an excited state at rest at x_1 and atom B is in its ground state at x_2. At time t_1 atom A descends to the ground state and emits a tachyon in the direction of B. Let E_1 be this event at t_1, x_1. Subsequently, at $t_2 > t_1$ atom B absorbs the tachyon and ascends to an excited state; this is event E_2, at t_2, x_2. Then at $t_3 > t_2$ atom B is excited and A is in its ground state. For an observer traveling at an appropriate velocity, $v < c$ relative to the laboratory frame, the events E_1 and E_2 appear to occur in the opposite order in time. Feinberg describes the experiment by stating that at $t_2^{'}$ atom B spontaneously ascends from the ground state to an excited state, emitting a tachyon which travels toward A. Subsequently, at $t_1^{'}$, atom A absorbs the tachyon and drops to the ground state.

It is clear from this that what is absorption for one observer is spontaneous emission for another. But if quantum mechanics is to remain intact so that we are able to detect such particles, then there must be an observable difference between them: The first depends on a controllable density of tachyons, the second does not. In order to elucidate this point, we should repeat the above experiment many times over. The possibility of reversing the temporal order of causality, sometimes termed 'sending a signal backwards in time' must be addresses [8,41,42]. Is this cause-effect statistical in nature? In the case of Bell's Theorem, these correlations are extremely strong whether explained by $v > c$ or $v = c$ signaling.

In [44], Bilaniuk, et al. formulated the interpretation of the association of negative energy states with tachyonic signaling. From the different frames of reference, thus to one observer absorption is observed and to another emission is observed. These states do not violate special relativity. Acausal experiments in particle physics, such as for the S-Matrix, have been suggested by a number of researchers [45]. Another approach is through the detection of Cerenkov radiation, which is emitted by charged particles moving through a substance traveling at a velocity, $v > c$. For a tachyon traveling in free space with velocity, $v > c$ Cerenkov radiation may occur in a vacuum cause the tachyon to lose energy and become a tardon [4]. See Figs. 2.8 and 2.12.

2.4 Velocity of Propagation in Complex 8-Space

In this section we utilize the Cauchy-Riemann relations to formulate the hyperdimensional velocities of propagation in the complex plane in various slices through the hyperdimensional complex 8-space. In this model finite limit velocities, $v > c$ can be considered. In some Lorentz frames of reference, instantaneous signaling can be considered. In Fig. 2.13 is displayed the velocity connection between remote nonlocal events, and in Fig. 2.14 is displayed temporal separated events or anticipatory and real time event relations.

It is important to define the complex derivative so that we can define the velocity, $v = v_{\text{Im}}$. In the *xit* or $x_{\text{Re}} i x_{\text{Im}}$ plane then, we define a velocity of $v = dx / d(it_{\text{Im}})$ We now examine in some detail the velocity of this expression, here $x = x_{\text{Re}}$. In defining the derivative of a complex function we have two cases in terms of a choice in terms of the differential increment considered. Consider the orthogonal coordinates x and it_{Im}; then we have the generalized function, $f(x, t_{\text{Im}}) = f(z)$ for $z = x + it_{\text{Im}}$ and $f(z)$ = $u(x, t_{\text{Im}}) + iv(x, t_{\text{Im}})$ where $u(x, t_{\text{Im}})$ and $v(x_{\text{Im}}, t_{\text{Im}})$ are real functions of the rectangular coordinates x and t_{Im} of a point in space, $P(x, t_{\text{Im}})$. Choose a case such as the origin $z_0 = x_0 + it_{0\text{Im}}$ and consider two cases , one for real increments $h = \Delta x$ and imaginary increments $h = i\Delta t_{\text{Im}}$. For the real increments $h = \Delta t_{\text{Im}}$ we form the derivative $f'(z_0) \equiv df(z)/dz_{z_0}$ which is evaluated at z_0 as

$$f' = \lim_{\Delta x \to 0} \left\{ \frac{\mu(x_0 + \Delta x, t_{0\text{Im}}) - \mu(x_0, t_{0\text{Im}})}{\Delta x} + i \frac{v(x_0 + \Delta x, t_{0\text{Im}}) - v(x_0, t_{0\text{Im}})}{\Delta x} \right\} \tag{2.24a}$$

or

$$f'(z_0) = u_x(x_0, t_{0\text{Im}}) + iv_x(x_0, t_{0\text{Im}}) \text{ for } u_x \equiv \frac{\partial u}{\partial x} \text{ and } v_x \equiv \frac{\partial v}{\partial x}. \tag{2.24b}$$

Again $x = x_{\text{Re}}$, $x_0 = x_{0\text{Re}}$ and $v_x = v_{x\text{Re}}$.

Now for the purely imaginary increment, $h = i\Delta t_{\text{Im}}$ we have

$$f'(z_0) \lim_{\Delta t_{\text{Im}} \to 0} \left\{ \frac{1}{i} \frac{\mu(x_0, t_{0\text{Im}} + \Delta t_{\text{Im}}) - \mu(x_0, t_{0\text{Im}})}{\Delta t_{\text{Im}}} + \frac{v(x_0, t_{0\text{Im}} + \Delta t_{\text{Im}}) - v(x_0, t_{0\text{Im}})}{\Delta t_{\text{Im}}} \right\} \tag{2.25a}$$

and $$f'(z_0) = -iu_{t\,\mathrm{Im}}(x_0, t_{0\,\mathrm{Im}}) + v_{t\,\mathrm{Im}}(x_0, t_{0\,\mathrm{Im}}) \tag{2.25b}$$

for $u_{\mathrm{Im}} = u_{t\,\mathrm{Im}}$ and $v_{\mathrm{Im}} = v_{t\,\mathrm{Im}}$ then

$$u_{t\,\mathrm{Im}} \equiv \frac{\partial u}{\partial t_{\mathrm{Im}}} \text{ and } v_{t\,\mathrm{Im}} \equiv \frac{\partial v}{\partial t_{\mathrm{Im}}}. \tag{2.25c}$$

Using the Cauchy-Riemann equations

$$\frac{\partial u}{\partial x} = \frac{\partial v}{\partial t_{\mathrm{Im}}} \text{ and } \frac{\partial u}{\partial t_{\mathrm{Im}}} = -\frac{\partial v}{\partial x} \tag{2.26}$$

and assuming all principle derivations are definable on the manifold and letting $h = \Delta x + i\Delta t_{\mathrm{Im}}$ we can use

$$f'(z_0)\lim_{h\to 0}\frac{f(z_0+h)-f(z_0)}{h} = \left.\frac{df(z)}{dz}\right|_{z_0} \tag{2.27a}$$

and

$$u_x(x_0, t_{0\,\mathrm{Im}}) + iv_x(x_0, t_{0\,\mathrm{Im}}) - \frac{\partial u(x_0, t_{0\,\mathrm{Im}})}{\partial x} + i\frac{\partial v}{\partial x}(x_0, t_{0\,\mathrm{Im}}) \tag{2.27b}$$

with v_x for x and t_{Re} that is $u_{\mathrm{Re}} = u_{x\mathrm{Re}}$, with the derivative form of the charge of the real space increment with complex time, we can define a complex velocity as,

$$f'(z_0) = \frac{dx}{d(it_{\mathrm{Im}})} = \frac{1}{i}\frac{dx}{dt_{\mathrm{Im}}} \tag{2.28a}$$

we can have $x(t_{\mathrm{Im}})$ where x_{Re} is a function of t_{Im} and $f(z)$ and using $h = i\Delta t_{\mathrm{Im}}$, then

$$f'(z_0) = x'(t_{\mathrm{Im}}) = \frac{dx}{dh} = \frac{dx}{idt_{\mathrm{Im}}}. \tag{2.28b}$$

Then we can define a velocity where the differential increment is in terms of $h = i\Delta t_{\mathrm{Im}}$. Using the first case as $u(x_0, t_{0\,\mathrm{Im}})$ and obtaining $dt_{0\,\mathrm{Im}}/\Delta x$ (with i's) we take the inverse. If u_x which is v_x in the $h \to i\Delta t_{\mathrm{Im}}$ case have both u_x and v_x, one can be zero.

In the next section, we present a brief discussion of n > 4D geometries. Like the complex 8D space, the 5D Kaluza-Klein geometries are subsets of the supersymmetry models. The complex 8-

space deals in extended dimensions, but like the TOE models, Kaluza-Klein models also treat n > 4D as compactified on the scale of the Planck length, 10^{-33} cm [1-6]. See Fig. 2.15, Chap. 13 and Fig. 13.2.

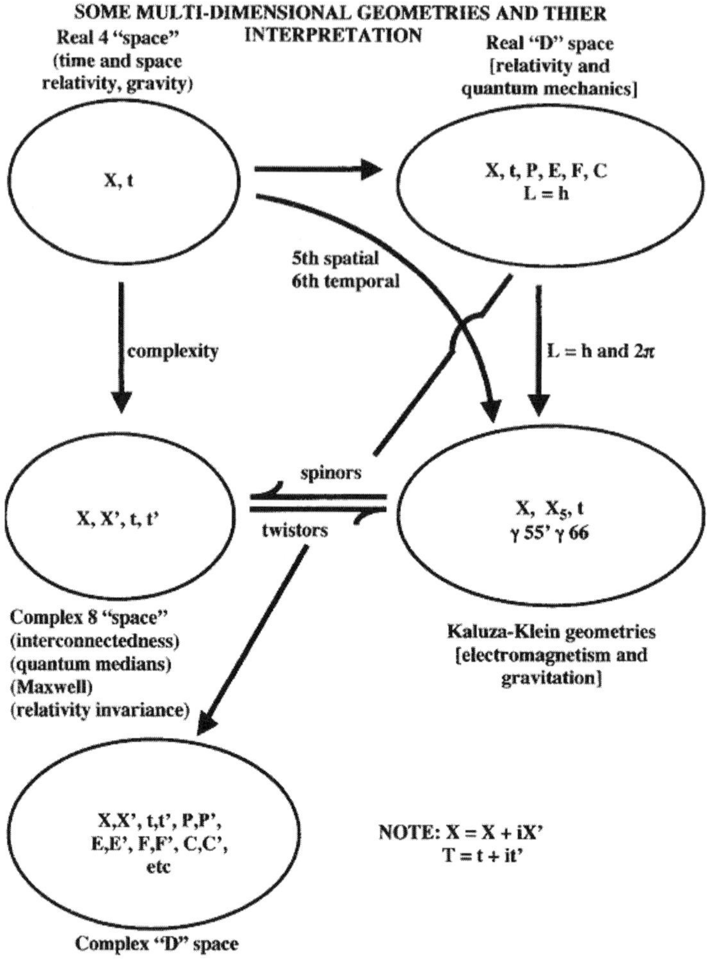

Figure 2.15. Representation of the usual 4-Space lightcone and four other multidimensional geometries. In the upper right is a representation of the 10D real Descartes geometry. Below and to the left of the usual 4D space is the complexified 8D space. To the lower right is the 5D and 6D Kaluza-Klein geometry and in the lower left is the complexified 10D Descartes space as a 20D complex Descartes space. Note that X represents x,y,z and P represents P_x, P_y, P_z in the upper right. The relationships of all these geometries are represented in this figure.

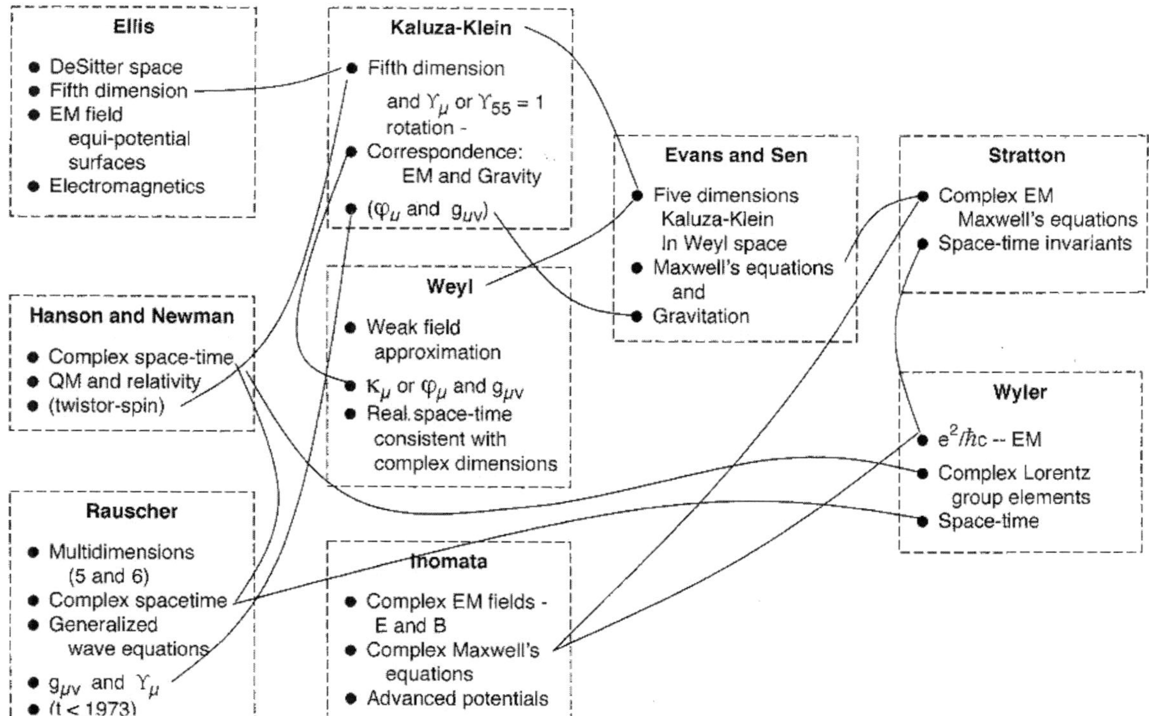

Figure 2.16. Relationship of multidimensional geometric theories. Comparing the differences between the concepts in these models is interesting because it may lead to unification electromagnetic and gravitational phenomena and model remote connectedness and nonlinear phenomena [7-12, 33, 34, 38, 39, 46, 50-54].

In 4D space (Fig. 2.9) event point, P_1 and P_2 are spatially separated on the real space axis as x_{0Re} at point P_1 and x_{1Re} at point P_2 with separation $\Delta x_{Re} = x_{1Re} - x_{0Re}$. From the event point P_3 on the t_{Im} axis we move in complex space from event P_1 to event P_3. From the origin, t_{0Im} we move to an imaginary temporal separation of t_{1Im} to t_{2Im} of $\Delta t_{Im} = t_{2Im} - t_{0Im}$. The distance in real space and imaginary time can be set so that measurement along the t_{Im} axis yields an imaginary temporal separation Δt_{Im} subtracts out, from the spacetime metric, the temporal separation Δx_{Re}. In this case occurrence of events P_1 and P_2 can occur simultaneous, that is, the apparent velocity of propagation is instantaneous.

For the example of Bell's Theorem, the two photons leave a source nearly simultaneously at time, t_{0Re} and their spin states are correlated at two real spatially separated locations, x_{1Re} and x_{2Re} separated by $\Delta x_{Re} = x_{2Re} - x_{1Re}$. This space-like separation, is forbidden by special relativity; however, in the complex space, the points x_{1Re} and x_{2Re} appear to be contiguous for the proper path 'traveled' to point at t_{1Im} along the imaginary axis. Because of the possibility of proper spacetime adjustment or transformation which is possible in the complex plane, separate spacetime locations can appear contiguous in the hyperdimensional 8-space. Hence the upper limit of velocity propagation is instantaneous. See Table 2.1 and Figs. 2.13, and 2.14. By adjusting our imaginary spatial and temporal advantage, v_{Re} and v_{Im} can be variously adjusted and effect apparent causal conditions from the 8D space to the 4D space [14].

Table 2.1
Multidimensional Models: Macrocosm and Consciousness

- Einstein-Minkowski 4D space expressed as three spatial and one temporal dimension. This is the usual observed 3-space modeled on the Minkowski lightcone diagram [7,8].
- 4-space can be expanded to a multidimensional quantum gravity space of 11D; spacetime, momentum, mass-energy, force, velocity, acceleration, power, pressure, and rotation which comprise Descartes 10D space [4,5].
- Complex 8D space is generated by complexifying the usual 4-space

using $\sqrt{-1}$, and has nonlocality and remote interconnectedness that can relate a twistor algebra to the spinor calculus of Kaluza-Klein 5D geometry and complexified Maxwell's equations and non-Hertzian phenomena.

- Kaluza-Klein geometry relates the Einstein-Minkowski 4-space of relativity to electromagnetic phenomena and complex 8-space.
- The Einstein-Minkowski 4-space, Kaluza-Klein 5-space, Rauscher 8-space, M-Theory 11-space and Amoroso 12-space relate topological geometries of modern particle physics to quantum theory and general and special relativity.
- These multidimensional models allow a domain to exist in the physical world for the action of local and nonlocal aspects of the reality of the observer.

2.5 Kaluza-Klein Geometries: A Possible Unification of Electro-Magnetic and Gravitational Phenomena

We will present a brief discussion of other multidimensional models and examine the manner in which they may relate to the complex 8D model which was presented in previous sections and in references [1-8]. In the last several decades there has been a great deal of interest in some specific types of 5D and 6D geometries. This revived interest is based on the work of two colleagues of Einstein, who received encouragement from him in the 1930's, Kaluza and Klein, who introduced a 5D covariant geometry which appears to have properties which suggest a method of unifying the electromagnetic, gauge-invariant field theories (Maxwell's equations) and the gravitational field [21,22] (gravitational potential). This particular multidimensional model appears to be useful to examine further because it not only demonstrates the relationship between electromagnetic phenomena and gravitational relativistic phenomena, but it appears to be consistent with the main body of physics [8,18]. The spinor calculus is an excellent framework for accounting for the coupling of the electromagnetic field to the gravitational field in a natural way rather than the usual phenomenological manner [8,37-39]. This approach is automatically accomplished by requiring periodicity of 5D spinor fields. The theory of spinors is used in unifying electromagnetic and gravitational phenomena based on the homomorphism between the group of Lorentz transformations in relativity and the group of unimodular linear transformations in Maxwell's theory [12]. It should be noted that this homomorphism is valid only in the weak Weyl field approximation for the gravitational field [46,47].

In addition to the general coordinate transformations of the four coordinates, x^μ, the preferred coordinate system permutation group is

$$x'^5 = x^5 + f\left(x^1, x^2, x^3, x^4\right). \tag{2.29}$$

Using this condition and the 5D cylindrical metric or $ds^2 = (\gamma_{ik} dx^i dx^k)$ yields the form

$$ds^2 = (dx^5 + \gamma_{\mu 5} dx^\mu)^2 + g_{\mu\nu} dx^\mu dx^\nu \tag{2.30}$$

where the second term is the usual 4-space metric. Greek indices μ, ν run 1 to 4 and Latin indices i, k run 1 to 5.

The quantity $\gamma_{\mu 5}$ in the above equation transforms like a gauge, *ds*

$$\gamma'_{\mu 5} = \gamma_{\mu 5} - \frac{\partial f}{\partial x^\mu}, \tag{2.31}$$

where the function *f* is introduced as an arbitrary function. Returning to our 5D metric form in its five compact form and 4D and 5D form gives

$$\gamma_{\mu\nu} = g_{\mu\nu} + \gamma_{\mu 5}\gamma_{\mu 5}. \tag{2.32}$$

Starting from the metric form in a five "cylindrical" space $ds^2 = \gamma_{ik} dx^i dx^k$ where indices *i,k* run 1 to 5, we introduce the condition of cylindricity that can be described in a coordinate system in which the γ_{ik} are independent of x^5, that is

$$\frac{\partial \gamma_{ik}}{\partial x^5} = 0. \tag{2.33}$$

Also, Kaluza-Klein assumed $\gamma_{55} = 1$ or the positive sign, $\gamma_{55} > 0$ for the condition of the fifth dimension for a 5D space, to ensure that the fifth dimension is metrically space-like [8]. We can also construct a 6D space for $\gamma_{66} = -1$ and $\gamma_{55} < 0$. Geometrically one can interpret x^5 as an angle variable so that all values of x^5 differ by an integral multiple of 2π corresponding to the same point of the 5D space, if the values of the x^μ- are the same. For this specific case, each point of the 5D spaces passes exactly one geodesic curve which returns to the same point. In this case, there always exists a perpendicular coordinate system in which $\gamma_{55} = 1$ and,

$$\frac{\partial \gamma_{5\mu}}{\partial x^5} = 0. \tag{2.34}$$

Other properties follow in which $g_{\mu\nu}$ and γ_{ik} are analogous and $\gamma_{\mu\nu} = g_{\mu\nu}$ and

$$\gamma_{55} = 1 + \gamma^{\mu\nu}\gamma_{\mu 5}\gamma_{\nu 5} \quad \text{also,} \quad \gamma^{\mu 5} = g^{\mu\nu}\gamma_{\nu 5} \tag{2.35}$$

The gauge-like form alone is analogous to the gauge group, which suggests the identification of $\gamma_{\mu 5}$ with the electromagnetic potential ϕ_μ. We can write an expression for an antisymmetric tensor

$$\frac{\partial \gamma_{\mu 5}}{\partial x^{\mu}} - \frac{\partial \gamma_{\mu 5}}{\partial x^{\nu}} = f_{\mu\nu} \tag{2.36}$$

which is an invariant with respect to the gauge transformation.

Using the independence of γ_{ik} of x_5 or $\partial \gamma_{ik} / \partial x^5 = 0$, the geodesics of the metric in five space can be interpreted by the expression

$$\frac{dx^5}{ds} + \gamma_{\mu 5} \frac{dx^{\nu}}{ds} = C \tag{2.37}$$

where C is a constant and s is a distance parameter. If we consider a generalized 5D curvature tensor, and using the form for $f_{\mu\nu}$ we can express it in terms of $f_{\mu\nu}$, the electromagnetic field strength

$$f_{\mu\nu} = \sqrt{\frac{16\pi G}{c4}} F_{\mu\nu} \tag{2.38a}$$

and then we can write

$$\gamma_{\mu 5} = \sqrt{\frac{16\pi G}{c4}} \phi_{\mu} \tag{2.38b}$$

The integration constant above can be identified as proportional to the ratio *e/m* of charge to mass of a particle traveling geodesics in the Kaluza-Klein space [37-39]; c is the velocity of light and G is the universal gravitational constant. The force term, $F = C^4 / G$ is found in Einstein's field equations in the stress energy tensor term and is identified as having cosmological significance by Rauscher [1].

Under specific conditions of the conformal mappings in the complex Minkowski space, one can represent twistors in terms of spinors. The spinor(s) will be said to "represent" the twistor. The twistor is described as a complex two-plane in the complex Minkowski space. References on twistor theory and the spinor calculus are cited in [8,48]. Twistors and spinors can easily be related by the general Lorentz conditions in such a manner as to retain the condition that all signals are luminal in the complex space. The conformal invariance of tensor fields (which can be Hermitian) can be defined in terms of twistors and these fields can be identified with particles. See Chap. 11.

We can represent twistors in terms of a pair of spinors, μ^A and π_A which are said to represent the twistor, $\mathbb{Z}^{\alpha}$. Conditions for this representation are

- the null infinity condition for a zero spin field, $\mathbb{Z}^{\alpha}\overline{\mathbb{Z}}_{\alpha}=0$,
- Conformal invariance, and
- Independence of the origin.

Twistors and spinors are related by the general Lorentz conditions in such a manner as to retain the fact that all signals are luminal in the real 4-space, which does not preclude superluminal signals in an XD space [18].

The twistor is described as a 2-plane in complex Minkowski space, M_4. Twistors define the conformal invariance of the tensor field which can be identified with spin or spinless particles. For particles with spin s we have $\mathbb{Z}^{\alpha}\overline{\mathbb{Z}}_{\alpha}=2s$. The twistor is derived from the imaginary part of the spinor field. The Kerr theorem comes out naturally. It is through the representation of spinors as twistors in complex Minkowski space that we can relate the complex 8-space model to the Kaluza-Klein geometries. In the 5D Kaluza-Klein geometries, the extra dimension, XD was considered to be a spatial rotational dimension in terms of $\gamma_{\mu 5}$.

The Hanson-Newman [33,34] and Rauscher [7-18] complex Minkowski space has introduced with it as an angular momentum, or helix or spiral dimension, called a twistor which is expressed in terms of spinors. We suggest that the problem of closed time-like loops may be resolved in terms of an additional dimension or dimensions which may, in one model, be represented by a helical world line in 5D and 6D space in such a manner that the world line does not collapse on itself and become multi-valued at a single spacetime point [39,44]. Note the twistor relates to the complex Schwarzschild metric yielding the Kerr (rotational) metric [32,45]. The Schwarzschild solution is seen as a "real slice" of a complex Minkowski space [32]. The complex Weyl tensor is viewed as a single complex field on the complex Minkowski space.

Some directions for further exploration of the relationship of our 8D model and the main body of physics may be made through the work of Hansen and Newman [33,34] and Kaluza and Klein [37-39]. Use of the Weyl weak field approximation may be used to examine the complex 8-space and electromagnetic phenomena [40] such as complex electric and magnetic fields which we explore further in the next section. Figure 2.12 presents a schematic of the relationship of some multidimensional geometries.

Basic to the Kaluza-Klein geometry is the series of papers published by Weyl [46,47] in which he forms a generalization of Riemannian geometry claiming to interpret all physical events in terms of gravitation and electromagnetism in terms of a "world metric" (note that this statement is much stronger than the Kaluza-Klein unification scheme, since it excludes strong and weak interactions). See Fig. 2.15. On Fig. 2.16 we present a comparison of some of the multi-dimensional $n > 4D$ geometries and complexification theories and their interconnectedness to each other.

The gauge transformation of the Weyl space is formulated in terms of a quantity, ϕ_μ, rather than $g_{\mu\nu}$ where ϕ_μ is the 4-vector potential. Rather than strict gauge-invariance of $g_{\mu\nu}dx^\mu dx^\nu = 0$ (usually where $\phi_\mu = 0$). In Weyl's theory uses complex wave mechanics for electrically charged matter for the wave function, ψ. Then for a gauge transformation we have

$$\phi_\mu' = \phi_\mu - \frac{i\hbar c}{\varepsilon}\frac{\partial f}{\partial x^\mu} \tag{2.39}$$

where $\psi' = \psi e^{if(x)}$ and where he considers the invariance conditions on the imaginary exponent in ψ instead of the real exponent in $g_{\mu\nu}$. Weyl modified certain inconsistencies that occurred with relativity [46]. The Weyl theory most likely set up the considerations for the Kaluza-Klein model in their attempt to unify gravity and electromagnetism by relating $g_{\mu\nu}$ and ϕ^μ.

P.A.M. Dirac [49] generalized the complex scalar field to a complex two-component field $\binom{\mu}{\nu}$ in order to express the Schrödinger wave equation in a relativistic invariant form The complex two component field is called a spinor because it relates to the spin degrees of freedom that were needed because of the Zeeman spectral splitting in atoms. The question arises; does the spinor field result from the conditions in quantum physics or relativity theory? Using the irreducible representation of the underlying groups in relativity theory, Einstein and Mayer discovered that the real 4D representation in relativity reduces to the direct product of two 2D complex representations. The complex two component functions that are the basis of these representations are the spinor variables that Dirac discovered earlier to describe the electron and anti-electron or positron. Therefore, the spinor variable is the most fundamental expression of the theory of relativistic invariance. In this form then, relativity theory can be quantized as formulated by Dirac.

The hypergeometric Schrödinger equation is second order in space and first order in time. The Klein-Gordon equation is second order in both space and time; whereas the Dirac equation is first order in space and time, which is like a square root of the Klein-Gordon equation that has two solutions. That is, the Dirac equation has both a positive and a negative solution. For other multidimensional perspectives see Fig. 2.13.

Since the spinor invariant is complex, it corresponds to two invariance conditions; one real and the other the imaginary part of the two components of the spinor. We have seen that by introducing a complex Minkowski space we may be able to achieve a reasonable interpretation of some of the apparent paradoxes in quantum physics, relativity theory and electromagnetism. Non-locality and superluminal signal propagation are precisely formulated. We will examine the implications of complex geometries for electromagnetic phenomena, Bell's theorem, and other remote connectedness phenomena in Chap. 4. Also in Chap. 11 we address the relationship of 5R and 6D geometries with spinors, twistors and quaternions.

2.6 Additional Thoughts on Current Physical Theory

The formalism of the complex 8D space and the 5D Kaluza-Klein space are incorporated into the current grand unification theories (GUT), supersymmetry models, with gravity, and string theory (M-Theory where matter is considered to be made of vibrating strings and branes instead of paint like particles), that describes the unifications of the four force fields in particle physics and current models of the universe. The four fundamental forces are the strong nuclear force mediated by quarks, electromagnetic force, weak nuclear decay force and the gravitational force of General Relativity..

The Kaluza-Klein model relates the electromagnetic and gravitational fields in which the photon (spin 1) mediates the electromagnetic field and the graviton (spin 2) mediates the gravitational field. This is why tensor analysis works. The electroweak force of the GUT model is mediated by $W^{\pm}, Z^0$ which are massive bosons for the electromagnetic and weak interactions. The mediators of the strong force are quarks and gluons. It becomes possible to relate the GUT theories (which only related the strong, electromagnetic and weak forces) to gravity via the use of the Kaluza-Klein geometry. These theories attempt to reduce “everything" to quarks and leptons mediated by the exchange of gauge bosons. This is currently termed the standard model.

References

[1] Rauscher, E.A. (1971) A Unifying Theory of Fundamental Processes, UCRL-20808 book, LBNL University of California Press.

[2] Rauscher, E.A. (1972) Closed Cosmological Solutions to Einstein’s Field Equations, Lett. Nuovo Cimento 3, 661.

[3] Rauscher, E.A. (1972) A Set of Generalized Heisenberg Relations and a New Form of Quantization, Lett. Nuovo Cimento 4, 757.

[4] Rauscher, E.A. (1972) A Possible Group Theoretical Representation of the Generalization Heisenberg Relations, Lett. Nuovo Cimento 5, 925.

[5] Rauscher, E.A. (1973) The Minkowski Metric for a Multidimesional Geometry, Lett. Nuovo Cimento 7, 361.

[6] Rauscher, E.A. (2005) Cosmogenises and quantum gravity, in R.L. Amoroso,

B.Lehnert & J.P. Vigier (eds.) Beyond the standard model, pp. 43-72, Oakland: The Noetic Press.
[7] Rauscher, E.A. (1979) Superluminal Transformations in Complex Minkowski Spaces, LBL-9752; and (1980) Found. of Phys. 10, 661.
[8] Rauscher, E.A. (1979) The Iceland Papers: Select Papers On Experimental And Theoretical Research on Physics of Consciousness, Foreward by B. Josephson, and A. Puharich, Amherst: Essentia Research Associates, P.A.C.E.
[9] Rauscher, E.A. (1978) Comples Coordiante Geometries in General Relativity and Electromagnetism, Bull. Amer. Phys. Soc. 23, 84.
[10] Rauscher, E.A. (1981) Conceptual changes in reality from new discoveries in physics, pp. 1407-1431, Proceeding of the 10th International Conferences on the Sciences, Vol. II, New York: ICF Press, E. Wigner comments on Rauscher, p. 1479.
[11] Ramon, C. & Rauscher, E.A. (1987) Remote connectedness in complex geometries, pp. 1423-1477, E.A. Rauscher, chair address pp. 1405-1407, Proceeding of the 11th Intl Conference on the Sciences, Volume II, New York: ICF Press.
[12] Rauscher, E.A. (1983) Electromagnetic Phenomena in Complex Geometries and Nonlinear Phenomena and Non-Hertzian Waves, Millbrae: Tesla Book Co.
[13] Rauscher, E.A. (1983) Electromagnetic and nonlinear phenomena in complex minkowski spaces, paper presented at the University of California, Los Angeles, and Bull. Am. Phys. Soc. 20, 351.
[14] Rauscher, E.A. and Targ, R. (2008) Integration of a complex spacetime and nonlocality, pp. 121-146, D.P. Sheehan (ed.) AIP proceedings 863.
[15] Rauscher, E.A. and Targ, R. (2002) Why only four dimensions will not explain non-locality, Journal of Scientific Explorations 16, 655.
[16] Rauscher, E.A. and Amoroso, R.L. (2009) Relativistic physics in complex Minkowski space, nonlocality, and quantum physics, in M.C. Duffy & J., Levey (eds.) Ether Spacetime and Cosmology, pp. 23-45, Montreal: Aperion.
[17] Rauscher, E.A. & Amoroso, R.L. (2008) The Schrödinger equation in complex Minkowski space, nonlocality and anticipatory systems, in D.M. Dubois (ed.) pp. 370-388, International Jour. Computing Anticipation Systems.
[18] Amoroso, R.L. & Rauscher, E.A. (2009) The Holographic Anthropic Multiverse, Singapore: World Scientific Pub.
[19] Haramein, N. & Rauscher, E.A. (2005) Collective coherent plasma modes in the media surrounding black holes, in R.L. Amoroso, B. Lehnert & J.P. Vigier (eds.) Searching for Unity in Physics, Oakland: The Noetic Press.
[20] Rauscher, E.A. (1968), Electron interactions and quantum plasma physics, J. Plasma Phys. 2, 517.
[21] Rauscher, E.A. and Van Bise, W. (1988) Non-superconducting Apparatus for Detecting Magnetic and Electromagnetic Fields, U.S. Patent, 4,724,390, February 9.
[22] Penrose, P. and Newman, E.J., (1978) Proc. Roy. Soc. 363, 445.
[23] Rauscher, E.A. (2010), NORA.
[24] Rauscher, E.A. and Van Bise, W. (1988) External Magnetic Field Impulse Pacemaker Non-Invasive Method and Apparatus for Modulating Brain Signals Through an External Magnetic or Electric Field to Pace the Heart and Reduce Pain, U.S. Patent Number 4,723,536, issued February 9.
[25] Rauscher, E.A. and Van Bise, W. Non-invasive Method and Apparatus for

Modulating Brain Signals Through an External Magnetic or Electric Field to Reduce Pain, U.S. Patent Number 4,889,526, issued December 26, 1989.
[26] Rauscher, E.A. and Van Bise, W. (1988) Magnetic field interaction with macro biological systems with applications to effects on physiology, in J.M. Srinivasan (ed.) Medicine Around the World, pp. 205-218, Phoenix: Gabrial Press, Print World, Nov., Proceedings from conference at the Madras Institute of Magnetobiology 1987).
[27] Amoroso, R.L. (2009), Universal quantum computing: anticipatory parameters predicting bulk implementation, in D.M. Dubois (ed.) pp. 254-282, Liege: IJCAS.
[28] Bell, J.S. (1964), Physics 1, 195.
[29] Aharonov, Y. and Bohm, D. (1959) Phys. Rev. 115,485.
[30] Silver Threads: 25 Years of Parapsychology Research (1993), eds. B. Kane, J. Millary and D. Brown and W. Harman Araeges, Conn.
[31] Prigogine, A. (1993) Chaotic Dynamics and Transport in Fluids and Plasmas: Research Trends in Physics Series, New York: American Institute of Physics.
[32] Rauscher, E.A. (1973) Superconductivity and Superfluidity in Multidimensional Geometries, LBNL, LBL-1572.
[33] Hansen, R.O. and Newman, E.T. (1975) Gen. Rel. and Grav. 6, 216.
[34] Newman, E.T. (1973) J. Math. Phys. 14, 774.
[35] Tipler, F.J. (1974) Phys. Rev. D9, 2203.
[36] B.S. DeWitt and N. Graham (eds.) (1973) The Many Worlds Interpretation of Quantum Mechanics, with papers by H,. Everett, N. Graham and J.A. Wheeler, Princeton: Pinceton Univ. Press.
[37] Kaluza, Th., (1921) *sitz. Berlin Press, A. Kad. Wis*. 1968.
[38] Klein, O. (1926) *Z. Phys*. 37, 805.
[39] Klein, O. (1927) *Z. Phys*. 41, 407.
[40] Newton, R.G. (1967) Phys. Rev. 162,1274.
[41] Feinberg, G. (1967) Phys. Rev. 159, 1089; Feinberg, G. (1970) Particles that go faster than light, Sci Am, 222, p. 68.
[42] Benford, G.A., Book, D.LO. and Newcomb, (1970), Tachyonic anti-telephone, Phys. Rev. D2, 263.
[43] Rauscher, E.A. and Amoroso, R.L. (2006) The physical implications of multidimensional geometries and measurement, pp. 161-177, in D.M. Dubois (ed.) Internatiobnal Journal of Computing Anticipation Systems 19.
[44] Bilaniuk, O.M.P, Sundersham, E.C.G, & Deshpande, V.K. (1962) Am J Physics, 30, 718.
[45] Chew, G. (1964) Analytic S-Matix, Benjamin Pub.
[46] Weyl, H. (1918), Math. Z. 20, 384.
[47] Weyl, H. (1957), Space-Time-Matter, Dover Pub.
[48] Penrose, R. (1967) J. Math. Phys. 8, 345.
[49] Dirac, P.A.M. (1928) The Quantum Theory of the Electron, Proceedings of the Royal Society of London, Series A, 117, 778, pp. 610-624.
[50] de Sitter, W. (1916) Monthly Notices, RAS 77,155.
[51] Evan, G.T. and Sen, D.K. (1973) J. Math. Phys. 14, 1668.
[52] Stratton, T.A. (1941) Electromagnetic Theory, McGraw Hill.
[53] Wyler, A. (1969) Acad. Sc. Paris, Series A, 269, 743.
[54] Wyler, A. (1971) Acad. Sc. Paris, Series A, 271, 186.

Chapter 3

Major Principles of Physics: Poincaré Invariance, Analyticity, Unitarity and Complex Minkowski Space

The principles of modern physics can be stated in terms of Poincaré invariance, or the homogeneity of spacetime analyzability or causality and unitarity or the conservation of probability. Essentially all theories of physics must obey these principles.

3.1 Major Principles of Physics

Major progress in physics was made with the realization of such principles of energy and mass conservation, cause – effect relations and the isotropy and homogeneity of spacetime. The concept of Lorentz invariance in which where and when i.e. in which coordinate system an experiment is conducted does not alter the laws of physics that the system obeys. In this chapter, we examine the major principles of physics and the manner in which they apply to the structure of the complexification of Minkowski 4-space [1]. Three major universal principles are used to determine the structure and nature of physical laws. These are Poincaré invariance and its corollary Lorentz invariance (which expresses the spacetime independence of scientific laws) [2-4] analyticity (which is a general statement of causality), and unitarity (which can be related to the conservation of physical qualities). These principles can apply to macroscopic as well as microscopic phenomena. Poincaré invariance has implications for both macroscopic and microscopic phenomena and unitarity is a condition on the wave function description in quantum physics. The quantum description of elementary particle physics has led to a detailed formation of the analyticity principle in the complex momentum plane. [5-9]

In Table 3.1 we list (top row) the major principles of physics, (second row) a brief statement of physical phenomena related to these principles, and (third row) the aspect of the theoretical model that applies to a particular category of remote, nonlocal phenomena. We illustrate the three principles of physics with brief explanations and with specific physical models such as Bell's theorem, complex coordinate model and the physics of vacuum state polarization. We also present a diagrammatic map of the relationships between the major principles of physics and nonlocality, anticipatory and complex multidimensional geometries. These geometries are fundamental to physics and to describing spacetime attributes on the manifold, (Fig. 3.1).

Table 3.1. The Principles of Physics and Their Suggested Relationship to Remote Connectedness Phenomenon

PRINCIPLE	Poincaré invariance	Analyticity	Unitarity
BRIEF STATEMENT OF THE PRINCIPLE	Homogeneity of spacetime	Causality	Conservation of probability
THEORY RELATED TO THE PRINCIPLE	Bell's theorem	Complex hyperdimensional geometry	Vacuum state polarization, Dirac states
REMOTE SPACETIME PHENOMENON	Non-Locality	Anticipation	Action at a distance

We suggest relationships of these principles to nonlocal, anticipatory systems [9-11]. We give details as to the manner in which we can utilize these physical theories to accommodate nonlocality at the macro as well as micro levels. In particular we consider a multidimensional geometrical model which appears to reconcile nonlocality and anticipation and causality in a self-consistent theoretical framework. Complex physical variables which can be tested for their consistency with the main body of physics also may demonstrate a fundamental relationship between relativity and quantum and electromagnetic phenomena. We also demonstrate that nonlocal and anticipatory phenomena is not denied by, but is compatible with Poincaré invariance, and the other major principles of the foundation of physics.

Poincaré invariance is the statement of the independence of physical laws and generalized coordinate transformations. The Poincaré invariance of the energy Hamiltonian implies the conservation of energy. Bell's inequality nonlocal interactions do not violate Poincaré or Lorentz invariance. In remote effects at a distance are allowed where not only information is distantly correlated but apparent energy or physical effects are transmitted, this could affect the Hamiltonian, which would no longer be Poincaré invariant. Local energy state changes from distant correlated informational events may act through the virtual Fermi-Dirac vacuum polarization and may conserve energy or unitarity and Poincaré invariance and analyticity. The analytic S-matrix can be seen as a matrix valued generalization of the Schrödinger probability amplitude, $\Psi^*\Psi = |\Psi|^2$, which is complexified but yields real measurable values. The zero energy analytic S-matrix can be formulated in terms of the Feynman formalism to the category of operators. The physically motivated hypothesis is that S has an expression $S = \sqrt{\rho}\, S_0$, such that S_0 is a universal unitary S-matrix and $\sqrt{\rho}$ is the square root of the state dependent density matrix. The S-matrix can be identified as a "square root" of the positive energy density matrix $S = \rho_x{}^{1/2} S_0$ where S_0 is a unitary matrix and ρ_x is the density matrix for positive energy part as the zero energy state. Then $SS^+ = \rho_+$ and $S^+S = \rho_-$ which gives the density matrix for the negative part of zero energy state. It is obvious that the S-matrix can be interpreted as a matrix valued generalization of the Schrödinger amplitude. The indices of the S-matrix correspond to configuration space spinors addressed in Chap. 11. The S-matrix is strongly associated to unitarity and the conservation of angular momentum energy and relevant quantum numbers such as charge, spin, etc. [9].

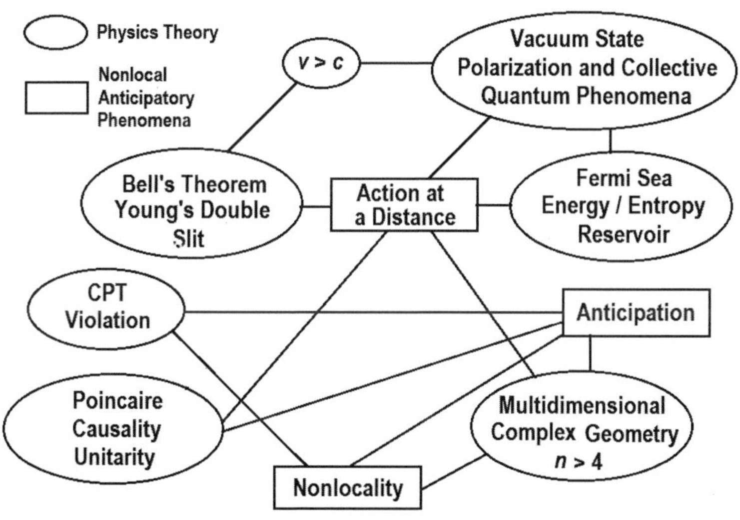

Figure 3.1. "Map" of Physics and the relationship to existing physical theory that accommodates the fundamental principles of nonlocal events in spacetime.

As applied to S-Matrix theory Poincaré theorem tells us that if a parameter of a differential equation such as ℓ or k appears only in functions which are holomorphic in some domain of the parameter, and if in some other domain, a solution of the equation is defined by a boundary condition which is independent of the parameter then this solution is holomorphic as a function of the parameter in the intersection of the two domains. Such parameters can be ℓ and k. In S-matrix theory, Argand plots in complex ℓ space where poles correspond to resonances or particles where ℓ is an angular momentum parameter. Also plots can be constructed in a complex energy space associated with the parameter, k, as $k = 1/\lambda$, the wave number and $p = \hbar/\lambda$, $p = mv = mc$, $E = mc^2$ so that $k \propto m \propto E$ for $c = 1$. These parameters are the independent variable of the differential equation which are hypergeometric and in non-relativistic form reduce to a time independent Schrödinger-like wave equation.

The ℓ, k variables are analogous to phase space variable (p,q) in momentum – distance. Other sets of complementary variables (E, t) and also others such as (p, E) and (x,t). See [12] on the generalization of the Heisenberg relations. Not the variable (p, E) act independent variables in the Lippman-Schwinger equation, which has an analogy to the Schrödinger equation. The independent variable (x,t) are those of the Schrödinger equation and most equations of physics.

An anticipatory system has the information, known and defined in the presence, to make an inference and discussion about the next action or inaction to be taken and hence, to make a change in the present to change the next or future states based on the predictions and fore knowledge about the relevant potential future states. Anticipation or "precognition" or to cognize a future even before the now on the light cone axis cannot be explained by superluminal signaling in 4-spacetime alone [10,11].

Tachyons or a superluminal signal alone will not explain anticipation precognition [13]. Feinberg states that tachyonic signals even at near the velocity of light will net one only a few nanoseconds/foot into the future on the light cone. If we choose a null light cone signal of v~c and for ~3 x 10^{10}cm/sec, then 1/c ~ 1/3 x 10^{-10} sec/cm and a nanosecond = 10^{-9} sec so that 1/c nanosecond x 1/30 cm for 2.54 cm = 1 inch, then 2.54 cm x 12^{11}/ft = 30.48 cm/ft or 30.48 cm = 1 ft ~ 30 cm therefore 1/c = 1 nanosecond/inch. In our consideration of anticipatory responses require the consideration of significant temporal advantages perhaps even hours. In the Gisin [14] test of Bell's theorem over km of distance [15-22]. One nanosecond / inch ≅ 3 nanosecond/cm and 10 km ~ 10^6 cm, then the Gisin experimental results require a factor of over 10^9 times over the 1 nanosecond/inch (for signaling transmission of the velocity of light)! If the time delay between the initial anticipation time at t_o and the verified result of anticipation or participation was t_1 then, for tachyonic signaling in $n = 4$ space would yield a requirement for a spatial separation of the events at t_0 and t_1 of 10^9 miles or greater (or about 10^{14} cm).

In order to accommodate precognition, anticipation or the results of Bell's theorem, one is required to address the issue and resolve the paradox by using on $n > 4$D space. As we stated before, the use of complex 8-space has the symmetry properties to satisfy the major principles of physics. The geometric approach to accommodate nonlocality is very consistent with Wheeler's statements that our understanding of physics will "come from the geometry, and not from the fields [23].

Hypothesis about the manner such anticipatory systems can exist are:

- An advanced wave, such as the Tachyon proposed by Feinberg [13]
- Heisenberg's quantum wave potentia model [24]
- Electromagnetic advanced and retarded waves [25]
- Cramer's advanced- Schrödinger retarded standing-wave transactional analysis [26].

In remote connected events, such as in Bell's theorem, the remote collapse of a wave function at one spacetime location Ψ_1 (x_1,t_1) determines the measured state collapse of the other spacetime, location Ψ_2 (x_2,t_2). In temporal separations of anticipatory systems between an initial event at t_1 as $\Psi_1(x_n,t_1)$ determines the state collapse of the wave function at t_2 for $\Psi_2(x_n,t_2)$. Note that for Ψ_1 and Ψ_2, x_n can have either both wave functions at $x_n = x_1$ or for both wave functions Ψ_1 at $x_n = x_1$ and Ψ_2 at $x_n = x_2$ where $\Delta x = x_2 - x_1$ can be an arbitrary spatial separation. Also temporal separation or anticipatory nonlocality occurs, which we reconcile in Chap. 2 and the following chapter [26].

References

[1] Minkowski, H. (1909) *Physikalische Zeitschrift* 10, 104, Lecture on space and time, Cologne, 21 September 1908, translated in Lorentz, Einstein, Minkowski and Weyl, the Principle of Relativity (1923) New York: Dover.
[2] Einstein, A., Lorentz, H.A.,Weyl, H. & Minkowski, H. (1923) The Principles of Relativity, New York: Dover.
[3] Lorentz, H.A. (1952) Theory of Electrons, New York: Dover.
[4] Einstein, A. (1905) *Annalen der Physik* 17, 891.
[5] Chew, G. (1964) The Analytic S-Matrix Frontiers of Physics, Benjamin; and Private communication, University of California, Berkeley.

[6] Rauscher, E.A. (1979) S-Matrix Theory and elastic and inelastic coupled channel alpha scattering theory and resonance decay widths, Lawrence Berkeley National Laboratory Report, LBL-8577.
[7] Rauscher, E.A. & Bruch, R. (1991) S-Matrix Approach to Fast – Projectile – Target Many Body Physics, University of Nevada, Reno, Department of Physics Accelerator group Report.
[8] Rauscher, E.A. (1994) An S-Matrix theory of Alpha's Decay, APS, Bull. Am. Phys. Soc., Nuclear Physics Meeting, Williamsberg, VA.
[9] Bruch, R., Rauscher, E.A., Fuelling, S., Schneider, D., Mannervik, S. & Larson, M. (1994) Collision processes of molecules and atoms, L. Byass (ed.) Encyclopedia of Applied Physics, AIP, 10, 437-470.
[10] Rauscher, E.A. & Amoroso, R.L. (2006) The physical implications of multidimensional geometries and measurement, Intl J. Comp. Anticipatory Sys, in D.M. Dubois, (ed.) CHA0S, Liege, Belgium, 19, pp. 169-177.
[11] Amoroso, R.L. & Rauscher, E.A. (2008) Emergence of generalized F-theory 2-branes from SUSY spacetime parameters of the discrete incursive oscillator, CASYS, Intl. J. Computing Anticipatory Systems 22, 283.
[12] Rauscher, E.A. (1979) Some physical models potentially applicable to nonlocality, in A. Puharich and B.D. Josephson (eds.) pp. 49-93, Essentia Research Associates reprint (1996) Ottawa: PACE Press.
[13] Feinberg, G. private communication, 1977.
[14] Gesin, N., Tittel, W., Brendel, J. & Zbinden, H. (1998) Phys. Rev. Lett. 81, 3563.
[15] Bell, J.S. (1964) Physics; 1, 195.
[16] Einstein, A. Podolsky, B. & Rosen, N. (1935) Phys. Rev. 47, 777.
[17] Bohm, D. (1952) Phys. Rev. 85, 166 and 180; and private communications, Birkbeck College, University of London, November 1977.
[18] Rauscher, E.A. (1977) Presentation: Univ. London, Nov 23, Physics Dep.
[19] Clauser, J.F. & Horne, W.A. (1971) Phys. Rev. 10D, 526 (1974), and private communication with J. Clauser and his presentation September 19, 1975 at the LBNL Fundamental "Fysiks" Group.
[20] Rauscher, E.A. (1978) Bull. Am Phys. Soc. 23, 84.
[21] Stapp, H. (1972) Theory of Reality, LBL-3837, April 29.
[22] Rauscher, E.A. (1976) Bull. Am. Phys. Soc. 21, 1305.
[23] Wheeler, J.A. (1978) private communication Austin, TX with EAR.
[24] Heisenberg, W. private communication with EAR.
[25] Ramon, C. & Rauscher, E.A. (1982) Remote connectedness in complex geometries and its application to the nature of the human mind, pp. 1423-1442, New York: ICF Publishing.
[25] Cramer, J.B. (1980) Phys. Rev. D22, 362.
[26] Rosen, R. (1985) Anticipatory Systems, New York: Pergamon Press.

Chapter 4

Nonlocal Interconnectedness as a Fundamental Principle of Reality

The world this appears as a complicated tissue of events in which connections of different kinds alternate or overlap or combine and thereby determine the texture of the whole - W. Heisenberg, 1938.

Two significant principles of the 20th century are the Heisenberg uncertainty principle and the nonlocality principles of the Einstein, Podolsky, Rosen paradox [1]. Both these elements of quantum theory have major physical and philosophical implication like Bell's Theorem and Space-Like Interconnectedness and Other Collective Coherent Phenomena Involve nonlocality. We also present a discussion of the EPR paradox and other experiments that may demand a nonlocal explanation of the phenomenon they display.

4.1 Bell's Theorem and Its Experimental Verification

One of the most significant theorem about the nature of physical systems is J.S. Bell's [2,3] formulation of the Einstein, Podolsky and Rosen (EPR) "completeness" formulation of quantum mechanics [1]. The EPR paper was written in response to Bohr's proposal the noncommuting operators which led to the Heisenberg uncertainty principle. This non Abelian algebra is said to comprise a complete theory of reality, at least at the quantum domain; this is the Copenhagen view. Einstein abhorred the uncertainty principle stating "God does not play dice with the universe". His vision was to determine the position and momentum of each particle in the universe and a unified field theory of the four force fields and explain all of reality. The hitch in this plan was that neither position-momentum or energy-time could not be exactly localized simultaneously by the Uncertainty Principle, $\Delta \times \Delta p \geq \hbar$ and $\Delta E \Delta t \geq \hbar$ [6]. Heisenberg's principle places restrictions on the absolute knowledge of the universe. The TOE (Theory of Everything) still holds the vision of Einstein's final theory of unification which, in Weinberg's view is, in a sense, a view of an absolutely complete final theory of everything [7].

Einstein, in his EPR paper defined a complete theory as one in which every element of the theory corresponds to an element of "reality" that is, for example, through the quantum principle, for every election, etc. there is an assignable wave future, Ψe. If the completeness principle holds, then the principle of nonlocality pervades the quantum world. The concept of non-locality was not well received by the physics community. Bohm introduced additional quantum non-observable variable or "hidden variables" in order to make the EPR quantum Bell quantified the EPR statement [8,9] and demonstrated mathematically that locality is incompatible with the statistical predictions of quantum mechanics.

The locality or separability assumption states that the result of a measurement on one system is unaffected by operations on a distant system with which it may have previously interacted or had become correlated, that is a lack of quantum entanglement. Bell states that "no theory of reality, compatible with quantum theory can require spatially separate events to be independent". That is, the measurement in the Clauser et al. experiment, of the polarization of one photon determines the polarization of the other photon at its respective measurement site. Bell discusses a specific experiment, Stern-Gerlach measurements of two spin one-half particles in the singlet spin state moving freely in opposite directions. If the spins are called s_1 and s_2 we can make our component spin measurements remote from each other at position (1) and (2), such that the Stern-Gerlach magnet at (1) does not affect another one at (2) and vice versa. Since we can predict, in advance, the result of measuring any chosen component of s_2 at (2) by previously measuring the same component spin of s_1 and (1), this implies that the result of the second measurement must actually be predetermined by the result at the first (1) remote from (2) measurement. In Bell's proof, he introduces a more complete specification of the parameters of a system by introducing parameters which in essence are hidden variables. Bell's proof is most eloquent and clear. He calculates the conditions on the correlation function for measurements at (1) and (2), as an inequality [9].

Bell's precise statement in his theorem made it possible for Clauser and Horne [10] to test the predicted statistical distribution of quantum processes and demonstrate a laboratory instance of quantum connectedness or nonlocality. Indeed, in Clauser's two photon system for spin 1 particles, two photodetectors remote from each other are each preceded by independent, randomly-oriented polarizers. The statistical predictions of quantum mechanics is borne out in the measurements made at the two photomultiplyer tubes (PMT); see Fig. 4.1.

In Bell's words "there must be a mechanism whereby the setting of one measuring device can influence the reading of another instrument, however remote" (they remain quantum mechanically entangled). Moreover the signal involved must propagate instantaneously so that a theory could not be Lorentz invariant. Lorentz invariance in the usual sense, implies $v \leq c$ [11]. Feinberg [12] discusses the relationship between Lorentz invariance and superluminal signals which he found not to be incompatible. It is not clear that superluminal signals must be invoked to derive Bell's theorem [12,14] but we believe that Bell's theorem demands $v > c$ or simultaneity.

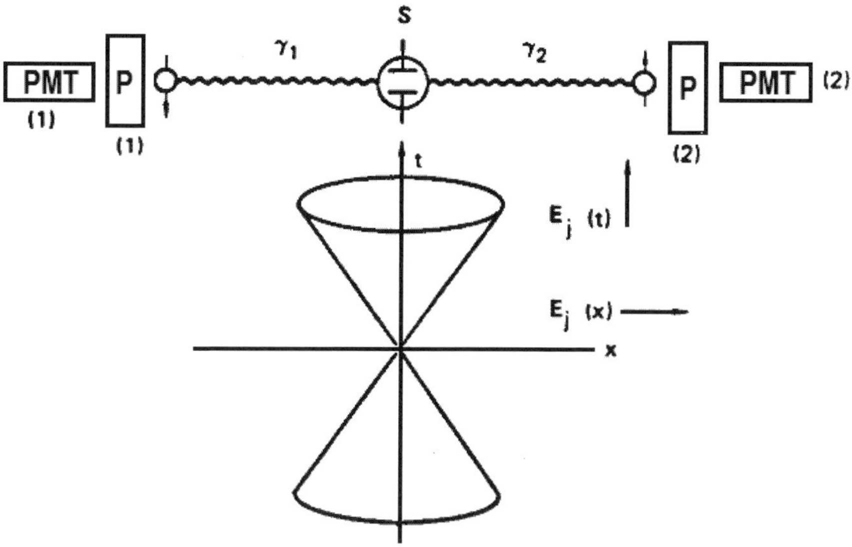

Figure 4.1. Schematic Diagram of the Design of the Clauser Bell's Theorem Correlation Function Experiment: The two detectors at positions (1) and (2) are Photomultiplier tubes (PMT) and P(1) and P(2) and polarizers for photons, Υ_1 and Υ_2 produced by the laser-stimulated radiative atomic cascade of a Calcium source, *S* that emits entangled photon pairs. The detectors of photon polarization at (1) and (2) appear to be outside each other's light cones; events $E_j(t)$ are purely time-like and events $E_k(x)$ are purely space-like.

Then the conclusion from Bell's theorem is that any hidden variable theory that reproduces all statistical predictions of quantum mechanics must be nonlocal, which implies remote connectedness. Of course thus far all these formulations involve microproperties only, but recent formulations seem to imply possible macroscopic consequences of Bell's theorem as well. It is believed that the key lies in formulating the correlation function representing the interconnectedness of previously correlated events, see Fig. 4.1.

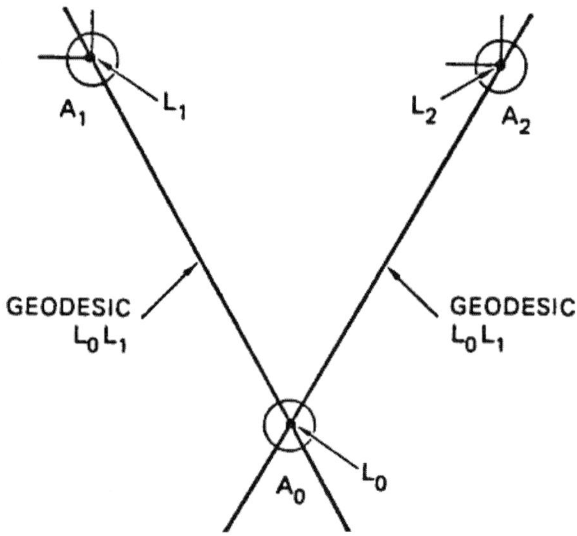

Figure 4.2. Common Point of Origin of Two Events Connected by a Light Signal: H. Stapp represents the non-local connection of two events E_1 at A_1 and E_2 at A_2 as connected by geodesics to a prior event E_0 at A_0.

Bell's theorem is formulated only in terms of a microscopic spin correlation function, usually for photons (Bosons) or electrons (Fermions). There may be some macroscopic non quantum remote correlated effects. A number of remote macroscopic effects do occur over kilometer distances for the Bell's theorem experiment discussed in Section 4.2. Non-quantum remote correlations exist. One such example of a Bell's theorem-like correlation is the Brown-Twiss effect [15] which involve long distance correlation. Stapp recently has expanded the pragmatic view of Bell's theorem and discusses the role of the macroscopic detection apparatus as well as the possible role of superluminal signals. He explores both cases for superluminal propagation or subluminal connection issuing from the points in common to the backward light cones coming from the two regions in Fig. 4.2.

We can write a general correlation function $C(\Theta)$ for example for an angle Θ between polarization vectors in two polarizers as $C(\Theta) = (1/2 + 1/2)\cos 2\Theta = \cos^2 \Theta$ for Clauser's experiment, or for odd integers we can write $nC(\Theta) - C(n\Theta) - (n-1) \leq 0$ which is Bell's inequality, specifically for $n = 3$ then $3C(\Theta) - C(3\Theta) - 2 \leq 0$. We can write in general $C(\Theta) = (1/2) + g\cos 2\Theta$ where g is determined by the particular experiment under consideration. See Fig. 4.3. The magnitude of correlation function constant, g relates to the type of non-local correlation experiment. For $g = ½$, we have the Bell's theorem photon-photon correlation. For $g \sim 0.25$ is the value of g related to the Furry experiment and $g \sim 0.15$ is the value of g related to the Brown-Twiss experiment. Both of these latter experiments relate to macroscopic correlation [16,17]. For example, the Brown-Twiss effect involves the macroscopic process of the small angle subtended in observing light from distant stars producing parallel rays of light such that their wave fronts are linear which is able to be described as a coherence function or correlation function. Although the photons appear correlated one cannot use the Brown-Twiss effect to demonstrate nonlocality. For $0.361 \leq g \leq 0.5$ (the shadowed region in Fig. 4.3) we have the only region in which one can experimentally demonstrate nonlocality [17]. It is important to note that the macrosystem phenomena of the Brown-Twiss effect and Furry experiment lie outside this region. It may be possible to calculate a macroscopic correlation function in a framework which will allow us to test nonlocality.

4.2 More Recent Long Distance Confirmations of Bell's Nonlocality

The physics of nonlocality has been repeatedly verified even over hundreds of kilometers. This research verifying nonlocality covers the period from 1971 to 1998 when one of us (EAR) met John Clauser at a meeting with David Bohm in the 1970's at Berkeley LBNL, at University of London Birkbeck College meetings, with Alan Aspect at Orsay, France and the more recent long distance measurements of Gisin, et al. in Italy.

In the Clauser experiments the position of the polarizers are set before the photons leave their source to reach the photomultiplyers. Aspect added a delayed chaise component to the experiment in which the polarizers are randomly set after the two photons leave the source. The photon's spin remain correlated in both cases. One of us (EAR) observed both experimental set ups and was extremely impressed with the Clauser and Aspect experimental designs and implementations. This most exciting research in current quantum physics is the investigation of what Bohm calls quantum-interconnectedness or nonlocal correlations. As we stated, first proposed by Einstein, Podolsky, and Rosen (EPR) in 1935, as evidence of a defect in quantum theory, and later formulated as a mathematical proof by Bell. It has now been repeatedly experimentally demonstrated that two quanta of light emitted from a single source, and traveling at the speed of light, in opposite directions maintain their connection to one another, so that each photon is affected by what happens to its "twin" many kilometers away, (Aspect et al.; [18] Bell, [3]; Friedman & Clauser [19]; Gisin et al. [20,21]).

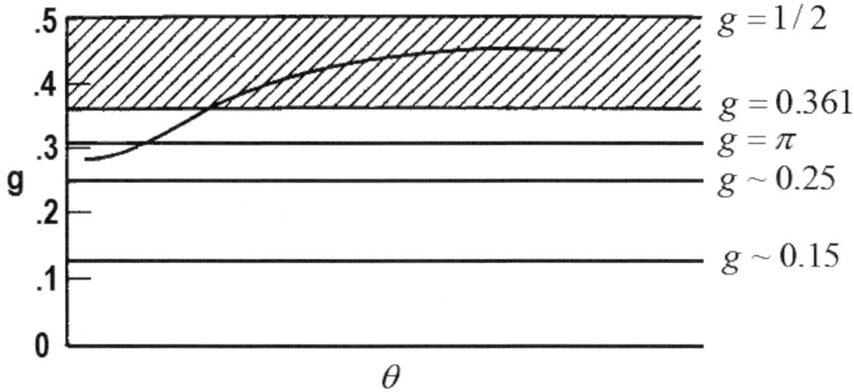

Figure 4.3. Relationship of the correlation function constant in various nonlocal correlation experiments. The curved line represents a plot of $g(\Theta)$ for n odd in the Bell's theorem correlation function $C(\Theta)$.

Clauser recently described his impressions of these nonlocality experiments to EAR. He said that quantum experiments have been carried out with photons, electrons, atoms, and even 60-carbon-atom Buckyballs. He said that "it may be impossible to keep anything in a box anymore." Bell emphasizes, "no theory of reality compatible with quantum theory can require spatially separate events to be independent." This is to say, the measurement of the polarization of one photon determines the polarization of the other photon at their respective measurement sites. This surprising coherence between distant entities is fundamental to the basis of nonlocality which is a property of both space and time. In writing on the philosophical implications of nonlocality, Stapp at LBNL and the Fundamental "Fysiks" Group states that these quantum connections could be the "most profound discovery in all of science" [9,22].

Bohm argues that we greatly misunderstand the illusion of separation in space and time. In his physics book, The Undivided Universe (Bohm and Hiley [23]), he discusses this illusion as he writes about the quantum-interconnectedness of all things. Bohm says "The essential features of the implicate order are, that the whole universe is in some way enfolded in everything, and that each thing is enfolded in the whole." This is the fundamental statement of the metaphor of the holographic ordering of the universe. It says that, like a hologram, each region of spacetime contains information about every other point in spacetime. This model was inspired by the indications of nonlocality in Bell's theorem. Bohm continues:

...all of this implies a thorough going wholeness, in which mental and physical sides participate very closely in each other. Likewise, intellect, emotion, and the whole state of the body are in a similar flux of fundamental participation. Thus, there is no real division between mind and matter, *psyche* and *soma*. The common term psychosomatic is in this way seen to be misleading, as it suggests the Cartesian notion of two distinct substances in some kind of interaction.

In the holographic universe of Bohm, there is a unity of consciousness, a "greater collective mind," with no boundaries of space or time. Bohm goes on to describe the famous Wheeler delayed choice experiment. He writes that experiments "can be designed to show that, according to quantum theory, the choice to measure one or another of a pair of complementary variables at a given time can apparently affect the physical state of things for considerable periods of time before such a decision is made". Such complementary variables are typically momentum and distance, or phase space variables, or in Wheeler's experiment they refer to the dual wave and particle nature of light, as observed in a two slit interference apparatus [24]. We discuss the design of Wheeler's developed chase experiment in Section 4.3.2. The Bell's theorem correlation of distant events and the principle of nonlocality is one of many forms of nonlocal interaction. See Chap. 2. It is clear that this principle of nonlocality has profound implications about the nature of nonlocality. The fundamental nature of nonlocality supersedes either just microscopic or the macroscopic phenomena and may occupy one point of commonality. The cover space, of which the quantum domain is a subset, is expressed in a complex Minkowski 8D and 12D space. In Section 4.2 we describe some of the possible implications and interpretations of Bell's theorem and its verification.

Bohm and Hiley express their assessment of the fundamental nature of reality based upon nonlocality as an acting principle of the universe. They state the following: "Our attitude is that we can sooner or later drop the notion of the quantum potential (as we can drop the scaffolding when a building is ready) and go on to radically new concepts, which incorporate the wholeness of form which we feel to be the essential significance of quantum descriptions. This implies that we have to go deeply into all our basic notions of space, time, and the nature of matter, which are at present inseparably intertwined with the idea of localizability, i.e. that the basic form of existence is that of entities that are located in well-defined regions of space and time. We have instead to start from nonlocality as the basic concept, and to obtain locality as a special and limiting case, applicable when there is relative functional independence of the various "elements" appearing in our descriptions. This means that our notions of space and time will have to change in a fundamental way [24]. The complex 8-space, see Chap. 2, is intrinsically a nonlocal spacetime geometry. Locality becomes a condensed approximation to an exact complete nonlocality.

4.3 Implications of Bell's Nonlocality Theorem

In this section, we explore some of the physical interpretations of Bell's theorem as well as the ontological and epistemological, philosophical and possible metaphysical implications of the theorem. The experimental verification of nonlocality and hence the completeness of the quantum theory leads to the conclusion of the fundamental existence of nonlocal interactions. In this sense is there a super-wave function, Ψ that was the origin of quantum entanglement at the Big Bang? Did this Ψ function lead to everything remaining correlated throughout cosmic evolution? In [22], Stapp and others discuss current physical theory and nonlocality. He states that "...the universal on a very basic level could be a vast web of particles, which remain in contact with one another over any distance and in no time".

The Fundamental "Fysiks" Group also called the Fundamental Physics Group, was started, organized and chaired by E.A. Rauscher for three years at LBNL [9,25]. Stapp stated in the F"F"G that the confirmation of the nonlocality of Bell's theorem is one of the most fundamental discoveries of the 20th century along with the Heisenberg Uncertainty Principle [6,22,26].

4.4 Conceptual and Philosophical Implications of Bell's Theorem

4.4.1 *Bell's Theorem*

If the statistical predictions of the quantum theory are correct, then principle of local causes is false. A Tacit assumption is that the photon counter efficiencies are not limited, in principle [25].

4.4.2 *Principle of Local Causes*

What happens in a spacetime region, "A" does not depend on variables subject to the control of an experimenter in distant space-like-separated region, "B". Stapp term this contra-factual definiteness and in addition to locality, CFD involves assigning reality to the quantum state wave function, Ψ, whether it is the state measured and observed or not [25]. See Fig. 2.1 in Chap. 2.

4.4.3 *Some Possible Conclusions About Bell's Theorem*

- Counter efficiencies are limited in principle.
- Statistical prediction of the quantum theory is not always correct.
- Pragmatic Philosophy termed the Copenhagen View of Quantum mechanism should be accepted according to Clauser [27]. We should concern ourselves with relationships between observations and practicality and not with models of external reality.

Arguments for the Copenhagen view

- Limitation on the Mind of Man: Our minds are probably geared to the problem of human survival by forming expectations about future experiences on the basis of past ones. (Pragmatist/Mechanist)
- Utility: To be useful science should concern itself with only experimental consequences.
- Verifiability: We can know the "truth" only through experiments. (Wheeler, "practical ontology" [28].)

4.4.4 *Contra-Factual Definiteness Fails*

The concept "does not depend on hidden variable..." used in theories which involve "contra-factual definiteness," the assumption that what would have happened if the experimenter had done something that he in fact did not do, is assumed to have some definite state which is an unknowable thing (epistemology) [29].

- Does quantum philosophy rule out contra-factual definiteness? In the double-slit and similar interference experiments "quantum philosophy" Copenhagen View dictates that we not think simultaneously about "what did happen and what would have happened" if some alternative experiment had been performed. However, quantum philosophy denies neither that the experimenter could have conducted another experiment, to quote Bohr: "...our possibilities of handling the measuring instruments allow us to make a choice between the different complementary types of phenomena we want to study" nor that "the other experiment would have had some definite result if it had been performed." It denies, rather, the metaphysical interpretation that the particle always goes definitely through one slit or the other. Pragmatic quantum philosophy yields economy that is particles and waves become "unified" ...but at a price: no description or reality is then possible from this approach, hence the quantum theory says nothing fundamental about the nature of reality.

- Models of Reality that violate contra-factual definiteness consideration of models of reality encompassing quantum phenomena is contrary to the "wisdom of elders of quantum theory", i.e. the Copenhagen View [29].

- One possible world: hence no hidden variable as Bohm hypothesized [30]. Ordinarily one thinks that either the experimenter has a choice, or if he/she has no choice, i.e. if everything is strictly deterministic then at least one can conceive of a world in which the "other" possible experiments were performed. Bell's theorem then implies that it is not possible to even conceive of these other worlds, if they are required to conform to quantum theory and the results in "A" not "B" do not depend on which experiment is conducted in "B" not "A".

- All possible worlds exist: via Everett-Graham-Wheeler (EGW) [31]. At each experiment, Ψ_i , the world breaks into, for example, 16 different worlds, each with an appropriate "weight" (this model is suggested by the Everett-Graham-Wheeler many-world interpretation of quantum theory. Note: Wheeler told this author (EAR) in 1978 that he no longer subscribes to the EGW model.

4.4.5 *Possible Interpretations of the Wave Function,* Ψ

- The wave function represents one to one mapping to the real world or to the probabilities of possible states of occurances in the world for the Schrödinger interpretation of quantum mechanics, $\Psi^*\Psi$ or the probability as a fictional mathematical symbol such as in classical statistical or stochastic mechanisms.

- Quasi-Real Potential model of Heisenberg represented the possibilities of what could happen [26,32].

4.4.6 *Objections to the Reality of Quantum Theory*

- Which represents the mathematical properties of probability function.
- The idea that the wave function, Ψ represents reality originates in

misinterpretation of Copenhagen claim of completeness.

- There is no fundamental relativistic form Ψ outside of the Dirac equation.
- Chew put forward the concepts approximate completeness and objectivity [33].

4.4.7 *Locality Fails*

- Nonlocal collapse of the (real) or actual wave function.
- Collective coherent nonlinear term in the quantum wave equation [34].
- Psychokinetic Effects and telepathy (*Gedankenbertrangurg*) [32].
- Continuous Nonlocal Reality (Problem of time and space ordering and the nature and properties of causality).
- Discrete Nonlocal Reality or the Theory of Events via Stapp, for example [25].
- Bell's theorem and the Clauser, Aspect and Gisin experiment proves locality fails [10,18,21].

4.4.8 *Concluding Remarks*

- The authors hold the concept that the quantum theory holds true and locality fails.
- That the wave function, Ψ in the theory of mathematical representation of physical properties, events and practices is valid.
- That nonlocality is true of the physical world.
- That consciousness is behind and operates through every physical event and hence one can extrapolate to universal consciousness [35-38].
- The EGW model is only valid in a Wigner-Rauscher nonlinear quantum formalism. In a linear theory, neither the EGW model or Bohm's Hidden Variables can be proved and hence are not practical.

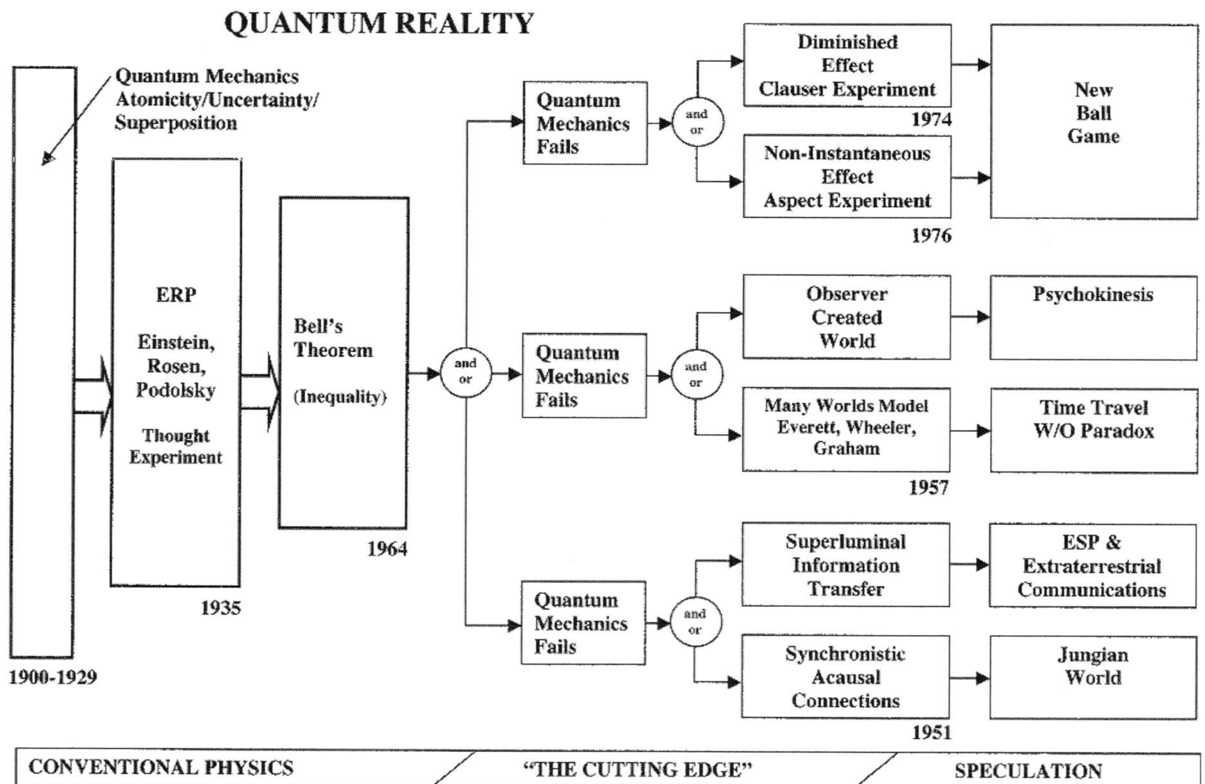

Figure 4.4. Historical development of quantum theory from turn of the last century through second half of that century to its current status. Collective construct of the members of the Fundamental "Fysiks Group" at LBNL organized by E.A. Rauscher of the Department of Theoretical Physics, LBNL.

- All measurement observation or interaction proves consciousness exists and the self-referential aspect of consciousness may imply that what is measured is aware of its change of state and is therefore conscious. If a system that is constructed in a suitable manner as to be sensitive to an intention interaction or a remote mentally effected system [6,39,40] is effected by human consciousness is it conscious as remote mentally effected system? Is such a system and perhaps all systems such as a cat, rat, or gnat conscious? It appears to be fitting to end this essay on a question. It is less destructive to utilize a rat or knot paradox experiment than a cat paradox experiment? A cat is aware it is alive, a rat also, what about the Ψ_{alive} state for a gnat? Is there a Ψ_{dead} state for a cat, rat or gnat, much less a human?! Certainly Bell's theorem and its test have lead us into a new age, where before for many centuries of the abhorrence of "action at a distance" has returned to us in a new form, not with Newton's gravity but at a more fundamental level of the quantum domain [41-43]. See Fig. 4.4.

4.5 Other Nonlocal Interactive Phenomenon and the Particle-Wave "Paradox" Resolved

In this section, and its subsections, we present a discussion of other theoretical constructs experiments that appear to exhibit nonlocality. Young's double slit experiment [44] and Wheeler's delayed choice experiments [45-47] not only elucidate some of the quantum properties of nonlocality, even though Young conceived and conducted his experiment before the development of the quantum theory, but some of the issues related to the wave-particle paradox, potential models and possible nonlocality. Whereas, light and even billiard balls via de Broglie waves, $p = \hbar / \lambda$ exhibit wave and particle-like properties. The issue of nonlocality is not an issue of locality and nonlocality but nonlocality exits as a fundamental regime and that is it. The Aharonov-Bohm experiment is discussed and can be interpreted as displaying the occurrences of nonlocality. Other interesting frame of reference and apparent nonlocality are considered such as Mach's principle.

4.5.1 *Young's Double Slit Experiment and its Extension, the Wheeler Delayed Choice Experiment*

The reason Young developed and conducted his research was to resolve whether light was a particle or a wave, a hot discussion of his time and also now [44]. Wheeler expanded this experiment in his delayed choice design, which more closely analogous to Aspect's experimental test of Bell's theorem. Young's double slit experiment of 1803 was designed to elucidate whether light was a particle or a wave. A variety of experiments and theories suggested wave like or particle like properties for light. Sir Issac Newton stuck with the corpuscular-particle theory of light even though he conducted 15 years of optic experiments involving reflection and refraction. Christiaan Huygens and others thought that light was wave like in nature and showed that light considered as a wave could travel in slight lines and follow the laws of reflection and refraction. He interpreted light to be a longitudinal wave with oscillations taking place along the line of propagation. Thomas Young's double slit experiment, with both slits open, demonstrated the existence of interference patterns of a wave nature of light, whereas, with only one slit open, only a spot of light is observed on the screen. Only light as a wave phenomenon would be consistent with a light source passing through two separate narrow slits that spread out and overlap to form light and dark interference bands at the screen [44].

In the experimental case in which the beam intensity of photons or electrons is so low as to allow the passage of only one single particle through one slit, a diffraction pattern will appear on the screen. If, in another distinct experiment, one slit is covered, no diffraction pattern occurs. The Young's double slit experiment is schematically represented in Fig. 4.5a and 4.5b. The appearance of the pattern on the screen when both slits are open and when a particle passes through one slit seems to imply that the particle or photon appears to "knew" or carries information to the screen that contained information about what would have happened had the particle gone through the other slit concurrently or simultaneously. The so termed "knowing" the other slit is open or closed by the single photon or electron appears to demand a form of nonlocality. A pilot wave or advanced potential appears to be an attempt to find a mechanism for this nonlocality just as the hidden variable hypothesis of Bohm is an attempt to explain Bell's nonlocality. Augustine Jean Fresnel furthered Young's work, which led to the construction of a mathematical basis of a wave theory of light. Young and Fresnel adopted the transverse theory of light.

Newton's great influence before Young and others, led to many years of the acceptance of the corpuscular nature of light, which he proposed. In fact, many years later, after Young, Huygens, Fresnel, et al. proposed the wave theory of the nature of light, Einstein presented the corpuscular-particle quantum nature of light having an energy $E = h\nu$ where ν is the frequency of the light. In 1905 Einstein published five papers, (his *annum mirabilis*) three of which were of major importance, one dealt with the photoelectric effect. This effect involves light shown on certain metals was found to stimulate the emission of electrons. Einstein applied Planck quantum, momentum $P = \hbar / \lambda$ where $\hbar$ is Planck's constant and λ is the wave length of light to the photoelectric process.

Experimental work had been conducted earlier in 1902 by Lenard and earlier by Hertz. Experimental determination showed that the intensity of the emitted electrons does not depend on the intensity of light but on its frequency. Higher intensity causes more electrons to be emitted. Einstein's theory predicted the experimental results precisely. So is light a particle or a wave? What is the fundamental nature of light? This history led to the particle wave paradox. However, nature does not admit of a paradox and paradox is caused by our lack of understanding of how nature works. Both waves and particles however, obey quantum nonlocality. This and other paradoxes may require moving beyond Aristotelian logic of an either–or concept. At its most basic nature, light may be neither or both (4-logic) [38] a particle or a wave but display particular attributes depending on what experiment is performed to examine its nature.

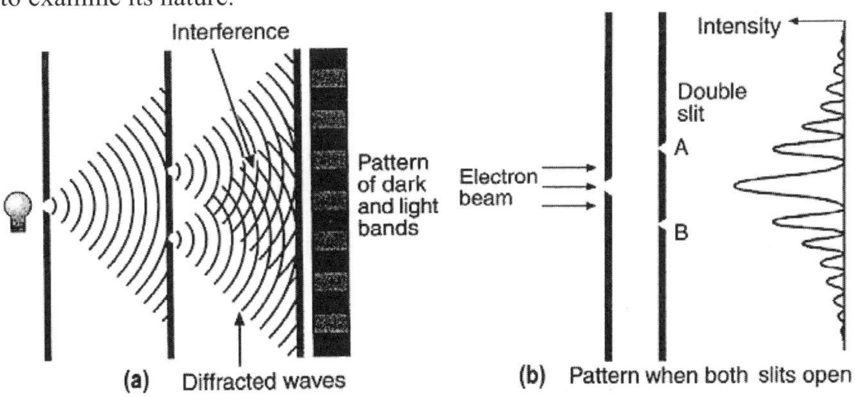

Figure 4.5. Double-slit experiment 4.5a) with light. When a beam of pure light passes through the experiment with the two holes open, the diffracted waves interfere to produce a characteristic pattern of light and shaded regions. 4.5b) with electrons. If one fires an electron beam through the experiment with two holes, one gets an interference pattern, as if the electrons were waves (de Broglie waves). The brightest part of the pattern is midway between the two holes. One does not get the pattern one would expect by adding up the two patterns corresponding to particles going through each of the two holes independently, which would give two bright peaks, one behind each hole for pure particle like properties of the electrons.

Huygens and Young first assumed light was a longitudinal wave. Then the double refraction of calcite or Island spar was carefully observed by Erasmus Barthalin. Objects observed through the crystal are refracted through two different angles. Fresnel's explanation of this phenomena was that one ray could be considered as a wave oscillating in one particular plane, the primary ray, and the other wave, the secondary ray, oscillated in a plane perpendicular to the first plane. These observations led to the transverse mode of light propagation. Young changed his mind and went with the transverse model of light. There is an analogy between the particle wave paradox and the Hertzian – non-Hertzian wave paradox, that is it depends on what experiment one conducts and the corresponding relevant formalism.

However, the argument continues over the wave-particle paradox. Just as water supported water waves, light in vacuum was considered to be supported by the lumeniferous or light-carrying ether, sometimes spelled aether. If light was longitudinal in oscillatory nature, the aether could be considered a fine gas like substance, but transverse waves can be transmitted through solids and hence because the velocity of light is so great, a very rigid solid at that. Some physicist of this era returned to the particle concept of light.

The aether abandonment came with Michelson's and later the Michelson-Morley experiment that used interferometry to measure the earth's movement through the aether, conducted with Edward Morley. The concept was that the aether was motionless, comprising an absolute frame of reference, and the earth traveled through it. It was expected from the Michelson-Morley experiment that interference would be observed in right angle light beams measured in parallel and perpendicular to the motion of the earth through the aether. No or few interference fringes were found – no aether? The vote was yes. Einstein's special theory is definably aether free. However, the aether model is not dead, as other aether models have arisen, some primarily mathematical in nature [50].

4.5.2 *Delayed Choice as an Extension of Young's Double Slit Experiment*

In the Aspect experiments the choice of the position of the two polarizers is made after the photons leave the source; the results of this experimental set up also obey the inequality and nonlocality of Bell's theorem [18]. A modification of Young's [44] double set experiment is proposed by Wheeler [45-47] termed the delayed choice experiment. The double slit experiment did much to clarify certain aspects of the Bohr-Einstein debates on "does a God play dice with the Universe?" [4]. That is, in the world of quantum theory, no elementary phenomena is a phenomena until it is recorded (and analyzed). This is the issue proposed by Wheeler who considered the Universe as a participatory Universe (which is counter to Bohr's Copenhagen view). It appears to us that the back cloth of nature is fundamental in science to deducing the nature of reality, not just the process of analysis of experimental data [49,50].

The double-slit experiment is considered both in the familiar Young's version and in the "delayed-choice" version. [46,47] The familiar experiment includes the source of photons at the bottom left, the entering slit, the first lens, the double slitted metal screen that covers it, and the photographic plate that registers interference fringes. In the delayed choice version, the continuous source of illumination on the bottom left by a source that gives off one photon per timed flash. The photographic plate is replaced to make it like a Venetian blind. We perform a last-minute choice, after the photon has already traversed the double-slitted screen, whether to open this blind or close it. Closed, it registers on a blackened grain of silver halide emulsion the arrival of that photon "through both slits" along both paths. Opened, it allows the light to be focused by the second lens on the two photon counters. Since there is only one photon, only one counter is activated and tells "through which slit the photon has passed through the screen, whether it shall have passed through only one slit or both. All the features of the photographic plate at the right and the slices of that plate are what convert the slats into a venetian blind like structure. See Fig. 4.6.

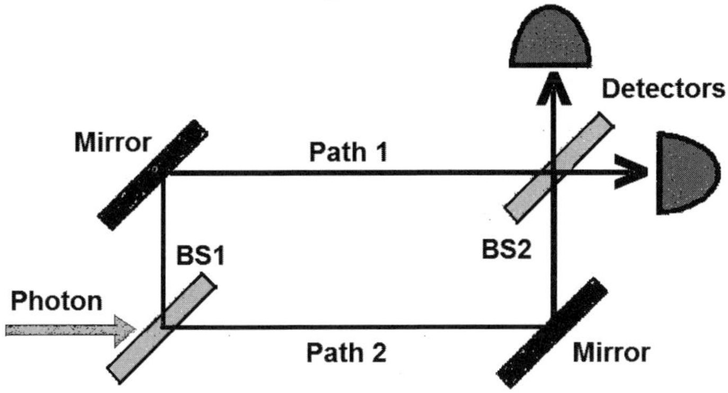

Figure 4.6. Wheeler's delayed choice experimental set up. Displayed is the photon source, then originally double slited metal screen on a first lens, BS1 to the second lens, BS2 where the photographic plate was made into a Venetian blind-like configuration which could be open or closed after the photon leaves the source and before it enters the photosensitive detectors.

As in Wheeler's description all the features to the right of the photographic plate, including the slicing of the plate into venetian blind like slats are fundamental to the delayed chaise experiment. A photon enters from the left and is recorded on the photographic plate by the blackening of a grain of silver bromide emulsion or silver halide in general. No matter how great the spacing in time between one photon and the next, the record of arrivals shows the standard two-slit interference pattern, basis for deducing that each photon has "gone through both slits", a divided photon? Not possible. One can also determine "through which slit" each quantum goes, Einstein argued, by measuring the vertical component of the kick that the photon imparts to the photographic plate. If it comes from the upper hole it kicks the plate down and from the lower hole, it kicks the plate up.

Einstein objected, stating that, through which slit did the photon go and through both slits is a logical inconsistency of the quantum theory. Bohr responded that we have conducted two separate experiments, not one. We can fasten the photographic plate to the apparatus so it will not move up and down. Then we can register the interference fringes. Or we can free it to slide up and down in a slot, not shown in figure. We can then measure the vertical kick of the photon. We cannot perform both experiments at the same time according to his complementarity principle [4]. The delayed choice experiment further exemplifies the property of nonlocality.

An obvious experiment is the triple or multiple slit experiment. What does adding another slit do to the interference pattern from the eight possible combinations for the photon to go through of open and closed slits. Max Born, in the 1920's, proposed that only pairs of photons can interfere and that adding one or more slits would not contribute any changes to the two slit interference pattern on the screen. There is no clear reason why quantum interference stops at two slits.

The test of the three or more slits experiment seemed an obvious one to us. It is only recently that U. Sinha et al. of the University of Waterloo, Canada conducted experiment using three parallel slits in a stainless steel plate, each 3×10^{-3} cm wide and 3×10^{-2} cm tall [48]. Various combinations of the three slits can be open or closed. The results demonstrated that the three slit interference pattern is the same as from a single or double slit interference pattern, that is no new fringes were observed [51]. More verification is of interest with electrons and other particles from the source. It appears, through that the Born hypothesis holds. Let us now examine some aspects of nonlocality that are macroscopic in nature and may relate to the nonlocal nature of consciousness [14,49-51].

4.5.3 *The Aharanov-Bohm Experiment, Fields and Potentials as Mechanisms of Non-Local Interactions*

Another interesting experiment and theory is the phase shift observed in the coherence of two electron beams in the Aharanov-Bohm effect. Changes in interference patterns are produced outside the actions of the fields of $\underline{E}$ and $\underline{B}$ and are ascribed to the action of the vector and scalar potentials $\underline{A}$ and ϕ. In a sense, the interferometry effect of these two beams, once correlated maintains a specific phase shift through the nonlocal interaction of the fields. The $\underline{E}$ and

$\underline{B}$ fields are regarded as primary because the field energy transfer is expressed in terms of them as the Poynting vector and the momentum transfer or Lorentz forces is also expressed in terms of $\underline{E}$ and $\underline{B}$. The potentials were introduced to obtain the canonical formalism. Aharanov and Bohm [52] theoretically formulated conditions, using a solenoid for conditions where $\underline{A}$ and ϕ have physical consequences where both $\underline{E}$ and $\underline{B}$ are zero. A number of experiments have been performed to confirm the existence of this effect such as in [53]. Experimental tests demonstrate that outside radiation fields pure potentials can exist without their associated E and B fields. See Fig. 4.7.

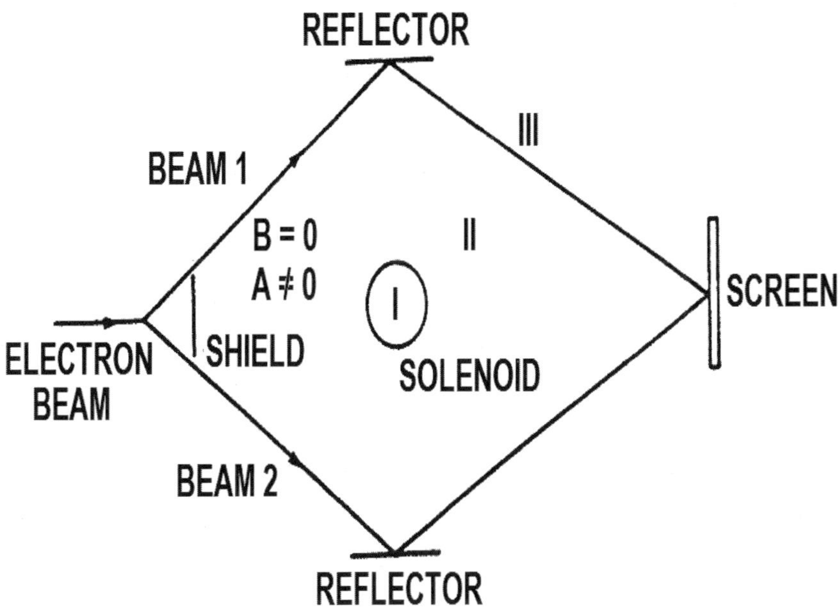

Figure 4.7. The schematic of the Aharonov-Bohm interferometry experiment with time – independent vector potential.

In the Aharanov-Bohm experiment [52], a coherent beam of electrons is split into two parts by a metal foil and each beam going to opposite sides of a centrally located solenoid. See Fig. 4.7. The solenoid is shielded by a thin plate. The two electron beams flow very close to the cylindrical solenoid of radius R and a length, l and N the number of windings N with a current flow of i. The magnetic field inside at the center of the solenoid is $B = \mu_0(Ni / l)$ and zero at its surface. Dependence on the $\cos\Theta$, where Θ is the angle from the center to the surface of the solenoid. After being reflected, the two electron interference fringes are measured at the screen. This set up insures that no fields exist outside the region of the solenoid. It was also assumed without $\underline{E}$ and $\underline{B}$ fields, no $\underline{A}$ and ϕ effects would be observed. However, the theoretical approach of Aharanov and Bohm and experimental tests determine a phase shift between the two beams of electrons at the screen demonstrating the effect of $\underline{A}$ and ϕ outside of $\underline{E}$ and $\underline{B}$. The phase shift is given as

$$\Delta\Phi = -\frac{e}{c\hbar}\oint \underline{A}.dx \text{ where } \oint \underline{A}\cdot d\underline{x} = \int H\cdot ds = \phi \qquad (4.1)$$

or the total magnetic flux inside the circuit. The vector potential cannot be zero outside the solenoid because the total flux through every circuit containing the origin is equal to a constant

$$\phi_0 = \int \underline{H}\cdot d\underline{s} = \int \underline{A}\cdot d\underline{x} \text{ for } B = \mu_0 H\ . \qquad (4.2)$$

For a singly connected region, $\underline{H} = \nabla\times\underline{A} = 0$ so that solution $\psi = \psi_0 e^{-\Phi/n}$ is the solution when $\underline{A} = 0$ so that $\nabla\Phi/\hbar = (e/c)\underline{A}$. But in the Aharanov-Bohm experiment, we have multiple connected regions outside the solenoid and ψ is no longer a single valued function so that the electron wave function splits into two parts $\psi = \psi_1 + \psi_2$ where ψ_1 represents the beam on one side and ψ_2 the beam on the other side so that the beams stay in a simply connected region so that we can write $\psi_1 = \psi_1^0 \exp -i\Phi_1/\hbar$ and $\psi_2 = \psi_2^0 \exp -i\Phi_2/\hbar$ so that Φ_1 and Φ_2 are equal to $e/c\int A\cdot dx$ along the paths of the first and second beam, respectively. The interference between the two beams depends on the phase difference.

$$(\Phi_1 - \Phi_2)/\hbar = \int \underline{A}\cdot dx = \frac{e}{\hbar c}\phi_0\ . \qquad (4.3)$$

Hence, the vector potential influences the electron interference pattern but may also influence their momentum. An experiment involving an array of solenoids is described in [59]. Experimental tests of the Aharanov-Bohm experiments have been performed which display electron interference patterns using various experimental set ups [53].

4.5.4 *Some Topics for Interference Experiments*

Figure 4.8, represents the relationship between the Aharanov-Bohm experiment, the Young's double slit experiment, Bells's theorem set up and a dual laser experiment. In the latter case, we examine dual path interferences between two lasers as correlated coherent source.

- The relation of the remote connectedness properties of Bell's theorem, Young's double slit experiment and laser interferometry from independent beams.
- The relationship between the advanced potential models and complex multidimensional geometries formulation of remote connectedness properties of the manifold.
- Interference effects produced by the superposition of light beams from two independent single-mode lasers for low beam intensities.

The central purpose is the experimental test using the dual laser source system to test the theoretical hypothesis developed in points 1 and 2. Positive results from such an experiment would have strong implication for the nature of the quantum measurement problem.

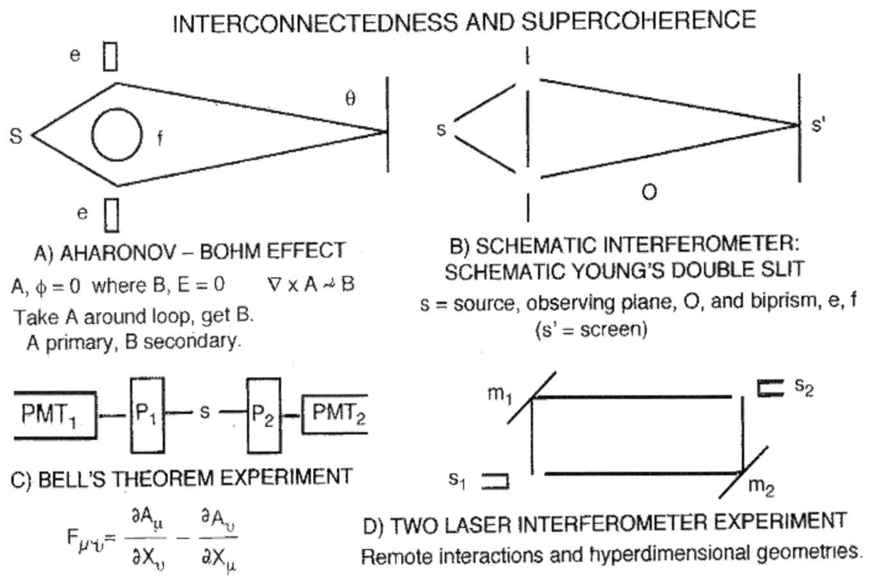

Figure 4.8. Schematic representations of experiments which involve nonlocal interconnectedness and supercoherence phenomena for four experimental set ups.

4.5.5 *Ernst Mach, Frames of Reference and Nonlocality*

Mach's principle states that the inertia of a body is due to the action of forces produced by all other bodies in the Universe. In a basic sense, such a hypothesis appears to imply a nonlocal connection of a remote inertial frame of reference. In a dynamic universe of special relativistic inertial frames and general relativistic non inertial frames how does Mach's principle apply? Is there a counter part of quantum nonlocality for longer scale correlations or other types of nonlocality even possibly a classical nonlocality? Mach's principle appears to imply nonlocal classical connections of remote events. Mach's philosophy influenced Einstein [54]. Mach's principle as we stated, appears to speak of a form of nonlocality in that local influences are produced by forces produced from other localities in the universe [55]. Rauscher and others have demonstrated unequivocally that, although Mach's principle relates to an absolute frame of reference in the Universe, it is not inconsistent with the mathematical formulism of relativity for both inertial and non-inertial frames of reference.

For a rotating and at rest bucket of water experiment was conducted by Sir Isaac Newton who hung the bucket of water at the end of a twisted rope and then let go of the rope. What one observes is that, as the bucket began to spin rapidly, the water's surface remained flat until viscous drag makes the water rotate and its surface became concave. When Newton stopped the bucket suddenly, the water continued to rotate and its surface remained curved until it stopped and the surface returned to being flat. Newton concluded that it was not the rotation of the water relative to the bucket that was important, since this relative rotation was associated with a flat water surface initially and finally with a curved surface, he interpreted that results to imply that one could state that there is rotation in relationship to an absolute space. Does the more massive body effecting a less massive one to a greater extent somehow represent an asymmetry in what is considered an inertial frame of reference? This is an interesting question from the point of view of Newton's law of Universal gravitation $F = \frac{4\pi}{G}\frac{m_1 m_2}{r^2}$ where G is the universal gravitational constant. In this sense, if $m_1 >> m_2$ then does m_1 have more inertial frame effect on m_2 where, for example, m_1 is like mass of the earth and m_2 is the mass of the water in the bucket.

Another experiment that appears to imply an absolute frame of reference for the rotation of the earth, or the fixed star system, is the behavior of the Foucault pendulum. In the mid 1850's, Jean Bernard Foucault conducted pendulum experiments on a pendulum with a heavy bob and long cord or wire. He noted that the pendulum remained in the same plane of oscillation, no matter how the point of attachment was twisted or rotated, that is a large pendulum maintained its plane of oscillation while the earth twisted under it. If the pendulum was at the north pole, the earth will make a complete twist under it in 24 hours. At the equator there is no twist at all, where the earth's rotation and velocity is the greatest, about one thousand miles per hour. To an observer on the earth's surface, it appears that the pendulum slowly rotates. Also the bulge at the earth's equator is a related phenomenon.

Mach suggests the logical question "How do we measure the inertial mass of a body?" From Newton's second law $m_i = F / a$ that is inertial mass is measured by the ratio of the sum of applied forces to it to the acceleration if produces. But the measurement of absolute acceleration requires the measurement of absolute displacement, whereas what we can really measure is the displacement of that body relative to other bodies. Therefore, according to Mach it is only by virtue of the presence of other bodies that a given body can be said to have inertial mass. Furthermore, Newton's bucket experiment and Foucault's pendulum appear to demonstrate that large masses at great (nonlocal) distances are more important than small masses nearby in defining inertial frames of reference. In what manner does this issue relate to Newton's universal law of gravitation and the universal law of equivalence of gravity?

If can be demonstrated that, although the tenet of special relativity is that there are no preferred reference frames, the structure of the theory may not preclude an absolute reference frame which is constant with Mach's principle. Mach's principle relates to the motion of material particles, separated from the close proximity to other bodies, as relative motion to the center of all other masses in the Universe. [56,57] Mach's explanation is that the difference between a nonrotating bucket of water with a flat, equipotential surface, and one rotating with a parabolic meniscus is created relative to the fixed stars. The stars were considered fixed at the turn of the last century; this concept changed with the Hubble expansion in 1921 for the analysis of stellar red shift data. The proportionality of rate of expansion of the rate of expansion to distance is Hubble's constant, $H = \dot{R} / R$. This expansion yields multiple frame of reference from which the expansion appears to be the same but it may not preclude another form of so termed fixed frame. The relative frames of the basket and what it is rotating to may represent a large scale, at least earth size, of nonlocal influence. Sciama [58] developed an interesting analogy between gravity and electromagnetism by forming a gravitational analogue of electromagnetic the acceleration dependent inductive force which produces photon emission. His attempt was to reconcile general relativity and Mach's principle including, in the context of this principle, to explain redshift, which is an interesting approach. Rowlands discusses Sciama's approach and make a Machian analogy of the so termed all pervasive Higgs field [55]. In

[59] is given a generalized discussion of nonlocality and the complex 8 space.

In [14] we have reinterpreted the meaning of Hubble's expansion law, $H = \dot{R}/R$ to derive the fundamental basis of the continuous-state principle, a key element in developing the Holographic Anthropic Multiverse (HAM) cosmological paradigm. HAM cosmology allows an infinitesimal photon mass, m_γ as described by the Proca equation which in a covariant polarized Dirac vacuum leads to the 'tired light' interpretation of cosmological redshift and a Cavity-QED spacetime exiplex model of the Cosmic Microwave Background Radiation (CMBR) as blackbody equilibrium conditions of emission and redshift as absorption. Thus redshift is virtual, a continuous-state inherent free-fall motion of the M-Theory backcloth illustrated in the HAM mantra 'continuous-state spin-exchange dimensional reduction compactification process' of symmetry breaking in Calabi-Yau mirror symmetry conditions. This, and the 'rest of the story' is a lot to chew on; which is developed to the degree possible in [14] which we will update and refine as possible.

4.6 Conclusion

The argument and attempted explanation concerning the particle-wave nature of light during the last five centuries or so has led to intense debate including some broken lifelong friendships. This also true of the debate over the fundamental existence of locality and nonlocality. The attempt in studying physics is to strive toward a more basic knowledge of the nature of reality. It is assumed that physics is the most fundamental of all sciences and its perhaps the basis for all human knowledge, using the precise and logical language of mathematics. Our current understanding of physics grows out of our attempt to understand the natural world and has been the result of accumulated knowledge by a succession of inductive and deductive inferences derived from observation and theoretical hypothesis and theory explanation and prediction.

The concept of a unified theory of physics or a theory of everything (TOE) assumes there is a point at which the origin of everything is explained and also that the origin of everything can be explained in terms of a single obvious source. It is clear that in every fundamental theory, should one exist, must accommodate the fundamental nature of nonlocality in both micro and macro systems. Therefore, it is imperative that a theoretical framework be constructed to accommodate nonlocality at a very basic level. Such a theory is exemplified by the formulation of the complex Minkowski space which has deeply imbedded in its structure nonlocality in space and time.

References and Notes

[1] Einstein, A. Podolsky, B. and Rosen, N. (1935) Can a quantum mechanical description of physical reality be considered complete?, Phys. Rev., 47, 777.
[2] Bell, J.S. (1964) On the Einstein, Podolsky, Rosen Paradox. Physics, 1, 195.
[3] Bell, J.S. (1966) On the problem of hidden variables in quantum theory, Rev. Mod. Phys., 38, 447.
[4] Jammer, M. (1973) The Philosophy of Quantum Mechanics, New York: Wiley; The Bohr-Einstein debates are covered in detail in this book.
[5] Schilpp, P.A. (1949) Albert Einstein: Philosopher-Scientist, Evanston: Library of Living Philosophers; Covers Bohr-Einstein epistemology in atomic physics.
[6] Rauscher, E.A. (1972) Generalized Heisenberg relations, Lett. N. Cim., 4, 747.
[7] Weinberg, S. (1968) Phys. Rev 166, 1568; and private communication at LBNL (1965) on EAR's concept of multiply charged elementary particles and the identification of the so termed 3,3 resonance of the doubly charged, Δ^{++}. Until that time, all elementary particles were thought to be singly charged.
[8] Bohm, D. (1952) Phys. Rev. 85, 166 and 180; and private communication at LBNL when he spoke to the Fundamental Physics ("Fysiks") Group and at Birkbeck College, London during EAR's teaching summer 1977 and EAR's discussion on the physics of hyperdimensional geometrics, University of London presentation, 1977.
[9]The Fundamental Physics ("Fysiks") Group founded and run by EAR (1963-1979) at LBNL to discuss the formulation of quantum theory, consciousness, nonlocality of Bell's theorem and the general foundations of physics.
[10] Clauser, J.F. & Horne, W.A. (1974) Phys. Rev. 100, 526; private communication with J. Clauser UCB, 1971 and presentations to the Fundamental Physics Group, 1975 and 1976.
[11] Herbert, N. (1975) Am. J. Physics 43, 315; and Bell's Theorem and Superluminal Signals, presentations to the Fundamental Physics Group of EAR at LBNL, 1975, 1976, 1977.
[12] Feinberg, G. (1967) Phys. Rev. 159, 1089; and private communications with EAR and H. Puthoff at the University of Columbia, 1975.
[13] Ramon, C. & Rauscher, E.A. (1980) Found. of Phys. 10, 661, demonstrates the compatability of superluminal signals and multidimensional geometries.
[14] Amoroso, R.L. & Rauscher, E.A. (2009) The Holographic Anthropic Multiverse: Formalizing the Complex Geometry of Reality, Singapore: World Scientific.
[15] Brown, R.H. & Twiss, R. (1956) A test of a new type og stellar interferometer on Sirius, Nature, 178, 1046c.
[16] Herbert, N. (1988) Faster Then Light: Superluminal Loopholes in Physics, American Library.
[17] Scaul, D. (1968) Phys. Rev. 137, 166; appreciation is given to S.P. Sirag for bringing this reference to EAR's attention.
[18] Aspect, A., Grainger, P. & Roger, G. (1992) Experimental tests of Bell's inequalities using time-varying analyzers, Phys. Rev. Lett., 49, 1804; and private communication with A. Aspect, Orsay, France, 1977.
[19] Freedman, S. & Clauser, J. (1972) Experimental test of local hidden variable theories, Phys. Rev. Lett., 28, 934-941.

[20] Gisin, N., Tittel, W., Brendel, J. & Zbinden, H. (1998) Violation of Bell inequalities by photons more than 10 km apart, Phys. Rev. Lett., 81, 3563.
[21] Gisin, N., Tittel, W., Brendel, J. & Zbinden, H. (1998) Quantum correlation over more than 10 km. Optics and Photonics News, 9, 41.
[22] Stapp, H., Nadeau, R. and Kafatos, M. (1999) The Nonlocal Universe: The New Physics and Matters of the Mind, Oxford: Oxford University Press; private communication, LBNL 1964-1973, EAR, and Kafatos with RLA, 2000.
[23] Bohm, D. and Hiley, B. (1993) The Undivided Universe London: Routledge; and private communication with D. Bohm and B. Hiley, 1977.
[24] Bohm, D. and Hiley, B. (1975) On the understanding of nonlocality as implied by quantum theory, Found. of Phys. 5 93.
[25] Stapp, H. Presentation on The Interpretation of Bell's Theorem presented to the Fundamental Physics Group, chaired by E.A. Rauscher at LBNL, October 31, 1975 and June 25, 1976.
[26] Heisenberg, W. (1938) Physics and Philosophy, New York: Harper-Collins; and private communication, 1972 with EAR.
[27] Clauser, J.F. (1975) Counter Efficiencies and Bell's Theorem, presented to the Fundamental Physics Group, Chaired by E.A. Rauscher at LBNL October 31.
[28] Wheeler, J.A. (1978) private communication University of Texas, Auston, June.
[29] (1973) The Many-Worlds Interpretation of Quantum Mechanics, B.S. DeWitt and N. Graham (eds.) includes papers by H. Everett, J.A. Wheeler, N. Graham and B.S. DeWitt, Princeton: Princeton University Press; and private communication with J.A. Wheeler and B.S. DeWitt and EAR, 1967 to 1979.
[30] Bohm, D. (1977) Bell's Theorem and Hidden Variables, presentation to the Fundamental Physics Group, LBNL chaired by E.A. Rauscher April 7.
[31] Eberhard, P. (1978) Nuovo Cimento, B46, 392; Lamport, L., Eberhard, P. and Rauscher, E.A. (1976) Everett-Graham-Wheeler many worlds interpretation of Quantum Theory, Fundamental Physics Group, LBNL, Chaired by E.A. Rauscher.
[32] Sirag, S.P. & Rauscher, E.A. (1975) Spacetime quantum signals and Heisenberg's potentia theory for Bell's theorem and the Young double slit experiment, Presentation to the Fundamental Physics Group, LBNL, chaired by EAR.
[33] Chew, G. (1968) Science 161, 762; and presentation S-Matrix bootstrap and models of physical reality, the Fundamental Physics Group Chaired by E.A. Rauscher at the LBNL, February 13, 1976.
[34] Rauscher, E.A. (1981) Conceptual changes in reality models from new discoveries in physics, pp. 1407-1431, in N.A. Salonen (ed.) The Search for Absolute Values and Creation of a New World, Vol. II, November 9-13, Seoul, Korea, New York: ICF Press.
[35] Rauscher, E.A. (2005) The unity of consciousness experience and current physical theory, pp. 87-116, Journal of ISSSEEM, 15,23.
[36] Rauscher, E.A. (2007) Quantum and the role for consciousness in the physical world, pp. 1-42, Journal of ISSSEEM, 16, 1.
[37] Van Bise, W. & Rauscher, E.A. (2004) Quantum mechanics and the role of intentional will, p. 73, Towards a Science of Consciousness, Proceedings, Tucson: University of Arizona.
[38] Penrose, R. (1994) Shadows of the Mind: A Search for the Missing Science of

Consciousness, Oxford: Oxford University Press.
[39] Schmidt, H. (1976) Psychokinetic effect on prerecorded targets, J. Am. Soc. Psychic Research 70, 267; and private communication.
[40] Van Bise, W.L. & Rauscher, E.A. (1988) Theoretical models and experimental data concerning the spacetime properties of consciousness and remote mental influence, pp. 1-222, Buck Fine Arts and Science Grant, PSRL-60788, revised 1990.
[41] Peat, F.D. (1990) Einstein's Moon: Bell's Theorem and the Curious Quest for Quantum Reality, Chicago: Contemporary Books, Inc.
[42] Bell, J.S. (1987) Speakable and Unspeakable in Quantum Mechanics, Cambridge: Cambridge University Press.
[43] Kaiser, D. (2011) How the Hippies Saved Physics, covers topics, experiences and interactions of the Fundamental "Fysiks" Group organized and chaired by EAR at LBNL and other research groups the S.F. Bay Area (1972-77), NY: W.W. Norton.
[44] Young, T. (1804) Experiments and calculations relative to physical optics. Philosophical Trans of the Roy. Soc. of London, 94, 1-16.
[45] Wheeler, J.A. (1979) Beyond the Black Hole, Center for Theoretical Physics, Austin: University of Texas.
[46] Wheeler, J.A. (1979) Frontiers of time, Center for Theoretical Physics, Austin: University of Texas.
[47] Wheeler, J.A. (1978) The "past" and the "delayed choice" double slit experiment, A.R. Marlow (ed.) Mathematical Foundations of Quantum Theory, New York: Academic Press.
[48] Sinha, U. et al (2010) Ruling out multi-order interference in quantum mechanics, Science. Vol. 329, p. 418.
[49] Rauscher, E.A. & Targ, R. (2001) The speed of thought: Investigation of a complex spacetime metric to describe psychic phenomena, J. Sci. Expl. 15, 331.
[50] Millay, J. (ed.) (2010) Radiant Minds: Scientists Explore the Dimensions of Consciousness, in PRG Anthology.
[51] Rauscher, E.A. & Amoroso, R.L. (2009) Relativitic physics in complex Minkowski space, nonlocality, ether model and quantum physics, in M.C. Duffy & J. Levy (eds.) The Ether Spacetime and Cosmology, Vol. 3, Physical Vacuum, Relativity and Quantum Physics, Montreal: Apeiron.
[52] Aharonov, Y. & Bohm, D. (1959) Significance of electromagntic potential in the quantum theory, Phys. Rev. 115, 485.
[53] Olariu, S. & Popesca, I.I. (1985) The quantum effect of electromagnetic fluxes, Rev. Mod. Phys. 57, 3390.
[54] Mach, E. (1959) The Analysis of Sensations, New York: Dover Pub.
[55] Rowlands, P. (2007) Zero to Infinity: The Foundation of Physics, Singapore: World Scientific.
[56] Rauscher, E.A. (1971) A unifying theory of fundamental processes, LBL/UCB, UCRL-20808.
[57] Rosen, N. (1965) Ann. Phys. 35, 426.
[58] Sciama, D.W. (1969) The Physical Foundation of General Relativity, New York: Doubleday.
[59] Renshaw, K. (2009) Beyond Einstein's Horizon: Science, Remote Viewing and ESP, Cambria: Constellation Press, and references therein.

Chapter 5

The Complexification of Maxwell's Equations

Should we not then use the equations of motion in high-energy as well as low energy physics? I say we should. A theory with mathematical beauty is more likely to be correct than an ugly one that fits some experimental data. – Albert Einstein

In this chapter we demonstrate that complex electric and magnetic fields are consistent with a geometry consisting of complex spacetime. We thus demonstrate that complex spacetime coordinates are not inconsistent with electromagnetic phenomena and may point to a direction for its unification with gravitational phenomena, in the weak Weyl field limit. The particular case we examine in detail is for an electron in a field where we derive Coulomb's equation. We examine this unification using the Weyl geometry in the linear approximation of the gravitational field.

5.1 Complex Electromagnetic Fields

The linear approximation of Weyl geometry [1-4] for the gravitational field is consistent with the conditions of the 5D Kaluza-Klein geometry [5,6]. We present the formalism for the complexification of the electric and magnetic fields in this approach. We obtain additional symmetry conditions on the classical form of Maxwell's equations; and we obtain a non-zero divergence condition for the magnetic field which may be identifiable with a magnetic monopole term.

The relationship of the geodesic world lines and the electromagnetic field lines involve the definition of the field line structure. The field lines represent equipotential surfaces or they are lines connecting equipotential surfaces on a field map. For the gravitational tensor potential, $g_{\mu\nu}$ this map is the geodesic path on the light cone, i.e., the path that a photon will take according to the least action principle. We can similarly define an electromagnetic vector potential in analogy to $g_{\mu\nu}$ which we denote, A_μ. We use the formalism of Weyl to describe the manner in which we can derive Maxwell's equations, and in particular, Coulomb's law from the properties of A_μ. We then expand this formalism to include electromagnetic field components with real and imaginary parts and discuss the implications of this formalism. We also relate this formalism into our complex spacetime multidimensional geometry and then demonstrate that a complex "space" can be represented as a multidimensional real space with complex rotation represented by a generalized Lorentz transformation, Θ. It is likely that the transformation Θ includes all the affine connections. See Fig. 5.1.

Inomata [7] and Rauscher [8-13] introduce a simple but elegant concept - complex components to the electric and magnetic field vectors. He starts from Maxwell's equations in their usual form for an electromagnetic media for electric charge, ρ_E and electric current, J_E. Then we write Maxwell's equations in their usual form [14] which build on the extensive work of Faraday and others [15]:

$$\nabla\cdot\underline{E}=4\pi\rho_E \qquad \nabla\times\underline{E}=-\frac{1}{c}\frac{\partial\underline{B}}{\partial t}$$
$$\nabla\cdot\underline{B}=0 \qquad \nabla\times\underline{B}=-\frac{1}{c}\frac{\partial\underline{E}}{\partial t}+\underline{J}_E \tag{5.1}$$

To introduce symmetry to Maxwell's equations one can introduce an imaginary "magnetic" charge, $i\rho$ or $i\rho_{\text{Im}}$ and imaginary "magnetic" current, iJ_M or iJ_{Im}, where again $i\equiv\sqrt{-1}$ and ρ_M and J_M are real quantities. Upon substitution into Maxwell's equations, we have

$$\nabla\cdot\underline{E}=4\pi\rho_E \qquad \nabla\times\underline{E}=-\frac{1}{c}\frac{\partial\underline{B}}{\partial t}+i\underline{J}_M$$
$$\nabla\cdot\underline{B}=4\pi i\rho_M \qquad \nabla\times\underline{B}=-\frac{1}{c}\frac{\partial\underline{E}}{\partial t}+\underline{J}_E \tag{5.2}$$

In this form we see that there are no real terms for the magnetic charge or current in terms such as $4\pi i\rho_M$ and iJ_{m}. Now we can derive real forms of Maxwell's equations by introducing complex E and B fields and separating real and imaginary parts of the equations.

Consider both the electric and magnetic fields to be complex quantities, that is

$$\underline{E}=\underline{E}_{\text{Re}}+i\underline{E}_{\text{Im}}, \qquad \underline{B}=\underline{B}_{\text{Re}}+i\underline{B}_{\text{Im}} \tag{5.3}$$

where $E_{\text{Re}}, E_{\text{Im}}, B_{\text{Re}}$ and B_{Im} are real quantities, then substitution of these two equations into the complex form of Maxwell's equations above yields, upon separation of real and imaginary parts, two sets of Maxwell-like equations where the real parts are the usual Maxwell's equations:

$$\begin{aligned} \nabla \cdot \underline{E}_{\mathrm{Re}} &= 4\pi\rho_E \qquad & \nabla \times \underline{E}_{\mathrm{Re}} &= -\frac{1}{c}\frac{\partial \underline{B}_{\mathrm{Re}}}{\partial t} \\ \nabla \cdot \underline{B}_{\mathrm{Re}} &= 0 \qquad & \nabla \times \underline{B}_{\mathrm{Re}} &= \frac{1}{c}\frac{\partial \underline{E}_{\mathrm{Re}}}{\partial t} + J_E \end{aligned} \tag{5.4}$$

where $\rho_E = \rho_{\mathrm{Re}}$ and $\rho_M = \rho_{\mathrm{Im}}$ and $J_E = J_{\mathrm{Re}}$ -and $J_M = J_{\mathrm{Im}}$ for the imaginary parts:

$$\begin{aligned} \nabla \cdot \left(i\underline{E}_{\mathrm{Im}}\right) &= 0 \qquad & \nabla \times \left(i\underline{E}_{\mathrm{Im}}\right) &= -\frac{1}{c}\frac{\partial\left(i\underline{B}_{\mathrm{Im}}\right)}{\partial t} + i\underline{J}_M \\ \nabla \cdot \left(i\underline{B}_{\mathrm{Im}}\right) &= 4\pi i\rho_M \qquad & \nabla \times \left(i\underline{B}_{\mathrm{Im}}\right) &= \frac{1}{c}\frac{\partial\left(i\underline{E}_{\mathrm{Im}}\right)}{\partial t} \end{aligned} \tag{5.5}$$

Note that the *i* drops from both sides of each equation, giving real equations in all cases.

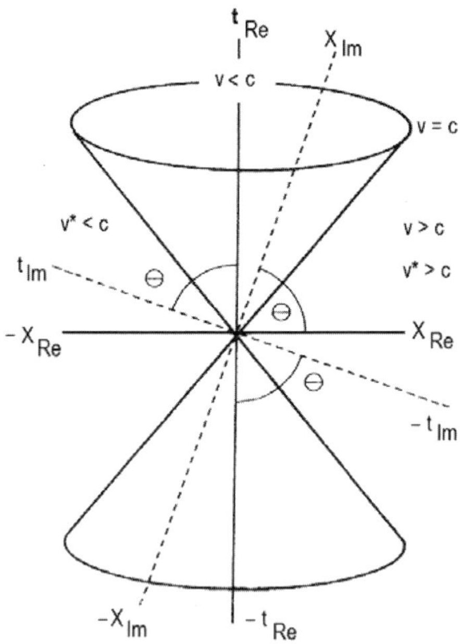

Figure 5.1. In the complex multidimensional space model we introduce, in addition to the usual orthogonal 4-space, four imaginary components, three spatial and one temporal. This is necessary in order to model remote connectedness and to retain the physical causality and symmetry conditions of conventional complex numbers. We can consider the eight orthogonal dimensions to be constituents of two intersecting light cones, one axis of real (*x*,*t*) and the other axis of imaginary (*y*,*t*) coordinates.

The real part of the electric and magnetic fields yield the usual Maxwell equations and the complex parts generates a "mirror" set of equations; for example, the divergence of the real component of the magnetic field is zero but the divergence of the imaginary part of the electric field is zero, and so forth. The imaginary part of the equations, the imaginary electric term replaces the real magnetic term, and vice versa. The structure of the real and imaginary parts of the fields form a symmetry in which electric real components are substituted by the imaginary part of the magnetic fields and the real part of the magnetic field being substituted by the imaginary part of the electric field in the second set of the equations [7,16].

The charge density and current density are expressed as complex quantities based on the separation of Maxwell's equations above. The complex generalized form for charge density and current is given as,

$$\rho = \rho_E + i\rho_M = \rho_{\mathrm{Re}} + i\rho_{\mathrm{Im}}$$

and (5.6)

$$J = J_E + iJ_M = J_{\mathrm{Re}} + J_{\mathrm{Im}}$$

where it may be possible to associate the imaginary complex charge with the magnetic monopole and, conversely, the electric current has an associated imaginary magnetic current.

The above definitions for the complex form of ρ and J appear to be interesting, where we let $\rho_E = \rho_{\mathrm{Re}}$ and $\rho_M = \rho_{\mathrm{Im}}$ and also $J_E = J_{\mathrm{Re}}$ and $J_M = J_{\mathrm{Im}}$ as before. For some interpretations we may not necessarily identify ρ_{Re} and J_{Re} as electric terms and ρ_{Im} and J_{Im} as magnetic terms. See [7,16] as there are other ways to examine the complexification of the E and B fields.

By considering the "mirror" imaginary B_{Im} and E_{Im} fields of the real E_{Re} and B_{Re} field we may have an explanation of electrostatic cooling. Extensive research on this effect, and the theoretical approach to electromagnetic cooling has been conducted by Rauscher and Beal [17,18]. If J_{E} is neglected then we have the usual case where $\nabla \cdot E_{\mathrm{Re}} = 4\pi\rho_E$ and $\nabla \times E_{\mathrm{Im}} = 0$ so that no extra or anomalous terms appear. In [19], Dirac suggested a model similar to ours and to that of Inomata. Considering the imaginary part of Maxwell's equations in complex form we have $\nabla \cdot \underline{B}_{\mathrm{Re}} = 4\pi\rho_M$, where identification of $\rho_M = \rho_{\mathrm{Im}}$ is reasonable and where the i term is eliminated from both sides of the equation. Then B_{Re} and ρ_M are real and we consider only real derivatives in the del operation. Later we will examine the complex form of ∇ and perform complex derivatives where we use the transformations $x = x_{\mathrm{Re}} + x_{\mathrm{Im}}$ and $t = t_{\mathrm{Re}} + it_{\mathrm{Im}}$ and other complex metric forms.

If we take $\nabla \times \underline{B}_{\mathrm{Im}} = 0$ then we have $\partial \underline{E}_{\mathrm{Im}} / \partial t = 0$ and if also $\nabla \times \underline{E}_{\mathrm{Im}} = 0$ then we have $\partial \underline{B}_{\mathrm{Im}} / \partial t = c\underline{J}_{\mathrm{Im}}$. We identify the temporal change of the imaginary part of the magnetic field term. If we use the definition $\underline{B} = \underline{B}_{\mathrm{Im}} + i\underline{B}_{\mathrm{Im}}$ then we can take the total magnetic derivative as

$$\frac{\partial \underline{B}}{\partial t} = \frac{\partial \underline{B}_{\mathrm{Re}}}{\partial t} + \frac{\partial \underline{B}_{\mathrm{Im}}}{\partial t} = c\left(\underline{J}_{\mathrm{Re}} + i\underline{J}_{\mathrm{Im}}\right) \tag{5.7}$$

and again we find the association of $\partial \underline{B}_{\mathrm{Im}} / \partial t = c\underline{J}_{\mathrm{Im}}$ We may be able to identify J_{Im} with a magnetic current, J_M, and associate a putative magnetic monopole current having one sign with the imaginary "mirror" part of the magnetic field. Before we proceed further with a physical interpretation of the imaginary component of the magnetic field, let us examine two issues in detail.

This formulation will assist us in understanding the physical interpretation of the complex model of Maxwell's equation. Currently we consider are the relationship between the complex form of E and B to the complex spacetime geometry and also the consideration of complex ($\underline{A}, \phi$) as a more useful and perhaps more primary interpretation of electromagnetic phenomena, rather than E and B.

5.2 Complex Electromagnetic Variables in Complex Multidimensional Spaces

We proceed from our 8D geometry. In [8,9], we defined the notation for the transformations, $x' \to x + ix$ and $t' \to t + i\tau$ which we have denoted as $x \to x_{\mathrm{Re}} + ix_{\mathrm{Im}}$ and $t \to t_{\mathrm{Re}} + it_{\mathrm{Im}}$. We can also denote $x \to x_1 + ix_2$ and $t \to t_1 + it_2$ in analogy to $E = E_1 + iE_2$ and $B = B_1 + iB_2$. We denote $E = E_{\mathrm{Re}} + iE_{\mathrm{Im}}$ and $B = B_{\mathrm{Re}} + iB_{\mathrm{Im}}$ as before.

In [8,9] we define a method for taking complex derivatives and apply this method to our examination of the Schrödinger equation in a complex Minkowski space. See Chap. 2. Because of the linear superposition principle approximation [10], we can solve the real and imaginary parts of the equation separately and sum them in the approximation of a small deviation from linearity. In the case of our calculation of the curl and divergence terms in Maxwell's equations we can no longer, to first order, make the linear approximation assumption.

We can define the divergence operation in the complex multidimensional geometry for a general vector $\underline{A} = \underline{A}_{\mathrm{Re}} + i\underline{A}_{\mathrm{Im}}$ (not to be confused with the vector potential, A_μ) where we have a vector form *A(x,y,z)* $= \hat{x}A_x + \hat{y}A_y + \hat{z}A_z$ where each component A_x can be written as $A_x = A_{x\mathrm{Re}} + iA_{x\mathrm{Im}}$, etc. Then

$$\nabla \cdot \underline{A} = \left[\frac{\partial \underline{A}}{\partial x_{\mathrm{Re}}} + \frac{\partial \underline{A}}{\partial y_{\mathrm{Re}}} + \frac{\partial \underline{A}}{\partial z_{\mathrm{Re}}} + \frac{\partial \underline{A}}{\partial i x_{\mathrm{Im}}} + \frac{\partial \underline{A}}{\partial y_{\mathrm{Im}}} + \frac{\partial \underline{A}}{\partial z_{\mathrm{Im}}} \right]. \tag{5.8}$$

We have $\underline{A} = A_{x\mathrm{Re}} + A_{y\mathrm{Re}} + A_{z\mathrm{Re}} + iA_{x\mathrm{Im}} + iA_{y\mathrm{Im}} + iA_{z\mathrm{Im}}$. Upon substitution we have twelve terms, six are real and six are imaginary. For $\nabla \cdot \underline{A}$, we have for $x \to x_{\mathrm{Re}} + ix_{\mathrm{Im}}$

$$\begin{aligned} \nabla \cdot \underline{A} = &+ \frac{\partial A_{x\mathrm{Re}}}{\partial_{x\mathrm{Re}}} + \frac{\partial A_{y\mathrm{Re}}}{\partial_{x\mathrm{Re}}} + \frac{\partial A_{z\mathrm{Re}}}{\partial_{x\mathrm{Re}}} - \frac{i\partial A_{x\mathrm{Im}}}{\partial_{x\mathrm{Re}}} - \frac{i\partial A_{y\mathrm{Im}}}{\partial_{x\mathrm{Re}}} - \frac{i\partial A_{z\mathrm{Im}}}{\partial_{x\mathrm{Re}}} \\ &- i\frac{\partial A_{x\mathrm{Re}}}{\partial_{x\mathrm{Im}}} - i\frac{\partial A_{y\mathrm{Re}}}{\partial_{x\mathrm{Im}}} - i\frac{\partial A_{z\mathrm{Re}}}{\partial_{x\mathrm{Im}}} + \frac{\partial A_{x\mathrm{Im}}}{\partial_{x\mathrm{Im}}} + \frac{\partial A_{y\mathrm{Im}}}{\partial_{x\mathrm{Im}}} + \frac{\partial A_{z\mathrm{Im}}}{\partial_{x\mathrm{Im}}} \end{aligned}. \tag{5.9}$$

Also there are 12 terms for the partial derivatives in $y \to y_{\mathrm{Re}} + iy_{\mathrm{Im}}$ and another 12 in terms of the partial derivatives of $z \to z_{\mathrm{Re}} + iz_{\mathrm{Im}}$. We address the dependence of *A* and its components as $A(x_{\mathrm{Re}}, y_{\mathrm{Re}}, z_{\mathrm{Re}}, x_{\mathrm{Im}}, y_{\mathrm{Im}}, z_{\mathrm{Im}})$. Also we have dependence of *A* and its components on other components; for example, we can have $A_{x\mathrm{Re}}(x_{\mathrm{Re}},..., A_{y\mathrm{Re}}, A_{z\mathrm{Re}}, A_{x\mathrm{Im}},...)$. Let us assume that when we consider *A* as the general symbol for *E* and *B*, that they are dependent only on real and imaginary components of space and time. In such a case we also have another 12 terms for $t \to t_{\mathrm{Re}} + it_{\mathrm{Im}}$ totaling 48 terms.

We can use certain approximations to examine the forms of the complex electromagnetic fields in complex spacetime. We will see that more general forms are useful in examining energy transmission for transverse and longitudinal components. Consider the two divergent forms of Maxwell's equations, $\nabla \cdot \underline{E} = 4\pi\rho$ and $\nabla \cdot \underline{B} = 0$. If we then write $E = E_{\mathrm{Re}} + iE_{\mathrm{Im}}$ and also $B = B_{\mathrm{Re}} + iB_{\mathrm{Im}}$ we have $E(x_{\mathrm{Re}}, y_{\mathrm{Re}}, z_{\mathrm{Re}}, x_{\mathrm{Im}}, y_{\mathrm{Im}}, z_{\mathrm{Im}}, t_{\mathrm{Re}}, t_{\mathrm{Im}})$. However let us consider only that $E(x_{\mathrm{Re}}, x_{\mathrm{Im}})$, $B(x_{\mathrm{Re}}, x_{\mathrm{Im}})$ and $\rho(x_{\mathrm{Re}}, x_{\mathrm{Im}})$, or more specifically that $E_{x\mathrm{Re}}(x_{\mathrm{Re}}, x_{\mathrm{Im}})$ and $E_{\mathrm{Im}}(x_{\mathrm{Re}}, x_{\mathrm{Im}})$. Now we will have

$$\nabla \cdot E = \frac{\partial E_{x\mathrm{Re}}}{\partial x_{\mathrm{Re}}} + i\frac{\partial E_{x\mathrm{Im}}}{\partial x_{\mathrm{Re}}} - i\frac{\partial E_{x\mathrm{Re}}}{\partial x_{\mathrm{Im}}} + \frac{\partial E_{\mathrm{Im}}}{\partial x_{\mathrm{Im}}} \tag{5.10}$$

$$= 4\pi(\rho_{\mathrm{Re}} + i\rho_{\mathrm{Im}}).$$

Collecting real and imaginary terms, we have two equations:

$$\frac{\partial E_{x\mathrm{Re}}}{\partial x_{\mathrm{Re}}} + \frac{\partial E_{x\mathrm{Im}}}{\partial x_{\mathrm{Im}}} = 4\pi\rho_{\mathrm{Re}}, \tag{5.11}$$

and

$$\frac{\partial E_{x\mathrm{Im}}}{\partial x_{\mathrm{Re}}} - \frac{\partial E_{x\mathrm{Re}}}{\partial x_{\mathrm{Im}}} = \rho_{\mathrm{Im}}. \tag{5.12}$$

Note now that the real and imaginary components are mixed.

In a similar manner we can write two similar expressions for $\nabla \cdot B = 0$ for real and imaginary components in complex space as

$$\frac{\partial B_{x\mathrm{Re}}}{\partial x_{\mathrm{Re}}} + \frac{\partial B_{x\mathrm{Im}}}{\partial x_{\mathrm{Im}}} = 0 \tag{5.13}$$

and

$$\frac{B_{x\mathrm{Im}}}{x_{\mathrm{Re}}} - \frac{B_{x\mathrm{Re}}}{x_{\mathrm{Im}}} = 0. \tag{5.14}$$

Again real and imaginary components are mixed, but since i exists on both sides of the second above equation, all four of the above equations are completely real. These equations are very restrictive in terms of purely spatial, and not temporal, dependence, and that $A_{x\mathrm{Re}}$ and $A_{x\mathrm{Im}}$ are taken as dependent on x_{Re} and x_{Im} where we take the term A as either E or B. In general, other terms such as $E_{x\mathrm{Re}}(y_{\mathrm{Re}}, t_{\mathrm{Im}}, E_{\mathrm{Re}})$, etc. can come into effect and we can approximate these by terms such as ε in E and β in B in the above equations so that terms in $\varepsilon_{\mathrm{Re}}(x_{\mathrm{Re}}, x_{\mathrm{Im}})$, etc., which appear as additional terms which we can consider to be small compared to the terms in the previous four equations. Perhaps terms such as $\beta(x_{\mathrm{Re}}, x_{\mathrm{Im}})$ and others might also act as effective terms. For example, we could write

$$\frac{\partial B_{x\mathrm{Re}}}{\partial x_{\mathrm{Re}}} + \frac{\partial B_{x\mathrm{Im}}}{\partial x_{\mathrm{Im}}} \equiv \frac{\partial B}{\partial x_{\mathrm{Re}}} \geq 0. \tag{5.15}$$

The above formalism does not represent strictly a projective geometry but is related to the concept that 4-space is a slice through a complex multidimensional space. We will make certain approximations which simplify the equations but they still remain nonlinear and give E and B fields of the form of $\mathrm{Sinh}^2(x)$, for parameter x [19-24].

We will examine in more detail how a projective geometrical form of the complex E and B fields form Hertzian as well as non-Hertzian waves. Then $\partial B_{x\mathrm{Re}} / \partial x_{\mathrm{Re}} = -\partial B_{x\mathrm{Im}} / \partial x_{\mathrm{Im}}$. The term on the right may be associated with a term in ρ such as ρ_{Im} or ρ_M. For example, we may have a form $\partial B_{x\mathrm{Re}} / \partial x_{\mathrm{Re}} = \rho_M$ or, in general, for the consideration of all components, $\nabla \cdot B = \rho_M$. The shadow imaginary terms to the real usual terms may supply insight as to new ways of interpreting conventional as well as novel electromagnetic phenomena. We will consider these issues in more detail in the Higgs field approximation

We turn our attention to the full detailed consideration of the set of derivatives involving complex $\underline{E}$ and $\underline{B}$ in complex spacetime. We use $\underline{E} = \underline{E}_{\mathrm{Re}} + i\underline{E}_{\mathrm{Im}}$ and $\underline{B} = \underline{B}_{\mathrm{Re}} + i\underline{B}_{\mathrm{Im}}$ and $x = x_{\mathrm{Re}} + ix_{\mathrm{Im}}$ and $t = t_{\mathrm{Re}} + t_{\mathrm{Im}}$; all terms such as $\underline{E}_{\mathrm{Re}}, \underline{E}_{\mathrm{Im}}, \underline{B}_{\mathrm{Re}}, \underline{B}_{\mathrm{Im}}$ and $x_{\mathrm{Re}}, x_{\mathrm{Im}}$ are real.

We use the Cauchy-Riemann relations [8-10]: $f(z) = u(x, y) + iv(x, y)$ and

$$f'(z) = \frac{\partial f}{\partial z} = \frac{\partial \mu}{\partial x} + i\frac{\partial v}{\partial x} = \frac{\partial v}{\partial y} - i\frac{\partial \mu}{\partial y} \tag{5.16}$$

for $x = x + iy$

Now consider the definitions

$$\mu(x, y) \equiv E_{\mathrm{Re}}(x_{\mathrm{Re}}, x_{\mathrm{Im}}), \quad v(x, y) \equiv E_{\mathrm{Im}}(x_{\mathrm{Re}}, x_{\mathrm{Im}}). \tag{5.17}$$

Then

$$f(z) = E(X_{\mathrm{Re}} + iX_{\mathrm{Im}}) \tag{5.18}$$

for $z \equiv x_{\mathrm{Re}} + ix_{\mathrm{Im}}$.

We have the two equations for $f'(z)$:

$$f'(z) = \frac{df}{dz} = \frac{\partial E_{\mathrm{Re}}}{\partial x_{\mathrm{Re}}} + i\frac{\partial E_{\mathrm{Im}}}{\partial x_{\mathrm{Re}}} = \frac{\partial E_{\mathrm{Im}}}{\partial x_{\mathrm{Im}}} - \frac{i\partial E_{\mathrm{Re}}}{\partial x_{\mathrm{Im}}} \tag{5.19}$$

Returning to $\nabla \cdot \underline{E} = 4\pi\rho$, we have

$$\nabla \cdot \underline{E} = \frac{\partial E_x}{\partial x} + \frac{\partial E_y}{\partial y} + \frac{\partial E_z}{\partial z} = 4\pi\rho$$

or (5.20)

$$\nabla \cdot \underline{E} = \frac{\partial E_x}{\partial (x_{\mathrm{Re}} + ix_{\mathrm{Im}})} + \frac{\partial E_y}{\partial (y_{\mathrm{Re}} + iy_{\mathrm{Im}})} + \frac{\partial E_z}{\partial (z_{\mathrm{Re}} + iz_{\mathrm{Im}})} = 4\pi\rho .$$

Using the Cauchy-Riemann relations there are two equations for $\nabla \cdot \underline{E}$,

$$\frac{\partial E_{x\mathrm{Re}}}{\partial x_{\mathrm{Re}}}+i\frac{\partial E_{x\mathrm{Im}}}{\partial x_{\mathrm{Re}}}+\frac{\partial E_{y\mathrm{Re}}}{\partial y_{\mathrm{Re}}}+i\frac{\partial E_{y\mathrm{Im}}}{\partial y_{\mathrm{Re}}}+\frac{\partial E_{z\mathrm{Re}}}{\partial z_{\mathrm{Re}}}+i\frac{\partial E_{z\mathrm{Im}}}{\partial z_{\mathrm{Re}}} = 4\pi(\rho_{\mathrm{Re}}+i\rho_{\mathrm{Im}}) \tag{5.21}$$

and

$$\frac{\partial E_{\mathrm{Im}}}{\partial x_{\mathrm{Im}}}-i\frac{\partial E_{\mathrm{Re}}}{\partial x_{\mathrm{Im}}}+\frac{\partial E_{y\mathrm{Im}}}{\partial y_{\mathrm{Im}}}-\frac{i\partial E_{y\mathrm{Re}}}{\partial y_{\mathrm{Im}}}+\frac{i\partial E_{z\mathrm{Re}}}{\partial z_{\mathrm{Im}}}+\frac{i\partial E_{z\mathrm{Re}}}{\partial z_{\mathrm{Im}}} = 4\pi(\rho_{\mathrm{Re}}+i\rho_{\mathrm{Im}}). \tag{5.22}$$

The above equations in terms of real spatial derivatives can be separated into real and imaginary terms as

$$\frac{\partial E_{\mathrm{Re}}}{\partial x_{\mathrm{Re}}}+\frac{\partial E_{y\mathrm{Re}}}{\partial y_{\mathrm{Re}}}+\frac{\partial E_{z\mathrm{Re}}}{\partial z_{\mathrm{Re}}}=4\pi\rho_{\mathrm{Re}} \tag{5.23}$$

which is the usual Maxwell equation ($\nabla\cdot E_{\mathrm{Re}}=4\pi\rho_{\mathrm{Re}}$). We also have the "mirror" equation

$$\frac{\partial E_{x\mathrm{Im}}}{\partial x_{\mathrm{Re}}}+\frac{\partial E_{y\mathrm{Im}}}{\partial y_{\mathrm{Re}}}+\frac{\partial E_{z\mathrm{Im}}}{\partial z_{\mathrm{Re}}}=4\pi\rho_{\mathrm{Im}} \tag{5.24}$$

where the i is canceled. This equation appears to be $\nabla\cdot \underline{E}_{\mathrm{Im}}=4\pi\rho_{\mathrm{Im}}$ as before.

For the second equation for $\nabla\cdot E=4\pi\rho$ from the Cauchy-Riemann relation. We can write two equations in terms of the imaginary parts of space

$$\frac{\partial E_{\mathrm{Im}}}{\partial x_{\mathrm{Im}}}+\frac{\partial E_{y\mathrm{Im}}}{\partial y_{\mathrm{Im}}}+\frac{\partial E_{z\mathrm{Im}}}{\partial z_{\mathrm{Im}}}=4\pi\rho_{\mathrm{Re}}$$

and (5.25)

$$\frac{\partial E_{\mathrm{Re}}}{\partial x_{\mathrm{Im}}}+\frac{\partial E_{y\mathrm{Re}}}{\partial y_{\mathrm{Im}}}+\frac{\partial E_{z\mathrm{Re}}}{\partial z_{\mathrm{Im}}}=-4\pi\rho_{\mathrm{Im}}$$

in which we have multiplied through by -*i*.

Let us define a new *del* operator in terms of imaginary components of space. We define this as, ∇_{Im} and the usual *del* operator, ∇ interchangeably as ∇_{Re}. Then we have our latter two equations which become

$$\nabla_{\text{Im}} \cdot \underline{E}_{\text{Im}} = 4\pi\rho_{\text{Re}} \quad \text{and} \quad \nabla_{\text{Im}} \cdot \underline{E}_{\text{Re}} = -4\pi\rho_{\text{Im}} \tag{5.26}$$

giving us two more unique new equations. Note the minus sign in the density term in the above equation. Similarly, we can write a set of $\nabla \cdot \underline{B} = 0$ and have $\nabla \cdot \underline{B}_{\text{Re}} = 0$, $\nabla \cdot \underline{B}_{\text{Im}} = 0$, $\nabla_{\text{Im}} \cdot \underline{B}_{\text{Re}} = 0$ and $\nabla_{\text{Im}} \cdot \underline{B}_{\text{Im}} = 0$. We can write forms such as $\nabla \cdot \underline{B}_{\text{Re}} = \nabla \cdot \underline{B}_{\text{Im}}$ where we identify the term $\nabla \cdot B_{\text{Im}}$ as a monopole component. We discuss this further in terms of the Higgs solitons model.

The Higgs mechanism involves the carriers of the electroweak force, the $W^{\pm}, Z^0$ Bosons which are hypothesized in analogy to the massless or near massless photon whereas standard hadrons, leptons and pions have mass which requires an explanation. Higgs et al. [13,25,26] suggests that there was an undetected field, the Higgs field, filling the universe. The concept is that a massless Boson such as a photon could absorb a Higgs Boson and create a massive particle. Salam and Weinberg utilized the Higgs mechanism in a renormalized form to develop the electroweak theory [26]. It has been suggested that the CERN LHC Tevatron may produce enough energy to uncover the elusive Higgs particle. The question becomes, how does an all pervasive Higgs field filling the universe relate to the nature of the vacuum plenum?

Elsewhere we have given clear indications that a small photon mass, m_γ probably exists [13,25]. The physics community has thought this would interfere with Gauge Theory, but this is not the case because Gauge Theory is only an approximation. This is a key indicator of M-Theory where Planck's constant, $\hbar$ is no longer fundamental but must be modulated by string tension T_s [25]. M-Theory, is based essentially on one parameter, string tension, T_S

$$T_S = e / l = (2\pi\alpha')^{-1}; \tag{5.27}$$

where e is energy, l is length of the string and α the fine structure constant, $e^2/\hbar c$ where this e is the electron charge. It is well known that the gauge condition is an approximation suggesting Planck's constant, $\hbar$ needs to be recalculated to satisfy the parameters of M-Theory [25]. Since our HAM cosmology is aligned with an extension of Einstein's energy-dependent spacetime metric $\hat{M}^4$, (or the alternate (+++-) convention) the Stoney e^2/c, an electromagnetic precursor to Planck's constant, [25] is therefore the choice for studying the recalculation. The factor added to $\hbar$ is string tension T_S, where T_0 can increase the size of $\hbar$ to the Larmour radius of the hydrogen atom in the small scale and lead to infinite size additional dimensionality cosmologically. Thus the fine-tuned Stoney, λ and the cosmological constant, Λ adjust the microscopic and cosmological domain limits of H_R respectively. Equation (5.28) illustrates the initial historical basis for this distinction

$$l_P = \sqrt{\frac{\hbar}{mc} \cdot \frac{Gm}{c^2}} \quad or \quad l_S = \sqrt{\frac{e^2}{4\pi\varepsilon_0 mc^2} \cdot \frac{Gm}{c^2}}, \tag{5.28}$$

where l_P *and* l_S are the length of the Planck and Stoney respectively.

One example for rescaling Planck's constant comes from Wolf [25]

$$\Delta x \Delta \rho = h \rightarrow h_0 \pm \Delta h\,. \tag{5.29}$$

He then suggests that

$$\Delta h = \frac{hv^2}{c} \tau_0 L_0 \tag{5.30}$$

where τ_0 and L_0 are time uncertainty and a discrete spacetime correction respectively. Wolf is able to speculate that this Planck rescaling has application to Neutron stars, CMBR and black hole formation. Our approach for a time, τ_0 and spacetime corrections, L_0 are different [25].

What does this mean for the Higgs mechanism? There are new topological conditions in Calabi-Yau mirror symmetry. With the addition of the parameters of string tension and string coupling to the fundamental structural-phenomenology of the nature of matter, mass arises in the 'topological charge' associated with the annihilation-creation vectors of the wave structure of matter in an extended view of the de Broglie-Bohm interpretation of quantum field theory. See Chap. 12.

We examine the equations involving the curl operation. When we calculate the curl of complex E and B fields in a complex geometry we have vector components and the curl operation becomes much more complicated. This is because, for a specific vector component, we have partial derivative terms as functions of other independent variables. We proceed from the standard form of the curl for a general vector $\underline{A} = \hat{x}A_x + \hat{y}A_y + \hat{z}A_z$ as

$$\nabla \times A = \begin{vmatrix} \hat{x} & \hat{y} & \hat{z} \\ \frac{\partial}{\partial x} & \frac{\partial}{\partial y} & \frac{\partial}{\partial z} \\ A_x & A_y & A_z \end{vmatrix}$$

$$= \hat{x}\left[\frac{\partial}{\partial y}A_z - \frac{\partial}{\partial z}A_y\right] + \hat{y}\left[\frac{\partial}{\partial z}A_x - \frac{\partial}{\partial x}A_z\right] \tag{5.31}$$

$$= \hat{z}\left[\frac{\partial}{\partial x}A_y - \frac{\partial}{\partial y}A_x\right]$$

where $\hat{x}, \hat{y}, \hat{z}$ are unit vectors and $\nabla \times A$ is a vector quantity. This is the usual three spatial dimensional quantity. The *del* operation, ∇ can be formed as the D'Alembertian operator, $\square$ with $\square = \nabla - (1/c^2)(\partial^2/\partial t^2)$ which includes ct terms. If we again write $\underline{A} = \underline{A}_{\text{Re}} + i\underline{A}_{\text{Im}}$ and also the complexified form of space and time, then we will have many more terms as part of the $\hat{x}_{\text{Re}}, \hat{y}_{\text{Re}}, \hat{z}_{\text{Re}}$ components as well as $\hat{x}_{\text{Im}}, \hat{y}_{\text{Im}}, \hat{z}_{\text{Im}}$ for $\hat{x}_{\text{Re}} \equiv \hat{x}$, etc. If we turn our attention to the curl expressions such as $\nabla \times E = -(1/c)(\partial B/\partial t)$ then we can consider E and B as cases of the general form of A (not to be confused with the vector potential of $(\underline{A}, \phi)$. The usual curl is derived for a 3 x 3 matrix. Consider the components $(\hat{x}_{\text{Re}}, \hat{y}_{\text{Re}}, \hat{z}_{\text{Re}}, \hat{x}_{\text{Im}}, \hat{y}_{\text{Im}}$ and $\hat{z}_{\text{Im}})$. Then we can write the generalized curl as

$$\nabla \times A = \begin{vmatrix} \hat{x}_{\text{Re}} & \hat{y}_{\text{Re}} & \hat{z}_{\text{Re}} & \hat{x}_{\text{Im}} & \hat{y}_{\text{Im}} & \hat{z}_{\text{Im}} \\ \frac{\partial}{\partial x_{\text{Re}}} & \frac{\partial}{\partial y_{\text{Re}}} & \frac{\partial}{\partial z_{\text{Re}}} & \frac{\partial}{\partial x_{\text{Im}}} & \frac{\partial}{\partial y_{\text{Im}}} & \frac{\partial}{\partial z_{\text{Im}}} \\ A_{x\text{Re}} & A_{y\text{Re}} & A_{z\text{Re}} & A_{x\text{Im}} & A_{y\text{Im}} & A_{z\text{Im}} \end{vmatrix} \tag{5.32}$$

which forms a 3 x 6 matrix.

This generalized form is necessary for analyzing $\nabla \times E = -(1/c)(\partial B/\partial t)$ and $\nabla \times \underline{B} = -\underline{1}(\partial \underline{E}/\partial t) + \underline{J}$ for complex $\underline{E}$ and $\underline{B}$. (Note: We can handle coupling to other terms or additional terms can be handled as coupling to the usual terms which we can define as the coupling term g^2, as in [10] and Chap. 10.)

Using the set of definitions, $E_x = E_{x\text{Re}} + iE_{xIM}$, $E_y = E_{y\text{Re}} + iE_{yIM}$, $E_z = E_{z\text{Re}} + iE_{zIM}$, $x = x_{\text{Re}} + ix_{\text{Im}}$, $y = y_{\text{Re}} + iy_{\text{Im}}$, $z = z_{\text{Re}} + iz_{\text{Im}}$, and also

$\hat{e}_x = \hat{e}_{x\mathrm{Re}} + i\hat{e}_{x\mathrm{Im}}$, $\hat{e}_y = \hat{e}_{y\mathrm{Re}} + i\hat{e}_{y\mathrm{Im}}$, $\hat{e}_z = \hat{e}_{z\mathrm{Re}} + i\hat{e}_{z\mathrm{Im}}$. We formed a vector addition for the limit vector coordinates. We can also form the modulus length as $|\hat{e}_x|^2 = \hat{e}_{x\mathrm{Re}}^2 + \hat{e}_{x\mathrm{Im}}^2$. For unit dimensions, $|\hat{e}_x|^2 = 2$. Expressing the usual form of the curl of $\underline{E}$, we can use the above equations to calculate $\nabla \times \underline{E}$ as

$$\nabla \times E = \begin{vmatrix} \hat{e}_x & \hat{e}_y & \hat{e}_z \\ \dfrac{\partial}{\partial x} & \dfrac{\partial}{\partial y} & \dfrac{\partial}{\partial z} \\ E_x & E_y & E_z \end{vmatrix} = \tag{5.33}$$

$$\hat{e}_x\left[\frac{\partial E_z}{\partial y} - \frac{\partial E_y}{\partial z}\right] + \hat{e}_y\left[\frac{\partial E_x}{\partial z} - \frac{\partial E_z}{\partial x}\right] + \hat{e}_z\left[\frac{\partial E_y}{\partial x} - \frac{\partial E_x}{\partial y}\right].$$

Using the above expression for complex forms of $\underline{E}$ and $\underline{x}$ we can write

$$\begin{aligned}\nabla \times E = (\hat{e}_{x\mathrm{Re}} + i\hat{e}_{x\mathrm{Im}})\left(\frac{\partial E_z}{\partial y} - \frac{\partial E_y}{\partial z}\right) + (\hat{e}_{y\mathrm{Re}} + i\hat{e}_{y\mathrm{Im}})\left(\frac{\partial E_x}{\partial z} - \frac{\partial E_z}{\partial x}\right) \\ + (\hat{e}_{z\mathrm{Re}} + i\hat{e}_{z\mathrm{Im}})\left(\frac{\partial E_y}{\partial x} - \frac{\partial E_x}{\partial y}\right).\end{aligned} \tag{5.34}$$

We can express the term in e_x as term I_{ex},

$$\begin{aligned}= (\hat{e}_{x\mathrm{Re}} + i\hat{e}_{x\mathrm{Im}})\left(\frac{\partial E_z}{\partial y} - \frac{\partial E_y}{\partial z}\right) = \hat{e}_{x\mathrm{Re}}\left(\frac{\partial E_z}{\partial y} - \hat{e}_{x\mathrm{Re}}\frac{\partial E_y}{\partial z}\right) \\ + i\hat{e}_{x\mathrm{Im}}\left(\frac{\partial E_z}{\partial y} - i\hat{e}_{x\mathrm{Im}}\frac{\partial E_y}{\partial z}\right).\end{aligned} \tag{5.35}$$

Applying the Cauchy-Riemann relations to the terms in $\hat{e}_x$ we have

$$\begin{aligned}I\hat{e}_x = \hat{e}_{x\mathrm{Re}}\left(\frac{\partial E_{z\mathrm{Re}}}{\partial y_{\mathrm{Re}}} + i\frac{\partial E_{z\mathrm{Im}}}{\partial y_{\mathrm{Re}}}\right) - \hat{e}_{x\mathrm{Re}}\left(\frac{\partial E_{y\mathrm{Re}}}{\partial z_{\mathrm{Re}}} + i\frac{\partial E_{y\mathrm{Im}}}{\partial z_{\mathrm{Re}}}\right) \\ + i\hat{e}_{x\mathrm{Im}}\left(\frac{\partial E_{z\mathrm{Re}}}{\partial y_{\mathrm{Re}}} + i\frac{\partial E_{z\mathrm{Im}}}{\partial y_{\mathrm{Re}}}\right) - i\hat{e}_{x\mathrm{Im}}\left(\frac{\partial E_{y\mathrm{Re}}}{\partial z_{\mathrm{Re}}} + i\frac{\partial E_{y\mathrm{Im}}}{\partial z_{\mathrm{Re}}}\right).\end{aligned} \tag{5.36}$$

We also have another set of terms which we define as $I'_{\hat{e}x}$ from the other of the Cauchy-Riemann relations

$$I'_{\hat{e}x} = \hat{e}_{x\mathrm{Re}}\left(\frac{\partial E_{z\mathrm{Im}}}{\partial y_{\mathrm{Im}}} - i\frac{\partial E_{z\mathrm{Re}}}{\partial y_{\mathrm{Im}}}\right) - \hat{e}_{x\mathrm{Re}}\left(\frac{\partial E_{y\mathrm{Im}}}{\partial z_{\mathrm{Im}}} - i\frac{\partial E_{y\mathrm{Re}}}{\partial z_{\mathrm{Im}}}\right)$$
$$+ i\hat{e}_{x\mathrm{Im}}\left(\frac{\partial E_{z\mathrm{Im}}}{\partial y_{\mathrm{Im}}} - i\frac{\partial E_{z\mathrm{Re}}}{\partial y_{\mathrm{Im}}}\right) - i\hat{e}_{x\mathrm{Im}}\left(\frac{\partial E_{y\mathrm{Im}}}{\partial z_{\mathrm{Im}}} - i\frac{\partial E_{y\mathrm{Re}}}{\partial z_{\mathrm{Im}}}\right). \tag{5.37}$$

Separation into real and imaginary parts of $I_{\hat{e}x}$ and $I'_{\hat{e}x}$ can be per-formed. For $I_{\hat{e}x}$ we have

$$I_{\hat{e}x\mathrm{Re}} = \hat{e}_{x\mathrm{Re}}\left(\frac{\partial E_{z\mathrm{Re}}}{\partial y_{\mathrm{Re}}}\right) - \hat{e}_{x\mathrm{Re}}\left(\frac{\partial E_{y\mathrm{Re}}}{\partial z_{\mathrm{Re}}}\right) - \hat{e}_{x\mathrm{Im}}\left(\frac{\partial E_{z\mathrm{Im}}}{\partial y_{\mathrm{Re}}}\right) + \hat{e}_{x\mathrm{Im}}\left(\frac{\partial E_{y\mathrm{Im}}}{\partial z_{\mathrm{Re}}}\right)$$
$$= \hat{e}_{x\mathrm{Re}}\left(\frac{\partial E_{z\mathrm{Re}}}{\partial y_{\mathrm{Re}}} - \frac{\partial E_{y\mathrm{Re}}}{\partial z_{\mathrm{Re}}}\right) - \hat{e}_{x\mathrm{Im}}\left(\frac{\partial E_{z\mathrm{Im}}}{\partial y_{\mathrm{Re}}} + \frac{\partial E_{y\mathrm{Im}}}{\partial z_{\mathrm{Re}}}\right) \tag{5.38}$$

and for $I'_{\hat{e}x}$ we have

$$I'_{\hat{e}x\mathrm{Im}} = i\hat{e}_{x\mathrm{Re}}\left(\frac{\partial E_{z\mathrm{Im}}}{\partial y_{\mathrm{Re}}}\right) - i\hat{e}_{x\mathrm{Re}}\left(\frac{\partial E_{y\mathrm{Im}}}{\partial z_{\mathrm{Re}}}\right) + i\hat{e}_{x\mathrm{Im}}\left(\frac{\partial E_{z\mathrm{Re}}}{\partial y_{\mathrm{Re}}}\right) - i\hat{e}_{x\mathrm{Im}}\left(\frac{\partial E_{y\mathrm{Re}}}{\partial z_{\mathrm{Re}}}\right)$$
$$= i\hat{e}_{x\mathrm{Re}}\left(\frac{\partial E_{z\mathrm{Im}}}{\partial y_{\mathrm{Re}}} - \frac{\partial E_{y\mathrm{Im}}}{\partial z_{\mathrm{Re}}}\right) - i\hat{e}_{x\mathrm{Im}}\left(\frac{\partial E_{z\mathrm{Re}}}{\partial y_{\mathrm{Re}}} - \frac{\partial E_{y\mathrm{Re}}}{\partial z_{\mathrm{Re}}}\right). \tag{5.39}$$

We have eight terms for $I_{\hat{e}x}$ and also 8 terms for $I'_{\hat{e}x}$. Therefore there are 16 terms for the $\hat{e}_x$ term of $\nabla \times \underline{E}$.

For all 3 components ($\hat{e}_x, \hat{e}_y$ and $\hat{e}_z$) of the curl, we have a total of 48 terms. Returning to $\hat{e}_x$ terms only then, let us consider these terms only in $\nabla \times \underline{E} = -1/c\,(\partial B/\partial t)$. From $I_{\hat{e}\mathrm{Re}}$, we have, using the separation of $\underline{B}$ into real and imaginary parts and using the x component only,

$$\hat{e}_{x\mathrm{Re}}\left(\frac{\partial E_{z\mathrm{Re}}}{\partial y_{\mathrm{Re}}} + \frac{\partial E_{y\mathrm{Re}}}{\partial z_{\mathrm{Re}}}\right) - \hat{e}_{x\mathrm{Im}}\left(\frac{\partial E_{z\mathrm{Im}}}{\partial y_{\mathrm{Re}}} + \frac{\partial E_{y\mathrm{Im}}}{\partial z_{\mathrm{Re}}}\right)$$
$$= -\frac{1}{c}\frac{\partial B_{x\mathrm{Re}}}{\partial x_{\mathrm{Re}}} \tag{5.40}$$

where we use the expression as

$$\frac{\partial \underline{B}}{\partial t} = \frac{\partial B_x}{\partial t} + \frac{\partial B_y}{\partial t} + \frac{\partial B_z}{\partial t} \tag{5.41}$$

and applying the Cauchy-Riemann relations to the x component of B we have for the temporal element $t = t_{\text{Re}} + it_{\text{Im}}$, for $\partial B_x / \partial t$, then $(\partial B_{x\text{Re}} / \partial t_{\text{Re}}) + i(\partial B_{x\text{Im}} / \partial t_{\text{Re}})$.

For real parts we consider the $\partial B_{x\text{Re}} / \partial t_{\text{Re}}$ term only, which we use in the above equation. We can define a term in terms of the imaginary directed component $\hat{e}_{x\text{Im}}$; let

$$g^2 A_{\text{Im}}\left(x_{\text{Re}}, y_{\text{Re}}, z_{\text{Re}}\right) \equiv -\hat{e}_{x\text{Im}}\left(\frac{\partial E_{z\text{Im}}}{\partial y_{\text{Re}}} + \frac{\partial E_{y\text{Im}}}{\partial z_{\text{Re}}}\right) \tag{5.42}$$

so that the expression now reads

$$\hat{e}_{x\text{Re}}\left(\frac{\partial E_{z\text{Re}}}{\partial y_{\text{Re}}} + \frac{\partial E_{y\text{Re}}}{\partial z_{\text{Re}}}\right) + g^2 \text{A}_{\text{Im}}\left(x_{\text{Re}}, y_{\text{Re}}, z_{\text{Re}}\right) = -\frac{1}{c}\frac{\partial B_{x\text{Re}}}{\partial x_{\text{Re}}}. \tag{5.43}$$

Terms not incorporated into the $g^2\text{A}$ term comprise the usual Maxwell equation. We consider g^2 to be a coupling to a small order perturbation term given by $\text{A}_{\text{Im}}\left(x_{\text{Re}}, y_{\text{Re}}, z_{\text{Re}}\right)$, where our components projected from the imaginary components of E and B lie on the imaginary axis $\hat{e}_{x\text{Im}}$. Contributions from other imaginary dimensions of space and time can yield contributions that give rise to transverse components of the electromagnetic field and can contribute to energy transmission terms.

From $I_{\hat{e}x\text{Im}}$ we can also form the equation

$$\hat{e}_{x\text{Re}}\left(\frac{\partial E_{z\text{Re}}}{\partial y_{\text{Re}}} + \frac{\partial E_{y\text{Im}}}{\partial z_{\text{Re}}}\right) - \hat{e}_{x\text{Im}}\left(\frac{\partial E_{z\text{Re}}}{\partial y_{\text{Re}}} - \frac{\partial E_{y\text{Im}}}{\partial z_{\text{Re}}}\right) = -\frac{1}{c}\left(\frac{\partial B_{x\text{Im}}}{\partial t_{\text{Re}}}\right) \tag{5.44}$$

where the "i's" cancel from both sides. The terms in this equation are components of the $\hat{e}_{x\text{Im}}$ direction.

Separation into real and imaginary parts are made for terms in $I'_{\hat{e}x}$ from the second coupling relation. For the real part we have

$$I'_{\hat{e}_{x\text{Re}}} = \hat{e}_{x\text{Re}}\frac{\partial E_{y\text{Im}}}{\partial z_{\text{Im}}} - \hat{e}_{x\text{Re}}\frac{\partial E_{z\text{Im}}}{\partial y_{\text{Im}}} + \hat{e}_{x\text{Im}}\frac{\partial E_{y\text{Re}}}{\partial z_{\text{Im}}} - \hat{e}_{x\text{Im}}\frac{\partial E_{z\text{Re}}}{\partial y_{\text{Im}}} \tag{5.45}$$

and similarly for the imaginary parts $I'_{\hat{e}_{x\text{Im}}}$. All these terms are in $x_{\text{Im}}, y_{\text{Im}}$ and z_{Im}. A similar process can be done for $I_{\hat{e}y}, I'_{\hat{e}y}$ and $I_{\hat{e}z}, I'_{\hat{e}z}$. In general, we can write

$$\nabla\times\left(\underline{E}_{\text{Re}}+i\underline{E}_{\text{Im}}\right)=\nabla\times\underline{E}_{\text{Re}}+i\nabla\times\underline{E}_{\text{Im}}$$

and (5.46)

$$\nabla_{\text{Im}}\times\left(\underline{E}_{\text{Re}}+i\underline{E}_{\text{Im}}\right)=\nabla_{\text{Im}}\times\underline{E}_{\text{Re}}+i\nabla_{\text{Im}}\times\underline{E}_{\text{Im}}\,.$$

For current purposes, we will not explore terms in ∇_{Im} which involve $\partial/\partial x_{\text{Im}},\partial/\partial y_{\text{Im}},\partial/\partial z_{\text{Im}}$, etc. We will briefly discuss the relationship of the complex electric and magnetic fields, complex spacetime metrics [8,9] and the interpretation of models of the magnetic monopole.

5.3 Complex Electromagnetic Field Vectors, Virtual Energy States and Magnetic Monopole Interpretations

We will briefly discuss some issues related to magnetic monopole model interpretations. Let us start from the metric element measure for fields associated with electric and magnetic charge. Essentially, if monopoles exist they will fill in the zeroes in Maxwell's equations. Comparing $\nabla\cdot\underline{E}=4\pi\rho_R$ and $\nabla\cdot\underline{B}=0$ and $\nabla\times\underline{B}-(1/c)(\partial\underline{E}/\partial t)=(4\pi/c)\underline{J}_E$ to $\nabla\times\underline{E}-(1/c)(\partial\underline{B}/\partial t)=0$ indicates complete symmetry if the zeros on the right side were replaced by ρ_B and J_B respectively. In relativistic notation we have for the electric current $J^{\nu}_{(E)}$, $\partial F^{\mu\nu}/\partial x_{\mu}=(4\pi/c)J^{\nu}_{(E)}$ and $\partial F^{\mu\nu}/\partial x_{\mu}=0$. If monopole fields exist the right side of the second equation would be written in terms of a 4D magnetic current $J^{\nu}_{(B)}$.

Dirac hypothesizes that the pole strength of a magnetic monopole-like electric charge would be quantized and that a conservation principle for monopole strength would exist analogous to electric charge conservation principles. In [27] we examine the role of magnetic monopoles in a real multidimensional geometry. We demonstrate that the form of the quantized monopole introduced by Schwinger [27], in which the electric and magnetic charge is put on an equal footing, is consistent with the n-dimensional Descartes geometry [27,28]. If we have $e^2/\hbar c=\ \alpha\sim1/137.037$, where α is the fine structure constant, we can form an analogous expression: $em/\hbar c=n$ where n is an integer [28]. This expression defines a quantized form of the magnetic monopole.

In the Dirac monopole model [19], (where m is the 'magnetic charge' which is termed g in Schwinger's notation) if the product of the pole strengths are given as $em=n\hbar c$ and $n=\frac{1}{2}$ (the smallest quantum value), then this gives $m \cong 68.5$ times the value of e. In the Schwinger model, n is taken as unity so that $m \cong 137e$. The latter value is the one usually considered in experimental explorations.

The set of assumptions for the Schwinger monopole is one of the simplest there is; it is the monopole structure for which most experimental detectors are designed to determine if monopoles exist. This picture brings into question the whole issue of the nature of charge as a quantized entity. We discuss the possibility of a more complete expression of charge as a quantum number in [24,28,29].

Teller [30] suggests that monopole detection will be made only in very high energy experiments. Alvarez [31], and his group conducted extensive monopole detection studies. Silvers presents some theoretical formulations that are relevant to the experimental detection of magnetic monopoles. Attempts have been made to find monopoles in moon rocks [32] by looking at heavy ionized radiation damage tracks [33]. Wheeler [34] has developed expressions for quantized charge which may have relevance to monopole formulation and detection.

5.4 Higgs Field Magnetic Monopole

Our model of plasma instabilities and superconductivity are based on the field theoretic approach. Both Abelian and non-Abelian fields are considered. The Abelian Higgs field can be represented as a 3D kink soliton which acts like a bare point soliton. We might identify such a system as a "vortex." In four dimensions we can identify a non-Abelian soliton as a static monopole [35].

The common definition in the quantum solutions of the sine-Gordon equation is that the institon is a finite action entity in space and time which is associated with the content of the vacuum. In elementary particle physics this institon state could be identified with the quark-gluon states. The soliton solution is an entity of finite energy in space [36] and time and is associated with the quark states in elementary particle physics.

The 3D Abelian Higgs confined field soliton, in the same absence of symmetry breaking, defines quark confinement [37]. The Lagrangian, L for the Higgs field is given as

$$L = \frac{1}{2} F_{\eta\nu} F^{\mu\nu} + \left| \frac{\partial \phi}{\partial x} + ieA\phi \right|^2 - \frac{g^2}{4} \left(|\psi|^2 - h^2 \right)^2 \tag{5.47}$$

where g^2 is a coupling term (which acts like a potential) to a nonlinear field factor, and h^2 is an additional field term. The Higgs-Goldstone spontaneous splitting is due to the field shift term *h*.

The electromagnetic field $F_{\mu\nu}$ is given in terms of the four-vector potential by $F_{\eta\nu} = (\partial A_{\mu}/\partial x_{\mu}) - (\partial A_{\nu}/\partial x_{\nu})$ where the vector potential, A_{μ} transforms as a gauge $A_{\mu} \rightarrow A_{\mu} - (1/e)h$ which defines the quantity *h*. The indexes μ and ν runs 1 to 4. The phase ψ represents the kink in the Higgs field in 3D. The form of ψ is given by its periodic form $\psi \propto e^{in\theta}$. We define the $A_{\mu} = (\underline{A}_j, \phi)$, where the index *j* runs 1 to 3, in their usual four space form. We use ϕ to represent the temporal component of the potential field A_{μ} where $\underline{A}_j$ is the vector potential.

Let us consider photon activation of pair production of a retarded (forward in time) and advanced (backward in time) potential waves in an analogy to the Cramer Transactional model [13]. The usual physical gauge condition gives $\psi = 0$ but for our coupling soliton theory, the kink $\psi \propto e^{in\theta}$ cannot be transformed away. The stability of the vortex solutions depends on the finite value of *n*. The gauge condition in the space with kink solitons becomes

$$A_{\mu}^{'} \rightarrow A_{\mu} - \frac{1}{e} n \frac{\partial \theta}{\partial x_{\mu}}. \tag{5.48}$$

The Lagrangian, *L* gives the trajectories of the soliton where *A* is considered as the pair producing photon field. Solitons are coupled as a 1/*e* term and dominate as the coupling term g^2 becomes larger. See Chap. 8 for generalized extended Gauge conditions.

In [10,35] we discuss how soliton solutions to the nonlinear Schrödinger equation relate to the kink monopole soliton. It is actually through the relativistic formalism for the soliton solutions of the Dirac equation that we see that the kink soliton monopole is one such solution [35]. See Chap. 12. Both the Schrödinger [10] and Dirac equation are solved in the complex Minkowski space which contributes the nonlinear term leading to the soliton solutions. The soliton retains its identity in space and time and acts as a field particle that acts as a signal for remote connectedness events. The form of the soliton explains the source of the effect of the vacuum state virtual states. The exciton (pair production) couples to acoustic or acustiton modes giving rise to the soliton solution (Chaps. 10 and 12).

The Higgs field monopole relates to the symmetry term in the complex form of Maxwell equations. The current solution to the electromagnetic equations are of nonrelativistic form. The Higgs field method is a relativistic form. We will outline a relativistic complexification of Maxwell's equations.

5.5 Some Further Speculations on Monopole Structures

The relation $F_{\mu\nu} = (\partial A^{\mu}/\partial x_{\nu}) - (\partial A^{\nu}/\partial x_{\mu})$ insures that the divergence of the B field is zero. In the condition where monopoles are allowed the condition on the relationship of $F_{\mu\nu}$ in terms of A^{μ} is relaxed. We can write an expression in terms of a monopole, field strength, m. Then we can write a for

$$F_{\mu\nu} = \frac{\partial A^{\mu}}{\partial x_{\nu}} - \frac{\partial A^{\nu}}{\partial x_{\mu}} - m\delta(f) f_{\mu\nu} \tag{5.49}$$

where f is an arbitrary given function of space, x,y,z and $f_{\mu\nu} = \left\{(\partial f^{\mu}/\partial x_{\nu}) - (\partial f^{\nu}/\partial x_{\mu})\right\}$.

A number of tests for monopoles have been explored. Eberhard summarizes some of these, including the Price, Lexan controversial plate examination [38] Dirac has suggested a possible test using a soliton model. The form of the coupling constant, m, will then depend on the geometric form of the soliton. A quantum theory can be constructed for specific types of monopoles. We can define a form for m from the soliton model $A(\phi) = m/4\pi\rho$ where $\rho = \sqrt{x^2 + y^2}$ for an x directionally oriented solenoid axis and ϕ is the zenith angle of (ρ, θ, ϕ). Consider the flux Φ and we then use the monopole condition $\partial\Phi/\partial\phi = iem/2\pi$ [39,45]. If we consider the quantized flux condition in superconductivity vacua, such as $(\Psi) \neq 0$, then Ψ^{+} acts as a creation operator and Ψ as a destruction operator for magnetic charge. Asymptotically we have $(\partial\Phi/\partial x^{\mu}) - ieA_{\mu}\Phi = 0$ with solutions of the form

$$\Phi \propto \Phi_0 e^{ie\int dx^{\mu} A_{\mu}(x)} \tag{5.50}$$

with the quantized condition for a closed path,

$$e\oint dx^{\mu} A_{\mu}(x) = 2\pi n \tag{5.51}$$

where n is an integer. So the quantized flux can be considered to be obeying the condition $2\pi n/e$. This condition holds for an infinite solenoid on the z axis (Aharanov-Bohm experiment).

More detailed consideration along this line may be fruitful to design a test for a possible monopole utilizing a solenoid configuration [41]. See Chap. 4. A more detailed examination of this picture and the suggested experiment by the Eyring Research Group should be made in which they suggest a test of the issues connecting $\underline{E}$ and $\underline{B}$, and $\underline{A}$ and ϕ [42]. In a suggested experiment by Mandelstam [43], gauge invariance and Poincaré invariance conditions need evaluation. The complexification of Maxwell's equations give us a detailed manner in which to formulate the nonlinear coupled terms, g^2.

5.6 The Structure of Non-Hertzian Waves in Complex Geometries and Electromagnetic Energy Transmission

Heinreich Hertz made two contributions that had a major influence on the interpretation of the nature and structure of electromagnetic waves. Maxwell had already shown the intimate relationship between electric and magnetic phenomena which had drawn together many of the discoveries by Faraday [15]. One of the two issues that Hertz put forward was that radio and light waves were part of the same phenomena; i.e. part of the electromagnetic spectra. The other was that electromagnetic waves were composed of the continuous orthogonal oscillations of electric and magnetic vector components transverse to the direction of motion. These oscillations traveled at the velocity of light (Maxwell) and the velocity of light is a constant in all frames of Einstein.

The former proposition of Hertz led to a coherent picture of many phenomena (such as radio, light, x-rays, and γ-rays) as part of the electromagnetic spectra. The condition on the vector oscillations of $\underline{E}$ and $\underline{B}$ may have been too restrictive and also that longitudinal components may exist and may have most significant implications [44-47]. Because of the great success of the former issue the second consideration was readily accepted. There was also a lack of understanding of Tesla's energy transmission ideas in his cryptic patents and also he was unable to complete vital tests of his ideas due to loss of funding from J.P. Morgan and his family [44]. Therefore the issue of longitudinal components of $\underline{E}$ and $\underline{B}$ and their possible interpretation as effects on A_μ or (A_μ, ϕ) was summarily dismissed from classical electromagnetic theory. The Aharonov-Bohm experiment appears to show that the (A_μ, ϕ) fields are detectable outside of the action of the $\underline{E}$ and $\underline{B}$ fields.

The ground wave and the ionospheric wave are set up in such a manner as to produce the predicted 1.57 ratio to the velocity of light which was stated by Tesla in one of his patents [44,47]. In his model Tesla treated the Earth as a finite capacitative reactance component surrounded by an ion shell of variable altitude, beginning at about 50 km in height, which represents a system whereby a resonant ringing signal can be set up and transmitted. Although the system represents a leaky capacitor with a Q of about 4 to 5 it is possible to set up a resonant state that appears as though a signal is transmitted and received from any two points on the Earth's surface. In actuality, according to the Rauscher-Van Bise model, the signal is not 'transmitted and received' but represents a nonlocal global coherent state. Any event which can 'wiggle' the static Earth-ionosphere magnetic flux is transmitted as both a local and nonlocal influence.

We will discuss in this section one model of non-Hertzian waves and suggest that there may be more modest tests of longitudinal wave effects and energy transmission than the major energy transmission program Tesla envisioned [44-48]. Some possible considerations for experiments may involve a solenoid Aharonov-Bohm type experiment and certain antenna designs for transmission and reception of significant signal, energy information and perhaps polarization experiments. See Chap. 4.

If we consider the complex form of $\underline{E}$ and $\underline{B}$ then we can consider an orthogonal space in which the real components $\underline{E}_{\text{Re}}$ and $\underline{B}_{\text{Re}}$ are transverse projections to the direction of propagation of the wave and are the usual transverse components. The orthogonal components $\underline{E}_{\text{Im}}$ and $\underline{B}_{\text{Im}}$ (where $\underline{E}_{\text{Im}}$ and $\underline{B}_{\text{Im}}$ themselves are real) are projections on the direction of propagation of the wave and comprise the longitudinal components. These longitudinal components may act in an acoustic-like or acusticon motion $\underline{E}_{\text{Re}}$, $\underline{B}_{\text{Re}}$, $\underline{E}_{\text{Im}}$ and $\underline{B}_{\text{Im}}$ are all mutually orthogonal although models can be considered in which, although maxima of $\underline{E}_{\text{Re}}$ and $\underline{B}_{\text{Re}}$ are 90° out of phase, those of $\underline{E}_{\text{Im}}$ and $\underline{B}_{\text{Im}}$ can be in phase or 90° out of phase [49].

Longitudinal oscillations of $\underline{E}_{\text{Im}}$ and $\underline{B}_{\text{Im}}$ (See Fig. 5.2) appear as presence and absence of these fields varying from maximum projection of $\underline{E}$ and $\underline{B}$ to zero projection on the direction of propagation. The constraint conditions $\underline{E} = \underline{E}_{\text{Re}} + i\underline{E}_{\text{Im}}$ and $\underline{B} = \underline{B}_{\text{Re}} + i\underline{B}_{\text{Im}}$ but we can also express the relationship between transverse and longitudinal components as $\underline{E} = \underline{E}_{\text{Re}} + ie\underline{E}_{\text{Im}}$ and $\underline{B} = \underline{B}_{\text{Re}} + ib\underline{B}_{\text{Im}}$ where *e* and *b* can be chosen to be greater than or less than unity. This way we can determine the relationship between the magnitude of the transverse and longitudinal components. The existence of the imaginary components of $\underline{E}$ and $\underline{B}$ derive their existence from the imaginary components of space and time. Dependent relationships such as $\underline{E}_{\text{Im}}(x_{\text{Im}}, t_{\text{Im}})$ can be found as well as $\underline{E}_{\text{Im}}(x_{\text{Re}}, x_{\text{Im}}, t_{\text{Re}}, t_{\text{Im}})$ can be formed. Essentially though, $\underline{E}_{\text{Im}}$ and $\underline{B}_{\text{Im}}$ derive their meaning from the components $(x_{\text{Im}}, t_{\text{Im}})$ as previously discussed.

New issues to address with the new formalism are primarily related to the possibility of non-Hertzian wave activity and transmission either in space or in a dielectric media. Possible means of "lossless" energy transmission or communication would necessarily involve non-Hertzian wave phenomena which does not attenuate in the usual $1/r^2$ diffusion mode. Of course laser light does not attenuate significantly in free space and is Hertzian and coherent, but a great amount of energy is not transmitted, nor can lasers be utilized (in their current form) to communicate with higher efficiency with undersea systems [50].

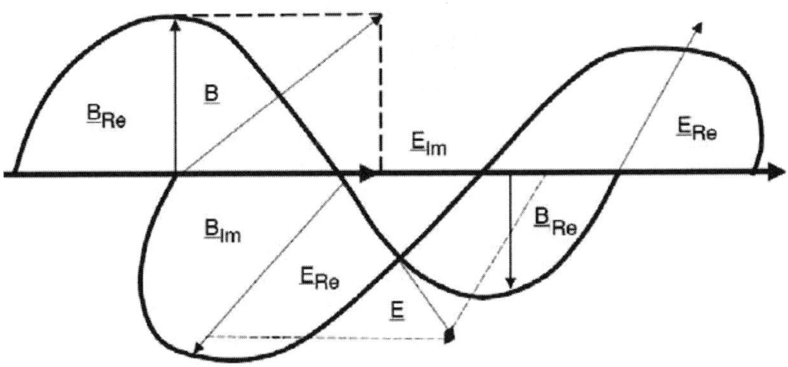

Figure 5.2. Hertzian and non-Hertzian waves. Where E and B are decoupled into two components $\underline{E}_{\text{Re}}$ and $\underline{E}_{\text{Im}}$ and $\underline{B}_{\text{Re}}$ and $\underline{B}_{\text{Im}}$.

Certain properties of the lasing phenomena do hold some clues for us because of its coherence properties. The possibility exists for utilization of the laser system (Hertzian wave) for remote communication, which can be formulated in terms of the remote connectedness properties of multidimensional geometries [8]. As stated before, phenomena such as Bell's interconnectedness theorem, Young's double slit phenomena, supercoherence phenomena and plasma instabilities (or coherent electron-electron states) etc. derive their properties and structures from the properties of remote connectedness and apparent superluminal connection in the 4-space as a subset of HD geometries [9,13].

Planck in his 1922 book on electromagnetic theory expresses the concept that energy is key to an understanding of Maxwell's equations and therefore proceeds from the Poynting vector, $\underline{S}$ (perpendicular to the vectors $\underline{E}$ and

$\underline{B}$) which is in the direction of energy flow [50,51]. Note that S is also called the radiant vector. The electromagnetic energy field is calculated by the work to create the field on ergs

$$w = \frac{\varepsilon E^2 + \mu H^2}{8\pi} \tag{5.52}$$

where ε is the dielectric constant or permittivity of the media and μ is the permeability (for free space or matter that is isotropic and non-ferromagnetic) and $c = 1/\sqrt{\varepsilon\mu}$ in matter and $c = 1/\sqrt{\varepsilon_0\mu_0}$ in vacuum. Then $\underline{S} = (c/4\pi)\underline{E}\times\underline{H}$, where $\underline{B} = \mu\underline{H}$ and the velocity in the direction of the Poynting vector is $\underline{v} = \underline{S}/w$ where $v = c$ is usually the case and where c is the velocity of light in vacuo. If we assume that each erg of moving energy has a mass of $1/c^2$, using $E = mc^2$ or a mass of about $1/(9\times10^{20})$ gm.; the energy in a cubic centimeter will have momentum equal to

$$\left(w/c^2\right)v = s/c^2 \tag{5.53}$$

for $\underline{B} = \mu\underline{H}$, or similar expressions, depending on the media. In free space this is the magnitude of the momentum in unit volume so that the electromagnetic momentum in free space may be thought of as ordinary momentum possessed by the moving electromagnetic field.

The vectors $\underline{E}$ and $\underline{B}$ are represented as waves of electric and magnetic fields moving in a direction of propagation perpendicular to their amplitude variation. This variation is sinusoidal and transverse to the direction of propagation of the electromagnetic disturbance. For propagation in the x direction then, $E_y = f(x - vt)$, $E_x = E_z = 0$, and $H_z = \sqrt{\varepsilon/\mu}\ f(x - vt)$; $H_x = H_y = 0$ and $v = c/\sqrt{\varepsilon/\mu}$ so that the wave can be in a media or free space. Then we have a wave equation

$$-\sqrt{\frac{\varepsilon}{\mu}}\frac{\partial f}{\partial x} = \frac{\varepsilon}{c}\frac{\partial f}{\partial t} \tag{5.54}$$

for $E_y = A\sin 2\pi(\frac{x}{\lambda}vt)$ and $H_z = \sqrt{t/\mu}\ A\sin 2\pi(\frac{x}{\lambda} - vt)$ which are plane wave forms. Now let us briefly discuss possible longitudinal components.

It probably would not make sense to consider longitudinal vector modes along $\underline{S}$ but scalar modes may be perfectly acceptable. As indicated by other calculations, acoustic type collective excitations arise from coherent, collective, nonlinear phenomena. Consider the propagation of an acoustic type mode, which are described as a soliton, if interaction with a source term (or exciton term) exists. Such a mode will not involve a Poynting vector energy term and with a source term would not obey the usual $1/r^2$ dispersion. Actually the recoherence from the nonlinear term overcomes the dispersion loss and disturbances do not eventually "wipe out", such as by water waves from a rock tossed into a pond, but retain their amplitude as in the soliton case [51]. Water waves cause interatomic friction and loss converts to heat in the water media. Electromagnetic energy disperses by dielectric (displacement currents), excitation of a media, and $1/r^2$ dispersion. The Hertzian wave momentum "pushes" through space.

The energy relationship for non-Hertzian waves is not of the form $S=(\varepsilon E^2+\mu H^2)/8\pi$ does not fall off as $1/r^2$ with distance. The question then becomes, what is the energy content in standing and transmitted coherent non-dispersive waves such as solitons? Certain properties of ELF waves may not only depend on their extremely long wave lengths (~10^9 cm), but also on a possible mechanism for creating and transmitting extremely low frequency nonlinear waves. These may have some non-Hertzian properties particularly below 10 kHz. These properties may explain low loss (non-attenuation) of wave energy and lack of frequency shifts when observed from different spatial locations in recently observed (since 1976) ELF phenomena [45]. The energy content is assumed to be distributed throughout the field in the direction of the Poynting vector, which is perpendicular to $\underline{E}$ and $\underline{B}$ and has a magnitude

$$S=\frac{c}{4\pi}EB\sin\theta. \tag{5.55}$$

As before the velocity of propagation in the direction of $\underline{S}$ is given as $v=S/\mathrm{E}$ and θ is the angle between $\underline{E}$ and $\underline{B}$.

The transverse mode may be associated with an acoustic-like wave of energy transmission. We have explored the manner in which acoustic modes reflect coherent, collective, nonlinear processes and relate to the coherent state, as modeled in the soliton physics of [10,24,45]. The soliton mode is pictured as a coupling of a collective acoustic mode to exciton (electron-positron) modes in a media. See Fig. 5.2 and Chap. 10.

Let us briefly examine a possible interpretation of a more general form of the electromagnetic field, **E**. We can consider complexification of **E** as $\mathbf{E}=\mathrm{E}_{\mathrm{Re}}+i\mathrm{E}_{\mathrm{Im}}$. Consider the terms $\varepsilon=\varepsilon_{\mathrm{Re}}+i\varepsilon_{\mathrm{Im}}$; $\mu=\mu_{\mathrm{Re}}+i\mu_{\mathrm{Im}}$; $E^2=E_{\mathrm{Re}}^2+E_{\mathrm{Im}}^2$ and $B^2=B_{\mathrm{Re}}^2+B_{\mathrm{Im}}^2$. We use the modulus of a vector form as $|E|^2=EE^*$, for example. Then we can form E as

$$\mathrm{E}_{\text{total}} = \frac{1}{8\pi}\{\varepsilon_{\text{Re}}E_{\text{Re}}^2 + i\varepsilon_{\text{Im}}E_{\text{Re}}^2 + \mathrm{E}_{\text{Re}}E_{\text{Im}}^2 + i\mathrm{E}_{\text{Im}}E_{\text{Im}}^2 +$$

$$\mu_{\text{Re}}B_{\text{Re}}^2 + i\mu_{\text{Im}}B_{\text{Re}}^2 + \mu_{\text{Re}}B_{\text{Im}}^2 + i\mu_{\text{Im}}B_{\text{Im}}^2\}. \quad (5.56)$$

We collect the terms in E_{Re} and E_{Im} The usual terms in E are $\varepsilon_{\text{Re}}E_{\text{Re}}^2$ and $\mu_{\text{Re}}B_{\text{Re}}^2$. We also have real terms $\varepsilon_{\text{Re}}E_{\text{Im}}^2$ and $\mu_{\text{Re}}B_{\text{Im}}^2$ which comprise E_{Re}. The parts that comprise E_{Im} are given by $\varepsilon_{\text{Im}}E_{\text{Re}}^2$, $\varepsilon_{\text{Im}}E_{\text{Im}}^2$, etc., as

$$\mathrm{E}_{\text{Im}} = \frac{1}{4\pi}\{\varepsilon_{\text{Im}}E_{\text{Re}}^2 + \varepsilon_{\text{Im}}E_{\text{Im}}^2 + \mu_{\text{Re}}B_{\text{Re}}^2 + \mu_{\text{Im}}B_{\text{Im}}^2 + i\}. \quad (5.57)$$

The traditional terms in E_{Re} as the usual terms as

$$\mathrm{E}_{\text{Re}} = \frac{1}{4\pi}\{\varepsilon_{\text{Re}}E_{\text{Re}}^2 + \mu_{\text{Re}}B_{\text{Re}}^2 + \varepsilon_{\text{Re}}E_{\text{Im}}^2 + \mu_{\text{Re}}B_{\text{Im}}^2\}. \quad (5.58)$$

These latter two terms come from projected longitudinal components of the electromagnetic field. The usual components, $\varepsilon_{\text{Re}}E_{\text{Re}}^2$ and $\mu_{\text{Re}}B_{\text{Re}}^2$.

The corresponding longitudinal Poynting vector is given as

$$S' = \frac{c}{4\pi}EB\cos\theta. \quad (5.59)$$

To be more precise, we have the usual transverse Poynting vector

$$S_{\text{Re}} = \frac{c}{4\pi}E_{\text{Re}}B_{\text{Re}}\sin\theta \quad (5.60)$$

and the longitudinal Poynting vector

$$S_{\text{Im}} = \frac{c}{4\pi}E_{\text{Im}}B_{\text{Im}}\cos\theta. \quad (5.61)$$

In each case respectively the angle θ is defined between $\underline{E}_{\text{Re}}$ and $\underline{B}_{\text{Re}}$ or, in S_{Im} as between $\underline{E}_{\text{Im}}$ and $\underline{B}_{\text{Im}}$. These expressions depend on the assumption that both the transverse and longitudinal components are transmitted at the velocity of light, c, and that c retains its relationship with ε and μ.

The constraints on ε_{Im} and therefore on $\varepsilon_{\text{Im}}E_{\text{Im}}^2$, etc. terms, must be such as to retain the relationship between c and $\varepsilon_{\text{total}}$ and μ_{total}. If the acoustic (longitudinal acusticon) mode of transmission should occur at some other velocity, such as $v > c$, then we need to examine the whole issue of Lorentz invariance. See Chaps. 2 and 9. Feinberg [10] has demonstrated that $v > c$ signals for tachyonic particles with complex mass can occur [25] and arguments such as these have been demonstrated to be consistent with the complex Minkowski space [8,44]. See Chap. 2. In fact, the structure of the metric demands a superluminal signal. Note that Tesla described a non-Hertzian superluminal signal [44]. The form of the Poynting vector then reflects signaling, should it be detected, in which S_{Im} would depend on some general velocity $v > c$. The longitudinal acoustic mode then may require new considerations in experimental detection designs that involve some of the considerations in the concepts in tachyon detection. It may well be that the monopole is a tachyon and may therefore require similar approaches to those of attempted monopole detection [37] in which remote connection in the multidimensional Cartesian geometry is related to superluminal signals and magnetic monopoles. Also similar considerations are made for complex geometries [9].

Two main issues come to mind. First, can information be transmitted by a superluminal acoustic wave?, and second, can energy be transmitted by a superluminal acoustic wave? We have previously demonstrated that collective coherent acoustic modes occur in matter in complex Minkowski spaces [24] and that acoustic modes coupled with vacuum state polarization may account for a variety of coherent phenomena such as plasma instabilities and superconductivity. These phenomena appear to depend on the remote connectivity of the manifold which is well described by the complex geometry.

Orthogonality of $\underline{E}_{\text{Re}}, \underline{B}_{\text{Re}}$ and $\underline{E}_{\text{Im}}, \underline{B}_{\text{Im}}$ is insured. A frequency dependent interaction between transverse and longitudinal components could lead to a standing wave, configuration. A self-reinforcing configuration could develop which would allow remote information transfer and interaction. Essentially such a model would be analogous to the coherence configuration of a laser but also have properties of nonlocality; possibly of energy 'transmission' or simultaneous information effects such as Bell's Theorem. See Chap. 4. Precise geometric transmitters (antennas) which form a nonlinear geometric array would be necessary to transmit the "acoustic" longitudinal components of the field.

Possible biological effects from ELF radiation may be due to nonlinear tissue "windowing" [21-23,52-72]. Nonlinear properties of tissues in which lipoproteins may act as receiving antennas could explain biological activity to ELF or higher frequency electromagnetic fields [24] which are not explained by the usual thermal effects, where intensity is below the half degree threshold. Additional calculations and interpretations are in progress which relate to both the laser coherence remote information effects (communication) and possible models of nonlinear transmitter receivers for ELF radiation.

Maxwell and Hertz primarily respectively dealt theoretically and experimentally with radio frequencies (RF) and above. Light can be produced by the excitation from charged particles such as e^- and p^+ in the atmosphere such as from lightning in the visible and x-ray region such as the sprits and jets in the upper atmosphere and the aurora borealis which lies above them. These phenomena tend to perturb the Earths steady state fields, as well as from solar wind activity leading to ULF, VLF and ELF phenomena. Most research has been in the MHz and above frequency region and only recently studied in geophysics [22,44,45,73-81] and biological science [20,21,52-72]. Maxwell's equations are wave equations and well described phenomena down to the upper KHz region but not so well for the ELF and VLF region of the electromagnetic spectrum. Some of the principles of the applications to low frequency phenomena can be listed as follows: Note that the standard Maxwell's equations fail in this region below about 10KHz because not only are Hertzian waves involved but so are non-Hertzian waves as formulated in this chapter. Phenomena in geophysics and biology exhibit both Hertzian and non-Hertzian phenomena and apply to the low end of the electromagnetic spectrum. Particularly in biology collective neuronal processes in the brain and cardiac system exhibit non-Hertzian receiving and transmitting modalities. Antenna designs are based on Hertzian wave phenomena and hence are not designed to pick up and receive non-Hertzian systems. Rauscher and Van Bise have designed such a system [22,44,45,73-81] which detects ELF, VLF and ULF signals involving Hertzian and non-Hertzian waves up to 500 kHz. These researchers have also applied this research to understanding and developing interactive systems with biological tissue to enhance its function through resonance coupling [20,21,52-72].

5.7 Summary and Concluding Remarks

Complexifying and expanding Maxwell's equations allows us to understand and expand upon our knowledge of low intensity and low frequency phenomena which is consistent with collective resonant recoherence of electric and magnetic transverse and longitudinal phenomena which can accommodate nonlocal interactions. With our new formulation, a number of systems can be reexamined and new ones developed. Some of the areas of research that can be examined and are being explored for technological development are:

- Energy systems and energy and information transmission and designing non-Hertzian antennas.
- Better design and development and theoretical understanding of controlling and utilization of plasma energy.
- More complete understanding of nonlocality in quantum processes and the development of quantum computation [13].
- A deeper and more complete and comprehensive understanding of the Earth and Earth's ionosphere and magnetosphere, understanding

and data analysis using the T-1050 detection system (Rauscher-Van Bise design patent US 4,724,390) and analysis program for better solar wind, Earth and lunar processes [22,45] as well as design and science method and volcanic prediction as to location, approximate magnitude with warning cycles six weeks, two to three weeks and about two hours before the impending seismic event by deployment and operation of an array of earthquake detecting precursor sights [22,73-81].

- Emergence of new medical modalities which are non-invasive, more gentle and medically effective and cost efficient. These involve methods of complete external cardiac normalization (US Patent 4,723,536 - non-invasive heart pacemaker) and pain reduction (US Patent 4,724,390) and elimination and brain wave effects that correct abnormal brain functions involving highly specific resonances tuned to biological tissue by pulsed magnetic fields. Biological maladaptive conditions not treatable by current medical modalities can be effectively treated with long term enhanced biological functional outcomes [20,21,23,58-71].

Some additional implications of complexification of Maxwell's equations for low, intermediate and high frequency phenomena:

- Relaxation and modification of gauge invariance conditions in which non-Hertzian or longitudinal waves occur. $A'_\mu \rightarrow A_\mu - \frac{1}{e} n \frac{\partial \Lambda}{\partial x_\mu}$ is modified. See Chap. 8.
- Comparison to the Boltzman-Maxwell or Vlassov Magneto Hydrodynamic (MHD) equations which allows transverse and longitudinal components of $\underline{E}$ and $\underline{B}$ in a high temperature plasma around 10^6 K.
- The usual condition, $F_{\mu\nu} = \frac{\partial \Lambda}{\partial y_\mu} - \frac{\partial \Lambda}{\partial x_\mu} \Rightarrow \nabla \cdot \underline{B} = 0$. In complexified modified form Maxwell's equations, $\nabla \cdot \underline{B} \neq 0$ have a monopole term mentioned earlier.
- Possible advanced potential 'pilot' wave mode of remote connection and Bell's Theorem. See Chap. 4.
- Possible explanation of the Aharonov-Bohm experiment where $\underline{A}, \varphi$ outside the effect of $\underline{E}$ and $\underline{B}$. We have $\underline{B} = \underline{B}_0 + \nabla \times \underline{A}$ and $\underline{E} = \underline{B}_0 - \nabla \phi$ for tensor potential, $\underset{\sim}{A}$, vector potential, $\underline{A}$ and scalar potential, ϕ. See Chap. 4.

- Finite but very small mass of the photon, m_γ has been postulated [25] and the mass of the neutrino has been detected experimentally [27,29].

Some examples of modifications of Maxwell's equations for ULF, ELF, VLF and LF frequency range are listed as follows:

- Maxwell's equations and the Hertzian electromagnetic wave assumptions are primarily applicable to $\omega \geq 10$ kHz. For example in MHD oscillatory collections, electron motion produces electromagnetic waves that have both transverse and longitudinal components even in the RF region of the EM spectrum [45]. Also the 10 kHz region and below apply to biological systems [20,21,52-72].
- For frequencies below 10 kHz we can treat Maxwell's equations in the slowly varying soliton-like envelope approximation (SVEA) [20,21].
- We consider periodic variations of the magnetic field governed by nonlinear evolution equations with dispersion, diffusion and dissipative modes overcome by nonlinear recoherences, $\varphi_{xx} - \varphi_{tt} - \sin\varphi = \delta\varphi_t - \gamma\varphi_{xx}$ where the δ and γ terms represent wave dissipation losses and x and t are the usual independent spacetime variables and $\sin\varphi$ is the nonlinear term. Note φ_{xx} stands for $\partial^2\varphi / \partial x^2$ and φ_{tt} stands for $\partial^2\varphi / \partial t^2$.
- Analogy is made to the Korteweg-deVries equation in which nonlinear terms of the dispersive losses, φ_{xxx} yield soliton solutions [20,21].
- Both transverse and longitudinal modes of excitation are generated and a generalization of the usual gauge conditions are formulated such as to accommodate both Hertzian and non-Hertzian phenomena.
- The Lagrangian forms for the modified gauge conditions are of the form $L = 1/2(\varphi_{xx})^2 - 1/2(\varphi_{tt})^2 - \cos\varphi$ are made, which is written for a model of naturally occurring coherent time evolutionary soliton-like wave.
- Some forms of relaxation of the gauge invariance effect conditions on the divergence of the magnetic field, $\nabla \cdot \underline{B} = 0$ and hence relate to the possibility of a magnetic monopole.

- The separation of E and B for the ELF region of the spectrum represents what occurs in some types of biological tissue and atmospheric and ionospheric phenomena and other applications..

The E and B fields no longer primarily act in concert as an electromagnetic wave, but can act as electric and magnetic fields separately but in a coordinated manner which occurs in the detection of biological signaling as well as in the detection and analysis of ionospheric resonances, seismic and volcanic precursors and other low frequency, low intensity resonant phenomena and other applications.

References and Notes

[1] Weyl, H. (1918) *Math. Z.* 20, 384.
[2] Weyl, H. (1918*) Gravitation und Elektrizität, Sitzungsberichte d. Preuss. Akad. d. Wiss.*: 465.
[3] Weyl, H. (1920) *Phys. Z.* 21, 649.
[4] Weyl, H. (1957) Space-Time-Matter, Dover Publications, Inc.
[5] Kaluza, T. (1921) *Sitz. Berlin Preuss, Acad. Wiss.* 966, B. Hoffman, private communication, SRI, August 24, 1976 with EAR The Kaluza work was pointed out to Hoffman by A. Einstein in the 30's as a significant direction to investigage and Hoffman suggested these references to EAR at SRI International.
[6] Klein, O. (1926) *Z. Phys.* 37, 895; Additional constraints consistent with the nonlocal model in terms of a complex 8-space may be accomplished by introducing a Kaluza-Klein-like tempora 6D of the form $\gamma_{66} = -1$ suggested by EAR.
[7] Inomata, S. (1976) Consciousness and complex electromagnetic fields, Electrotechnical Laboratory, MITI, 5-4-1 Mukodai-cho, Tanaski-city, Tokyo.
[8] Ramon, C. & Rauscher, E.A. (1980) Found. Physics 10, 661.
[9] Rauscher, E.A. (1979) in The Iceland Papers, select papers on experimental and theoretical research on the physics of consciousness, Frontiers of Physics Conference, Amherst: Essentia Research Associates; and (1996) Ottawa: PACE.
[10] Rauscher, E.A. (1981) Conceptual changes in reality models from new discoveries in physics, Tecnic Research Laboratories publication PSRL-1076, M/B Conference, Palo Alto, Octobver 10-11; and (1981) Proceedings of the 1st International Symposium on Non-Conventional Energy Technology, October 23-24 1981 Ontario: University of Toronto; and Proceedings of the 10th International Conference on the Sciences, Vol II, pp. 1407-1431, New York: ICF Press.
[11] Rauscher, E.A. (1978) Bull. Amer. Phys. Soc. 23, 84.
[12] Rauscher, E.A. (1983) Electromagnetic Waves in Complex Geometries and Nonlinear Phenomena, Non-Hertzian Waves and Magnetic Monopoles, Millbrae: Tesla Books.
[13] Amoroso, R.L. & Rauscher, E.A. (2009) The Holographic Anthropic Multiverse: Formulating the Complex Geometry of Reality, Singapore: World Scientific.
[14] Maxwell, J.C. (1873) A Treatise on Electricity and Magnetism, Oxford:

Clarendon Press, London, Vols. I &II); and (1865) A dynamic theory of the electromagnetic field, Royal Society Transactions, Vol. CLV, pg. 459. Appreciation to E. Byrd for supplying the author with a copy of this paper.
[15] Faraday, M. (1885) Experimental Researches in Electricity, Vols. I, II, III, in R. Taylor & W. Francis, (eds.) London: University of London.
[16] The alternative of defining and using $E' = E_{Re} + iB_{Im}$ and $B' = B_{Re} + i/c\, E_{Im}$ would not yield a description of the magnetic monopole in terms of complex quantities, but yields, for example, $\nabla \cdot (iB_{\text{Im}}) = 0$ in the second set of equations. The proper gauge conditions are not well described by this approach.
[17] Rauscher, E.A. & Van Bise, W. (1984) Electrostatic cooling theoretical and experimental analysis and application ot heat transfer enhancement in 2219-T87 aluminum weldments, PSRL-1072P-1, TRL report for NASA Lockheed-Martin Aerospace Advanced Quality Technology Dept. New Orleans, Contract SC-W02846, December 1983, Part I, 84 pages, December 1984, Part II, 90 pages; This report received the US Air Force outstanding recognition.
[18] Beal, J. (1979) Space Shuttle Program, Lockheed-Martin, Martin Marietta, New Orleans; private communication with EAR
[19] Dirac, P.A.M. (1948) Phys. Rev. 74, 817.
[20] Rauscher, E.A. & Van Bise, W. (1988) External Magnetic Field Impulse Pacemaker Non-Invasive Method and Apparatus for Modulating Brain Signals through an External Magnetic or Electric Field to Pace the Heart and Reduce Pain, U.S. Patent Number 4.723.536 issued February 9.
[21] Rauscher, E.A. & Van Bise, W. 1989 Non-invasive Method and Apparatus for Modulating Brain Signals Through an External Magnetic or Electric Field to Reduce Pain. U.S. Patent Number 4.889.526 issued December 26.
[22] Rauscher, E.A. & Van Bise, W. (1988) Non-superconducting Apparatus for Detecting Magnetic and Electromagnetic Fields, U.S. Patent Number 4.724.390, Febraury 9.
[23] Rauscher, E.A., Van Bise, W. , & Kroening, R. (1989) Magnetic control of low back pain, Bull. Amer. Phys. Soc. 34, 109.
[24] Haramein, N. & Rauscher, E.A. (2005) Collective coherent oscillation plasma modes in surrounding media of black holes, in R.A. Amoroso, B. Lehnert & J.P. Vigier (eds.) Beyond the Standard Model: Searching for Unity in Physics, pp. 279-331, Oakland: The Noetic Press.
[25] Randall, L. (2005) Warped Passages: Unraveling the Mysteries of the Unvierse Hidden Deimensions, New York: Harper Collins.
[26] Veltman, M. (2003) Elementary Particle Physics, Singapore: World Scientific.
[27] Rauscher, E.A. (1971) A Unifying Theory of Fundamental Processes, UCRL-20808, UCB / LBNL report.
[28] Ford, K.W. (1963) Magnetic monopoles, Sci. Am., December, p. 122.
[29] Vigier, J.P. (2002) Selected papers, Jean-Piere Vigier and the Causal Stochastic Interpretation of Quantum Mechanics, S. Jeffers, B. Lehnert, N. Abramson & Chebotarov, (eds.) Montreal, Aperion; and private communications with RLA.
[30] Teller, E. (1966) Quarks and magnetic monopoles, Talk at the LLNL, January 1966; and private communication with EAR.
[31] Alvarez, L.W. et al., (1970) A magnetic monopole detector utilizing

superconducting elements, UCRL-19756; and private communication.
[32] Silvers, D. (1979) Possible bending of a magnetic monopole to a particle with electric charge and magnetic dipole moment, UCRL-19794.
[33] Price, B. (1978) private communication, University of California, Berkeley, February 1971, June 1978.
[34] Wheeler, J. (1962) Geometrodynamics, New York: Academic Press; and fruitfull private communication with EAR.
[35] Rauscher, E.A. & Amoroso, R.L. (2010) Solutions to the Dirac equation in complex Minkowski spaces, in progress.
[36] Maharana, J. (1982) Proc. Theoret. Phys. 68, 277.
[37] Nimanton, P. (1979) Nucl. Phys. B158, 141.
[38] Eberhard, P. (1975) Status of the search for magnetic monopoles, presented at the Division of Particles and Fields, August 27, 1975, and LBL-4289; private communication with EAR.
[39] Dirac, P.A.M. (1931) Proc. Roy. Soc. A133, 60.
[40] Alvarez, L.W. (1975) Talk at the Stanford lepton/photon symposium, August 1975; and LBL-4260; private communication with EAR.
[41] Mandelstam, S. (1976) Talk LBNL, March 1; and private communication EAR.
[42] Discussions and communications with R. Curtin & D.L. Faust, Eyring Research Institute, Inc., Provo, Utah.
[43] Hertz, H. (1955) The Principles of Mechanics, , p.1899, New York: Dover.
[44] Tesla, N. (1989) Nikola Tesla Complete Patents, Vols. I & II; and Nikola Tesla Selected Patent Wrappers, Vols. I-IV, compiled by J.T. Ratzlaff , Milbrae: Tesla Books; and Colorado Springs Notes as well as various papers and articles.
[45] Rauscher, E.A. & Van Bise, W. (2002) Fundamental excitatory modes of the earth and earth ionosphere resonant cavity, pp. 233-268, in Thomas Valone (ed.) Harnessing the Wheelwork of Nature, Tesla's Science of Energy, Kempton: Adventures Unlimited Press.
[46] Tesla, N. (1905) Art of Transmitting Electrical Energy through Natural Mediums, US Patent 787,412, April 18.
[47] In calculating the velocity ratio of air and ground waves, one approach is to consider an air (earth ionosphere) wave travelling at v_2 and a through-the-earth wave traveling at v_1. Consider two waves emitted from the same location on the earth's surface, one in the air and the other through the earth and both traversing paths in the same time so as to come back to the emission location as reinforced. The path length for the air wave is πD and the through-earth wave is 2D. For equal time of travel, the velocity becomes $v_2 / v_1 = \pi / 2 = 1.57$. In this analysis, the greater velocity wave, v_2, is the air wave. If v_1 is chosen to be the velocity, then the relative velocity (v_2) is $\pi / 2$=1.57 time the speed of light. We could also consider the velocity v_2 as the velocity of light and then v_1 is $2 / \pi$=0.64 smaller than the velocity of light.
In Tesla's patents he makes it clear that the ground wave is the more rapidly travelling wave and the air wave is an electromagnetic wave travelling at the velocity of light. The above analysis is therefore not consistent with Tesla's model. In fact, there would be a mixing and reinforcing of a phonon/earth wave and an electromagnetic wave in the rarefield air and interaction. Therefore the above simple geometric problem does not apply. The problem, in fact, invokes phonon

(longitudinal) and transverse electromagnetic wave interactions, as discussed in the next section.

[48] Rauscher, E.A. (1984) Multi-Dimensional Formalism of Maxwell's Equations and Solutions to Maxwell's Equation in Complex Geometries, Lecture presented to Colorado College. Colorado Springs, CO. August 11, 1984. Sponsored by the International Tesla Society and the Pikes Peak Section of IEEE, Colorado Springs. Meeting conceived, sponsored, and chaired by E.A. Rauscher.
[49] Rauscher, E.A. (1968) J. Plasma Phys. 2, 517.
[50] The laser system utilizes an external energy flux to produce stimulated emission, whereas coherence in super-conductivity is achieved through the effective Cooper pair interaction at low temperatures. The laser system is more like the I. Prigogine self-ordering system. See [15] for further details.
[51] Scott, A. (1980) The birth of a paradigm, in P.L.E. Uslenghi (ed.) Non-Linear Electromagnetics, pp. 3563, New York: Academic Press.
[52] Becker, R.O. & Selden, G. (1985) The Body Electric: Electromagnetics and the Foundation of Life, New York: William Marrow Press; and private communication between R.O. Becker, EAR & Van Bise, W. .
[53] Adey, W.R. (1980) Frequency and power windowing in tissue interactions with weak electromagnetic fields, Proceedings of the IEEE 68, 119C.
[54] Pressman, A.S., (1968) Electromagnetic Fields and Living Nature, Moscow, English translation (1970) Electromagnetic Fields and Life, New York: Plenum.
[55] Van Bise, W. (1928) Low power radio frequency and microwave effects on human electroencephalogram and behavior, Physiological Chem. and Physics 10 387.
[56] Stocklin, P.L. & Stocklin, B.F. (1981 low power microwave effects on the human electroencephalogram: supporting the results of Bise, Physical Chem and Physics 13, 175.
[57] Van Bise, W. (1977) Radiofrequency induced interference response in the human nervous system, pp. 1221, Radiation Health; and Hearings Before the committee on commerce, science, and transporatation, United States Senate, 95th congress, Serial No. 95-19 June.
[58] Rauscher, E.A. (2010) Mathematical Medicine: A New Medical Model, Treatment Modality and Methods for Non-Invadive Medicine, book in progress.
[59] Rauscher, E.A. (1988) Theories and Experiments in Healing II ICAT. International Congress on Medical Thherapies. Sao Paulo, Brazil, September.
[60] Rauscher, E.A. & Van Bise, W.L. (1987) Magnetic field interaction with macro biological systems with applications to effects on physiology, medicine around the world, in T.M. Srinivasan (ed.) pp. 205-218, Phoenix: Gabriel Press; Print World, November 1988, Proceeding from conference Madras Inst. of Magnetobiology 1987.
[61] Rauscher, E.A. & Van Bise, W. (1989) Magnetic flux control of low back pain, Proceedings of the Association for the Advancement of Medical Instrumentation (AAMI), St. Louis, pp. 76, May.
[62] Van Bise, W.L. & Rauscher, E.A. (2001) Magnetic field impulse cardiovascular stimulation for normalizing a arrhythmias and/or heart block BEMS, European Bioelectromagnetics Association, 23rd Annual Meeting, pp. 155, St. Paul, June 10-14.
[63] Rauscher, E.A. & Van Bise, W.L. (2001) Pulsed magnetic field treatment and chronic back pain, BEMS, European Bioelectromagnetic Association, 23rd Annual

Meeting, pp. 56, St. Paul, June 10-14.
[64] Rauscher, E.A. & Van Bise, W.L. (2002) Medical application of pulsed magnetic fields in medicine, Presentation talk to the Department of Bioelectromagnetism, Beijing Polytecnic Institute University, Beijing, October 4.
[65] Rauscher, E.A. & Van Bise, W. (1987) Magnetic field flux induction into the visual system of humans, Institute for Electrical and Electronic Engineers IEEE Annual Conference on Engineering in Medicine and Biology Society, pp. 1590, Boston, November.
[66] Van Bise, W. & Rauscher, E.A. (1988) multiple extremely low frequency magnetic and electromagnetic field effects on human electroenchphalogram and behavior, pp. 57-58, Annual Review of Research on Biological Effects of Electric and Magnetic Fields form the Generation, Delivery and use of Electricity, DOE. EPRI; and U.S. National Inst. of Environmental Health Sciences, Tucson, September.
[67] Rauscher, E.A. (1981) Probing into control mechanisms in cellular healing processes, A San Francisco Initiative, Doctors Research Group, PACE Canadian Journal p. 8 February.
[68] Rauscher, E.A. (1981) The Biophysics of Electrical Pathways and Magnetic Fluxes in the Human Heart Wall The Doctors Research Group. PACE Canadian Journal page 9, February.
[69] Rauscher, E.A. (1982) Theoretical and field studies of ELF radiation and coherent nonlinear phenomena, Intl. Symp. on the Biological Effects of Electromagnetic Radiations, Learned Societies Conf., University of Ottawa, June.
[70] Rauscher, E.A. (1985) The mathematical theory of the magnetic piezoelectric resonator, MPR-7944-T device, for treatment of pain, Tecnic Research Laboratory report PSRL-12764, April.
[71] Van Bise, W. & Rauscher, E.A. (1999) The relatinship of extremely low frequency electomagnetic and magnetic fields associated with seismic and volcanic activity and natural and artificial ionospheric disturbances, 2nd Magnetic, Electric and Electromagnetic Methods in Seismology and Vulcanology, Ministry of Interior, Public Administration and General Secretariat for Civil Protection, General Secretariat for Research and Technology. National Institute for Geophysics (NIG) and Institute of Advanced Methodologies for Environmental Analysis, IMAAA-CNR. Chania, Greece, September.
[72] Han, M., Yang, X., Lee, J., Alla C.H. & Muneoka, K. (2008) Development and regeneration in the neonatal digit tip in mice, Developmental Biology, 315 125.
[73] Rauscher, E.A. & Van Bise, W. (1987) Observations of local and global earth-ionospheric excitation for earthquake and volcanic prediction, Bull. Am. Phys. Soc. 32, 67.
[74] Van Bise, W. & Rauscher, E.A. (1987) Instrumentation and techniques for analysis of extremely low frequency (ELF) magnetic field impulses preceding geologic events, Bull. Am. Phys. Soc. 32, 67B.
[75] Van Bise, W. & Rauscher, E.A. (1988) Fundamental excitatory modes of the earth and earth-ionosphere resonant cavity, Field Physics Division. Tecnic Research Laboratories Report. PSRL-702C-II. July, Proceedings of the IEEE. Colorado Springs Section, pp. 3-34.
[76] Van Bise, W.L. & Rauscher, E.A. (2003) Detection and analysis of precursor

magnetic signatures, Preceding the Turkey and Greek Seismicity, Geophysical Research Abstracts 5, 14637.
[77] Van Bise, W.L. & Rauscher, E.A. (2002) Nonlinear coherent collective resonant plasma propagation modes in the ionosphere, Bull. Am. Phys. Soc. 47, 169.
[78] Rauscher, E.A. & Van Bise, W. (1999) The relationship of extremely low frequency electromagnetic and magnetic fields associated with seismic and volcanic natural activity and artificial ionospheric disturbances, pp. 459-487, in M. Hayakawa (ed.) Atmospheric and Ionospheric Electromagnetic Phenomena Associated with Earthquakes, Tokyo: Terra Scientific Pub. Co. (TERRAPUB).
[79] Van Bise, W. & Rauscher, E.A. (1989) Non-superconducting systems for detecting and analyzing low intensity pure magnetic fields, Tecnic Research Laboratories Report PSRL-5476B; and Bull. Am. Phys. Soc. 34 p. 109.
[80] Van Bise, W.L. & Rauscher, E.A. (1996) Ambient electromagnetic fields as possible seismic and volcanic precursors, pp. 221-242, I M. Hayakawa & Y. Fujinowa (eds.) Electromagntic Phenomena Related to Earthquake Prediction, Tokyo Terra Sci. Pub. Co. (TERRAPUB).
[81] Rauscher, E.A. (with W.L. Van Bise's ghost) The Living Earth: Forces that Shape Seismic and Volcanic Occurrences, in progress.

Chapter 6

Vector and Scalar Potentials, Advanced and Retarded Waves and Nonlocal Phenomena

The issue of whether Bell's theorem and other remote connectedness phenomena, such as Young's double slit experiment, demands superluminal or space-like signals or prior generated luminal signals is an area of hot debate. This also relates to the existence of advanced vs. retarded potentials and annihilation creation operators which are of interest in this regard. Using the complex model of A^{μ} we will examine the issue of the nonlocality of Bell's theorem as quantum mechanical 'transactions' providing a microscopic communication path between detectors across space-like intervals, which violate the EPR locality postulate [1]. See Chap. 4. This picture appears consistent with the remote connectedness properties of complex Minkowski space. Also there are implications for macroscopic communications channels; another area of debate. Detailed discussions of Bell's theorem are given in [2].

6.1 Vector and Scalar Potentials and Fields

We formulate fields in terms of A or $\underset{\sim}{A} = (A^j, \phi)$ where A^j is $\underline{A}$ rather than the tensor, $F_{\mu\nu}$ or $\underline{E}$ or $\underline{B}$. We proceed from the usual continuity equation $\nabla \cdot \underline{J} + \partial\rho / \partial t = 0$ and utilize the expression $F_{\mu\nu} = \partial A_{\nu} / \partial x_{\mu} - \partial A_{\mu} / \partial x_{\nu}$. For the usual retarded potentials then, we have the Lorentz condition

$$\nabla \cdot \underline{A} + \mu\varepsilon \frac{\partial \varphi}{\partial t} = 0 \quad \text{and} \quad \nabla^2 \underline{A} - \mu\varepsilon \frac{\partial^2 \underline{A}}{\partial t^2} = -\mu \underline{J} \tag{6.1}$$

We also derive

$$\nabla^2 \phi - \mu\varepsilon \frac{\partial^2 \phi}{\partial t^2} = -\frac{1}{\varepsilon}\rho . \tag{6.2}$$

Equations (6.1) and (6.2) are the usual retarded potential solutions. The radiation field in quantum electrodynamics (QED) is usually quantized in terms of (A, φ). Conversion back to the $\underline{E}$ and $\underline{B}$ fields can be performed using $\underline{E} = -\nabla\varphi - \partial\underline{A}/\partial t$ and $\underline{B} = \nabla\times\underline{A}$. Quantization of the field consists of regarding the phase space coordinates (*x, k*) or (*q, p*) as quantum mechanical coordinates of a set of equivalent harmonic oscillators using the variables of $p = E/c = \hbar\omega/c$ and $c = \lambda\!\!\!^{-}\omega$ so that $k = n\omega/c$. [3]. Using the second quantized method and treating k_r, q_r and A_r as quantum numbers then we have quantized allowable energy levels n_r and η_r such as $W = \sum_r (n_r + \eta_r)\hbar\omega_r$ for two quantum states, n and η. Solutions are given in the form

$$\Psi \propto \sum_{n_r} n_r \exp-\left[\frac{iW(n_r)}{\hbar}\right] \tag{6.3}$$

and we have a Hamiltonian equation of motion $\dot{p}_{ab} + (ck)^2 q_{ab} = 0$ or $\dot{q}_{ab} = p_{ab}$ and for its Hamiltonian

$$H = \tfrac{1}{2}\sum[p_{ab}^2 + (ck)^2 q^2 q_{ab}^2]. \tag{6.4}$$

The electromagnetic field energy of the volume integral $(E^2 + B^2)/8\pi$ is just equal to the Hamiltonian.

We examine such phenomena as absorption and polarization in terms of the complexification of $\underline{E}$ and $\underline{B}$ or $\underline{A}$ and ϕ. Defining the usual $D = \varepsilon E$ (for displacement field) and $B = \mu H$ are performed for a homogeneous isotopic media. If we introduce p_0 and m_0 as independent of $\underline{E}$ and $\underline{H}$ where the induced polarizations of the media are absorbed into the parameters ε and μ, we have

$$D = \varepsilon E + p_0 \quad \text{and} \quad H = \tfrac{1}{\mu}B - m_0\,. \tag{6.5}$$

Then we define a complex field as

$$\underline{\underline{Q}} \equiv \underline{B} + i\sqrt{\varepsilon\mu}\,\underline{E} \tag{6.6}$$

so that we have Maxwell's equations now written as

$$\nabla\times\underline{\underline{Q}} + i\sqrt{\varepsilon\mu}\frac{\partial\underline{\underline{Q}}}{\partial t} = \mu\underline{J} \quad \text{and} \quad \nabla\cdot\underline{\underline{Q}} = i\sqrt{\frac{\mu}{\varepsilon}}\rho\,. \tag{6.7}$$

Using vector identities [3] and resolving into real and imaginary parts, we have

$$\nabla^2\underline{H} - \varepsilon\mu\frac{\partial^2\underline{H}}{\partial t^2} = -\nabla\times\underline{J} \quad \text{and} \quad \nabla^2\underline{E} - \varepsilon\mu\frac{\partial\underline{E}}{\partial t^2} = \mu\frac{\partial\underline{J}}{\partial t} + \frac{1}{\varepsilon}\nabla\rho \tag{6.8}$$

for the magnetic and electric fields.

We define Q in terms of the complex vector potential that $A_{\mathrm{Re}} \to V_{\mathrm{complex}}$ and $\phi_{\mathrm{Re}} \to \varphi_{\mathrm{complex}}$ where V is the complex potential as a vector-like quantity. Then

$$\underline{Q} = \nabla \times \underline{V} - i\sqrt{\varepsilon\mu}\frac{\partial \underline{V}}{\partial t} - i\sqrt{\varepsilon\mu}\nabla\varphi \tag{6.9}$$

subject to the condition similar to before, $\nabla \cdot \underline{V} + \varepsilon\mu \, {\partial\varphi}/{\partial t} = 0$. Then

$$\nabla^2 \underline{V} - \varepsilon\mu \frac{\partial^2 \underline{V}}{\partial t^2} = -\mu \underline{J} \text{ and } \nabla^2\varphi - \varepsilon\mu\frac{\partial^2\varphi}{\partial t^2} = -\frac{1}{\varepsilon}\rho. \tag{6.10}$$

Separation into real and imaginary parts of these potentials, $\underline{V}$ and ϕ can be written as

$$\underline{V} = \underline{A}_{\mathrm{Re}} - i\sqrt{\frac{\mu}{\varepsilon}}\underline{A}_{\mathrm{Im}} \text{ and } \varphi = \phi_{\mathrm{Re}} - i\sqrt{\frac{\mu}{\varepsilon}}\phi_{\mathrm{Im}}. \tag{6.11}$$

Upon substitution into the equation for Q and separation into real and imaginary parts we have

$$\underline{B}_{\mathrm{Re}} = \nabla \times \underline{A}_{\mathrm{Re}} - \frac{\mu\partial \underline{A}_{\mathrm{Im}}}{\partial t} - \mu\nabla\phi_{\mathrm{Im}};$$

$$\underline{E}_{\mathrm{Re}} = -\nabla\phi_{\mathrm{Re}} - \frac{\partial \underline{A}_{\mathrm{Re}}}{\partial t} - \frac{1}{c}\nabla \times \underline{A}_{\mathrm{Im}}. \tag{6.12}$$

The usual equations for the fields result when A_{Im} and ϕ_{Im} are taken as zero.

If free currents and charges are everywhere zero in the region under consideration, then we have

$$\nabla \times \underline{Q} + i\sqrt{\varepsilon\mu}\frac{\partial \underline{Q}}{\partial t} = 0 \text{ and } \nabla \cdot \underline{Q} = 0 \tag{6.13}$$

and we can express the field in terms of a single complex Hertzian-like vector $\underline{L}$ as the solution of

$$\nabla^2 \underline{L} - \varepsilon\mu\left({\partial^2 \underline{L}}/{\partial t^2}\right) = 0. \tag{6.14}$$

We can define L by

$$L \equiv \xi_{\mathrm{Re}} - i\sqrt{\frac{\mu}{\varepsilon}}\xi_{\mathrm{Im}} \tag{6.15}$$

where $\phi_{\mathrm{Re}} = -\nabla \cdot \underline{\xi}$ and we can write such expressions as

$$A_{\mathrm{Im}} = \mu\varepsilon\frac{\partial \xi_{\mathrm{Im}}}{\partial t} \quad \text{and} \quad \phi_{\mathrm{Im}} = \nabla \cdot \underline{\xi}_{\mathrm{Im}}. \tag{6.16}$$

This formalism works for a dielectric media but if the media is conducting the field equations is no longer symmetric then the method fails. Symmetry is maintained by introducing a complex induced capacity $\varepsilon' = \varepsilon_{\text{Re}} \pm i(\sigma_{\text{Im}} / \omega)$. If the vector B is in a solenoid charge-free region then this method works. Calculation of states of polarization by the complex method demonstrates its usefulness and validity. Also, absorption can be considered in terms of complex fields. In the complex space, V may also contain non-Hertzian as well as Hertzian components, L. We will apply this method to solutions that can be described as retarded and advanced and may explain Bell's theorem and other nonlocal phenomena. Linear and circular polarization can be expressed in terms of complex vectors $A = A_{\text{Re}} + iA_{\text{Im}}$. The light quanta undergoing this polarization is given as, $\hbar\omega\hat{n} = \hbar\sigma = \hbar k$. Complex unit vectors are introduced so that real and imaginary components are considered orthogonal. We have a form such as $\underline{A} = (\underline{A} \cdot \hat{\ell}_{\text{Im}})\hat{\ell}_{\text{Re}} + (\underline{A} \cdot \hat{j}_{\text{Im}})\hat{j}_{\text{Re}}$. The linearly polarized wave at angle θ is

$$\underline{A} = \frac{A}{\sqrt{2}}(\hat{\ell}_{\text{Re}} e^{-i\theta} - i\underline{j}_{\text{Re}} e^{i\theta}). \tag{6.17}$$

Now let us consider use of this polarization formalism to describe the polarization-detection process in the calcium source photon experiment of Clauser, et al. [4], Aspect, et al. [5] and Gisin, et al. [6]. First we examine solutions to the field equations for time-like and space-like events. The non-locality of Bell's theorem appears to be related to the remote connectedness of the complex geometry and the stability of the soliton over space and time.

We will consider periodically varying fields which move along the x-axis later in this chapter. For source-free space, we can write

$$c^2 \nabla^2 \underline{F} = \varphi^2 \frac{\underline{F}}{\varphi t^2} \tag{6.18}$$

where $\underline{F}$ represents either $\underline{E}$ or $\underline{B}$. The two independent solutions for this equation are [7]

$$\underline{E}_{\pm}(x,t) = E_0 \sin(2\pi kx \pm vt)$$

and (6.19)

$$\underline{B}_{\pm}(x,t) = B_0 \sin 2\pi(kx \pm vt)$$

and k is the wave number and v the frequency of the wave. The $\pm$ sign refers to the two independent solutions to the above second order equation in space and time. The wave corresponding to E_+ and B_+ will exist only when $t_0 <$

0 (past lightcone) and the wave corresponding to $\underline{E}$ and $\underline{B}$ will exist for $t_0 > 0$ (future lightcone) where t_0 is at the origin of the lightcone or the moment "now". Then the E_- wave arrives at a point x in a time t after emission, while E_+ wave arrive at x in time, t before emission (like a tachyonic signal).

6.2 Advanced and Retarded Solutions

Using Maxwell's equations for one spatial dimension, x, and the Poynting vector which indicates the direction of energy and momentum flow of the electromagnetic wave, we find that E_+ and B_+ correspond to a wave emitted in the $+x$ direction but with energy flowing in the $-x$ direction. For example, $E_+(x,\ t)$ is a negatives-energy and negative-frequency solution. The wave signal will arrive $t = x/c$ before it is emitted, and is termed an advanced wave. The solution $\underline{E}(x,t)$ is the normal positive-energy solution and arrives at x in time, $t = x/c$, after the instant of emission and is called the retarded potential, which is the usual potential.

The negative energy solutions can be interpreted in the quantum picture in quantum electrodynamics as virtual quantum states such as vacuum states in the Fermi-Dirac sea model [8]. See Chap. 12. These virtual states are not fully realizable as a single real state but can definitely effect real physical processes to a significant testable extent [9]. The causality conditions in S-matrix theory, as expressed by analytic continuation in the complex plane, relate real and virtual states [10,11] and Chap. 4. Virtual states can operate as a polarizable media leading to modification of real physical states. In fact, coherent collective excitations of a real media can be explained through the operations in a underlying virtual media [9]. These virtual states in physical plasma operating through collective quantum electron states, effect the dielectric constant, conductivity and other electromagnetic properties of plasma which, experimentally differ from the classical properties and agree with theoretical quantum conditions which include the vacuum state [9,12].

Four solutions emerge: Two retarded (F_1 and F_2) connecting processes in the forward light cone and two advanced, (F_3 and F_4) connecting processes in the backward slight cone [13]. These four solutions are

$$\begin{aligned} F_1 &= F_0 e^{-i(-kx-\omega t)}, \quad F_2 = F_0 e^{i(kx-\omega t)}; \\ F_3 &= F_0 e^{i(-kx+\omega t)}, \quad F_4 = e^{i(kx+\omega t)} \end{aligned} \tag{6.20}$$

where F_1 is for a wave moving in the $(-x, +t)$ direction, F_2 is for a $(+x, +t)$

moving wave, F_3 is for a (-*x*, -*t*) moving wave, and F_4 is a (+*x*, -*t*) moving wave. F_1 and F_4 are complex conjugates of each other and F_2 and F_3, are complex conjugates of each other, so that $F_1^+ = F_4$ and $F_2^+ = F_3$; where the usual solutions to Maxwell's equations are then retarded plane wave solutions [3,13].

The proper formulation of nonlocal correlations, which appear to come out of complex geometries provides a conceptual framework for a number of quantum mechanical paradoxes and appear to be explained by Bell's nonlocality, Young's double slit experiment, the Schrödinger cat paradox, superconductivity, superfluidity, and plasma 'instabilities' or coherent collective states including Wheeler's 'delayed choice experiment'. (See Chap. 4.) A paradox is caused by a lack of under-standing of a physical observation and is resolved by a new and better comprehension of the interpretation of the observation and/or new observation. Interpretation of these phenomena is made in terms of their implications about the lack of locality and the decomposition of the wave function which arises from the action of advanced waves which 'verify' the quantum mechanical transactions or communications. See Fig. 6.1.

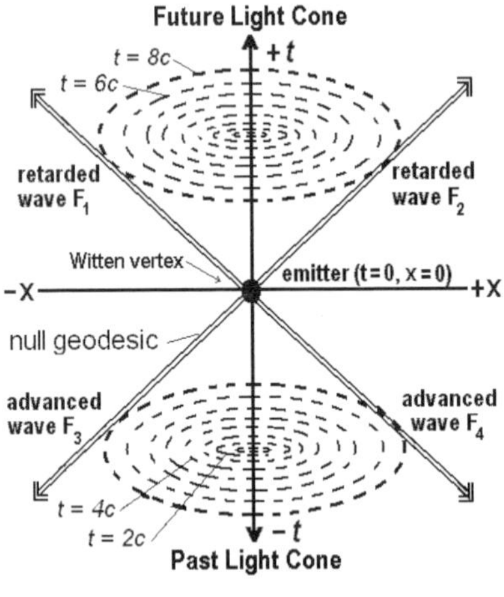

Figure 6.1. Adaptation of a complex Minkowski lightcone showing advanced-retarded future-past elements, $F_1 - F_4$, see Eq. (6.20), of a Cramer wavefront transaction with a central Witten model Ising lattice string vertex able to undergo continuous-state symmetry transformations of the Riemann sphere, $0 \leftrightarrow \infty$ rotation.

Cramer [13] demonstrated that the communication path between detectors in the Bell inequality experiments can be represented by space-like intervals that produce the quantum mechanical result by the addition of two time-like 4-vectors having time components of opposite signs, which demonstrate the locality violations of Bell's theorem; and are consistent with the Clauser, Fry, Aspect and Gisen experiments [4-6]. This model essentially is an 'action-at-a-distance' formalism [14].

One can think of the emitter (in Bell's or Young's experimental quantum condition) as sending out a pilot or probe 'wave' in various allowed directions to seek a 'transaction' or collapse of the wave function. A receiver or absorber detects or senses one of these probe waves, 'sets its state' and sends a 'verifying wave' back to the emitter confirming the transaction and arranging for the transfer of actual energy and momentum. This process comprises the nonlocal collapse of the wave function. De Broglie termed such a wave a pilot wave. The question becomes: does such a principle have macroscopic effects? The distance record for Bell's nonlocality theorem was 10km in 1997 [6], obtained by Nicolas Gisin and his team at the University of Geneva. Starting from a Geneva railway station they sent entangled photons along optical fibers through the city to destinations separated by 10km. They showed that observing the state of one member of the pair instantaneously determined the state of the other.

An attempt to examine such possible macroscopic effects over large distances has been made by Partridge [15]. Using 9.7GHz microwaves transmitted by a conical horn antenna so that waves were beamed in various directions. Partridge found that there was little evidence for decreased emission intensities in any direction for an accuracy of a few parts per 10^{9th}. Interpretation of such a process is made in terms of advanced potentials. Previously mentioned complex dimensional geometries give rise to solutions of equations forming subluminal and superluminal signal propagations or solitons. See Chaps. 9 and 10.

The possibility of a remote transmitter-absorber communicator now appears to be a possibility. The key to this end is an experiment by Pflelgov and Mandel [16]. Interference effects have been demonstrated, according to the authors, in the superposition of two light beams from two independent lasers. Intensity is kept so low that, to high probability, one photon is absorbed before the next is emitted. The analogy to Young's double slit experiment is enormous. See Chap. 9.

In Wheeler's work [17-19], he presents a detailed discussion of the physics of delayed choice photon interference and the double slit experiment (based on the Solvay conference Bohr-Einstein dialogue). Wheeler discusses the so-called Bohm 'hidden variables' as a possible determinant that nonlocality collapses the wave function [17]. Remote wave functions once entangled remain entangled over space-like separation, i.e. provide a possible solution to the Schrödinger cat paradox. Further theoretical and experimental investigation is indicated; but there appears to be a vast potential for remote non-local communication and perhaps even energy transfer [3]. In Chap. 9 we detail the forms of transformations of the vector and scalar potentials at rest and in moving frames, continuing our formulation in terms of $(\underline{A},\phi)$. The issues of sub and superluminal transformations of $\underline{A}$ and ϕ are given in a complex Minkowski space. Both damped and oscillatory solutions are found and conditions for advanced and restored potentials are given.

References

[1] Einstein A., Podolsky B., & Rosen M. (1935) Can a quantum-mechanical description of physical reality be considered complete? Phys. Rev. 47, 777.
[2] Bell, J.S., (1964) Physics 1, 195.
[3] Amoroso, R.L. & Rauscher, E.A. (2009) The Holographic Anthropic Multiverse: Formulating the Complex Geometry of Reality, Singapore: World Scientific.
[4] Clauser, J.F. and Horne, W.A. (1971) Phys. Rev. 10D, 526.
[5] Aspect, A. & Dalibard, G.R. (1982) Experimental test of Bell's inequalities using time-varying analyzers, Phys. Rev. 49, 804.
[6] Gisin, N., Tittel, W., Brendel, J. & Ebinden, H. (1998) Maximal violation of Bell's inequality for arbitrarily large spin, Phys. Rev. Lett. 81, 3563.
[7] Jackson, J.D. (1975) Classical Electrodynamics, New York: Wiley and Sons.
[8] Cufaro, N., Petroni, N. & Vigier, J-P (1983) Dirac ether in relativistic quantum mechanics, Found. Phys. 13, 253.
[9] Rauscher, E.A. (1968) J. Plasma Phys. 2, 519.
[10] Rauscher, E.A. & Amoroso, R.L. (2009) Relativistic physics in complex minkowski space, nonlocally ether model and quantum physics, in M.C. Duffy and J. Levy (eds.) Ether Space-Time and Cosmology, Vol. III, Physical Vacuum, Relativity and Quantum Physics, pp. 23-47, Montreal: Apeiron.
[11] Rauscher, E.A. & Amoroso, R.L. (2005) The Schrödinger equation in complex Minkowski space, and nonlocal anticipatory systems, 1st Unified Theories, Budapest, Hungary, in R.L. Amoroso, I. Dienes & C. Varges (eds.) Oakland: The Noetic Press.
[12] Rauscher, E.A. (1987) Soliton solutions to the Schrödinger equation in complex Minkowski space, pp. 89-105, Proc. of the 1st Intl Conference on Energy, Toronto.
[13] Cramer, J.G. (1980) Phys. Rev. D22, 362.
[14] Rauscher, E.A. (2010) Quantum mechanics and the role of consciousness in the physical world, in R.L. Amoroso (ed.) Complementarity of Mind and Body: Realizing the Dream of Descartes, Einstein and Eccles, New York: Nova Science.
[15] Partridge, R.B. (1973) Nature, 244, 263.
[16] Pflelgov, R.L. & Mandel, L. (1967) Phys. Rev. Lett. 24A, 766.
[17] Wheeler, J.A. (1982) Int. J. Theoret. Phys. 21, 557.
[18] Ciufolini, I. & Wheeler, J.A. (1995) Gravitation and Inertia, Princeton: Princeton University Press.
[19] Wheeler, J.A. (1978) Frontiers of Time, Austin: University of Texas Press.

Chapter 7

The Complex Form of Relativistic Maxwell's Equations

We outline the relativistic formalism which gives a more comprehensive explanation of the complexification scheme. Such issues as considering the Higgs Boson as a soliton depends on Lorentz invariance and relativistic causality constraints. We relate the complexification of Maxwell's equations to models of nonlocal micro and macro phenomena. In this chapter we relate the electromagnetic fields, $F^{\mu\nu}$ and A_μ, the potentials to the gravitational field, $G_{\mu\nu}$. We examine, for example, the manner in which advanced potentials may explain the remote connectedness which is indicated by the Clauser test of Bell's theorem. Similar arguments apply to Young's double slit experiment. The collective coherent phenomena of superconductivity is also explainable by considering the relativistic field theoretic approach in which wave equations are solved in the complex Minkowski space.

7.1 Relativistic Conditions for Maxwell's Equations in Complex Geometries and Invariance of the Line Element

This section introduces the relativistic form of Maxwell's equations. The fields $\underline{E}$ and $\underline{B}$ are defined in terms of ($\underline{A}, \phi$), the four vector potential; and the relativistic form of $\underline{E}$ and $\underline{B}$ is presented in terms of the tensor field, $F_{\mu\nu}$ (where indices μ and ν run 1 to 4). We then complexity $F_{\mu\nu}$ and determine the expression for the four vector potential $A_\mu = \left(\underline{A}_j, \phi\right)$ in terms of $F_{\mu\nu}$ (index j runs 1 to 3). Discussion of line element invariance is given in terms of the fields $F_{\mu\nu}$.

In Chap. 6 we describe the complex form of A_μ fields and through the formalism in this section we can relate this to the complex forms of $\underline{E}$ and $\underline{B}$. We utilize Weyl's action principle to demonstrate the validity of the use of the complex form of $F_{\mu\nu}$ [1]. Weyl relates the gravitational potential, $G_{\mu\nu}$, to the electromagnetic 'geometrizing' potential A_μ, or geometrical vector, using the principle of stationary action for all variations $\delta G_{\mu\nu}$ and δA_μ [2]. The quantity A_μ, or vector potential, which we identify with A^μ by symmetry relations on the complex conjugate of *A*, is related to $F_{\mu\nu}$, the electromagnetic force field, E by a set of gauge invariant relations. The electromagnetic force $F_{\mu\nu}$ is independent of the gauge system. The curl of A_μ has the important property

$$F_{\mu\nu} = \frac{\partial A_\mu}{\partial x_\nu} - \frac{\partial A_\nu}{\partial x_\mu} \tag{7.1}$$

where $F_{\mu\nu}$ is antisymmetric or $F_{\mu\nu} = -F_{\nu\mu}$, and changing A_μ to $A'_\mu = A_\mu + \partial\phi / \partial x_\mu$ is a typical gauge transformation where the intrinsic state of the world remains unchanged.

Defining the 4-vector potential as A_μ, which is written in terms of the three vector $\underline{A}_j$ and ϕ, where ϕ is the fourth or temporal component of the field. The indices μ, ν run 1 to 4 and *j* runs 1 to 3.

Then Maxwell's equations in compact notation in their usual tensor form in terms of $F_{\mu\nu}$, (for $c = 1$) are

$$F_{\mu\nu} = \begin{pmatrix} 0 & -B_z & B_y & E_x \\ B_z & 0 & -B_x & B_y \\ -B_y & B_x & 0 & E_z \\ -E_x & -E_y & -E_z & 0 \end{pmatrix} \tag{7.2}$$

then the equations $\nabla \times \underline{E} = -(1/c)(\partial \underline{B} / \partial t)$ and $\nabla \cdot \underline{B} = 0$ can be written as

$$\frac{\partial F_{\mu\nu}}{\partial x} + \frac{\partial F_{\mu\nu}}{\partial y} + \frac{\partial F_{\mu\nu}}{\partial z} = 0 \tag{7.3}$$

or $\nabla \times \underline{F}_{\mu\nu} = 0$ for $x^1 = x$, $x^2 = y$, $x^3 = z$, and $x^4 = t$.

To complexity the elements of $F_{\mu\nu}$ we can choose the conditions, for

$$\left(F_{41}, F_{42}, F_{43}\right) = i\underline{E} \text{ and } \left(F_{23}, F_{32}, F_{12}\right) = \underline{B},$$

or (7.4)

$$\left(E_x, E_y, E_z\right) = i\underline{E} \text{ and } \left(B_x, B_y, B_z\right) = \underline{B}.$$

The complex conjugate of the electric and magnetic fields are written in terms of the complex conjugate of $\underline{F}$ or $F^{*}_{\mu\nu} = -F^{\mu\nu}$. Tin this regard there is a useful theorem that states [3] $\nabla_{123} \times \underline{F} = \nabla^4 \cdot F^*$ or $(\nabla_{xyz} \times F = \nabla^t \cdot F^*)$. For $\left(F^{23*}, F^{31*}, F^{12*}\right) = i\underline{E}$ and $\left(F^{41*}, F^{42*}, F^{43*}\right) = -\underline{B}$ we will then obtain $\partial F^{\mu\nu*} / \partial x^\nu = 0$ or $\nabla \cdot \underline{F}^* = 0$ which gives the same symmetry between real and imaginary components as ours and Inomata's formalism [4].

The expressions for the other two Maxwell equations $\nabla \cdot \underline{E} = 4\pi\rho$ and $\nabla \times \underline{B} = \frac{1}{c}\frac{\partial \underline{E}}{\partial t} + \underline{J}_e$ can be obtained by introducing the concept of the vector potential in the Lorentz theory as first noticed by Minkowski [5]; we have the 4-vector forms $\left(\phi_1, \phi_2, \phi_3\right) = \underline{A}$ and $\phi_4 = i\phi$, so that $\underline{B} = \nabla \times \underline{A}$ and $\underline{E} = -\nabla\phi - \frac{1}{c}\frac{\partial \underline{A}}{\partial t}$. This yields $F_{\mu\nu} = \frac{\partial A_\nu}{\partial x^\mu} - \frac{\partial A_\mu}{\partial x^\nu}$ or $\underline{F} = \nabla \times \underline{A}$ for the vector and scalar potentials $\underset{\sim}{A} = \left(A_1, A_2, A_3, \phi\right)$. If A is a solution to $\underline{F} = \nabla \times \underline{A}$ then $\phi'_\mu + \frac{\partial \phi}{\partial x^\mu}$ also is also a solution by gauge invariance conditions and $\nabla \cdot \underline{A} + \frac{1}{c}\frac{\partial \phi_4}{\partial t} = 0$. We term the fourth component of A as ϕ or ϕ_4 inter-changeably. Then from Lorentz theory we have the 4D form as $\frac{\partial A^\mu}{\partial x^\mu} = 0$ or $\nabla \cdot \underline{A} = 0$. We now write the equations for $\nabla \cdot \underline{E} = 4\pi\rho$ and $\nabla \cdot \underline{B} = \frac{1}{c}\frac{\partial \underline{E}}{\partial t} + \underline{J}_e$ as

$$\frac{\partial F^{\mu\nu}}{\partial x^\nu} = s^\mu \text{ or } \nabla \cdot \underline{F} = s. \tag{7.5}$$

The most general covariant transformation group of electromagnetic field equations, which are more general than the Lorentz group, is formed by affine transformations which transform the equation of the light cone, $s^2 = 0$ into itself. The properties of the spacetime manifold are defined in terms of the constraints of the line element, which relate to the gravitational potential, $G_{\mu\nu}$. We also form an analogy of the metric space invariant to the electromagnetic source vector, s_μ [6]. The Lorentz group contains the Lorentz transformations as well as inversion with respect to a 4D sphere, or hyperboloid in real coordinates. Frank [7] discusses the Weyl theory and gives a proof that the Lorentz group together with the group of ordinary affine transformations, is the only group, in which Maxwell's equations are covariant [7]. Recall that an affine transformation acts as $x^\mu = \alpha^\mu_\nu x^\nu$ with an inverse $x^\nu = \alpha^{-1}_\mu{}' x'^\mu$. The affine group contains all linear transformations and the group of affine transformations transforms $s^2 = 0$ on the light cone into itself.

In the Weyl geometry, where we have from before, $\underline{F} = \nabla \times \phi$ and

$$\nabla \cdot \underline{F} = \frac{1}{\sqrt{g}} \frac{\partial \sqrt{g} F^\mu}{\partial x^\mu} \tag{7.6a}$$

and

$$\nabla^\mu \cdot F = \frac{1}{\sqrt{g}} \frac{\partial \sqrt{g} F^\nu}{\partial x^\nu} \tag{7.6b}$$

with the signature (+,+,+,-) and where $\sqrt{g}$ is the square root of the metric tensor representation of $g_{\mu\nu}$, which is proportional to *x,y,z*. Then using the theorem in Pauli [8],

$$\nabla_\mu \cdot \nabla \times \underline{F} = \nabla_\mu \nabla \cdot \underline{F} - \Box F_\mu \tag{7.7}$$

and from before, $\nabla \cdot \underline{F} = s$ and since $\nabla \cdot \phi = 0$ and then $\partial \phi^\mu / \partial x^\mu = 0$ and we have from

$$\nabla_\mu \cdot \nabla \times \underline{A} = \nabla_\mu \nabla \cdot \underline{A} - \Box \phi_\mu = s_\mu \tag{7.8}$$

or

$$\Box A_\mu = -s_\mu \tag{7.9}$$

for our potential equation, where $\Box$ is the D'Alembertian 4-space operator, and

$$\Box = \partial^{\mu}\partial_{\mu} = \eta^{\mu\nu}\partial_{\nu}\partial_{\mu} = \frac{\partial^2}{\partial x^2} + \frac{\partial^2}{\partial y^2} + \frac{\partial^2}{\partial z^2} - \frac{1}{c^2}\frac{\partial^2}{\partial t^2} = \nabla^2_{R^3} - \frac{\partial^2}{\partial t^2} \qquad (7.10)$$

where $\eta^{\mu\nu}$ is a metrical like transform.

The important aspect of this consideration [9] is our ability to relate the electromagnetic potential to a corresponding spacetime metric interval s or s^2. Hence we can construct the invariant relations for our fields in terms of our Lorentz invariance 4-space conditions where the $\sqrt{g}$ relates to s and $g_{\mu\nu}$ to s^2. We relate the introduction of a complex spacetime to the complex expansion of the electric and magnetic fields in this section and demonstrate their self-consistency. We examine this in more detail at the end of this section where we consider a generalized affine connection. We relate the electromagnetic potential, A_{μ} and ϕ_4 to $g_{\mu\nu}$ as $\sqrt{g}$ and also to the square root of the invariant, or s.

The key to the relationship of complex $F_{\mu\nu}$ and complex spacetime is the analogy between ϕ and $g_{\mu\nu}$. We can relate the electromagnetic scalar potential into the interval of time as in Eq. (7.9), $\Box A_{\mu} = \phi = -s_{\mu}$ and we make the analogy of the electromagnetic potential, A_{μ} to the gravitational potential, $G_{\mu\nu}$ which is related to the invariance conditions on s^2. Both potentials are related to space-like or time-like interval separation. Note that in the $\Box A_{\mu} = -s$ equation we have a $\sqrt{g}$ factor in order to form the invariant. In the equation for s^2, the invariant is found directly as $s^2 = g_{\mu\nu}x^{\mu}x^{\nu}$. We address the set of invariant relations for the case of complex $\underline{E}$ and $\underline{B}$ fields at the end of this section. We relate this then to the deSitter algebras and the complex Minkowski metric.

We associate the E_x component of $F_{\mu\nu}$ or $F_{41} = E_x$ with ϕ_4' as follows:

$$F_{41} = E_x = \phi_4' \frac{e}{r^2} \qquad (7.11)$$

in which $4\pi e$ or e is associated with electric charge on the electron. This approximation is made in the absence of a gravitational field. Maxwell's equations are intended to apply to the case in which no field of force is acting on the system or in the special system of Galilean coordinates $A^{\mu} = \left(A_x, A_y, A_z, \phi\right)$, where $A^j = \left(A_x, A_y, A_z\right)$ is the vector potential and ϕ is the scalar potential and A^{μ} is the covariant form. For the contravariant form, we have $A_\mu = \left(-A_x - A_y - A_z, \phi\right)$, and in empty space we have

$$\Box A_\mu = 0\,. \tag{7.12}$$

In non-empty space then

$$\Box A^\mu = J^\mu \tag{7.13}$$

or we can write this as

$$\nabla^2 \underline{A}^\mu - \mu\varepsilon \frac{\partial^2 \underline{A}^\mu}{\partial t^2} = -\underline{J}^\mu \tag{7.14}$$

which is true only approximately for the assumption of flat space for Galilean coordinate transformations. This is the condition which demands that we consider the weak Weyl limit of the gravitational field.

The invariant integral, $\mathfrak{I}$ for $F^{\mu\nu}$ is given by

$$\mathfrak{I} = \frac{1}{4}\int F^{\mu\nu} F_{\mu\nu} \sqrt{g}\, d\tau \tag{7.15}$$

where $d\tau$ stands for *dx,dy,dz*.

The quantity, $\mathfrak{L}$ is called the action integral of the electromagnetic field. Weyl [10] demonstrated that the action integral is a Lagrangian function, or

$$\mathfrak{L} = \int dt \iiint \frac{1}{2}\left(B_x^2 + B_y^2 + B_z^2 - E_x^2 - E_y^2 - E_z^2\right) dx dy dz \tag{7.16}$$

which is of the form $\mathfrak{L} = (T - V)dt$. Note the definition for the kinetic energy, T and the potential energy, V for the Hamiltonian is $H = T + V$ the Lagrangian, $\mathfrak{L} = T - V$. By describing an electron in a field by Weyl's formalism one has a more general, but more complicated, formalism than the usual Einstein-Galilean formalism [11]. We write a generalized Lagrangian, $\mathfrak{L}$ in terms of complex quantities. For example, we form a modulus of the complex vector $\underline{B}$ as $\left|B\right|^2 = \underline{B}\underline{B}^* = \underline{B}_{\mathrm{Re}}^2 + \underline{B}_{\mathrm{Im}}^2$. This is the Lagrangian form for the real components of $\underline{E}$ and $\underline{B}$ in 4-space with $E = E_{\mathrm{Re}} + iE_{\mathrm{Im}}$ and $B = B_{\mathrm{Re}} + iB_{\mathrm{Im}}$ for the complex forms of $\underline{E}$ and $\underline{B}$. The complex Lagrangian in complex 8-space becomes

$$\mathfrak{L} = \iint dt_{\mathrm{Re}} dt_{\mathrm{Im}} \iiint_{\mathrm{Re}} \iiint_{\mathrm{Im}} \frac{1}{2}\left(B_{\mathrm{Re}}^2 - E_{\mathrm{Re}}^2 + B_{\mathrm{Im}}^2 - E_{\mathrm{Im}}^2\right) \times$$
$$dx_{\mathrm{Re}} dy_{\mathrm{Re}} dz_{\mathrm{Re}} dx_{\mathrm{Im}} dy_{\mathrm{Im}} dz_{\mathrm{Im}}\,. \tag{7.17}$$

This is an 8D integral, six over space, two over time (not represented here) where all quantities of the integrand are real because they are squared quantities. We also write an expression for a generalized Poynting vector and energy relationship. There are two equations which define a vector quantity, A_μ in electromagnetic theory which corresponds to the gravitational potential, $G_{\mu\nu}$ (which relates to the metric, $g_{\mu\nu}$). We have

$$\frac{\partial}{\partial g_{\mu\nu}}\left(\frac{1}{4}F^{\mu\nu}F_{\mu\nu}\right)=\frac{1}{2}\mathrm{E}^{\mu\nu} \tag{7.18}$$

and

$$\frac{\partial}{\partial A_{\mu}}\left(\frac{1}{4}F^{\mu\nu}F_{\mu\nu}\right)=-\underline{J}^{\mu} \tag{7.19}$$

where $\mathrm{E}^{\mu\nu}$ is the energy tensor and J^μ is the charge and current vector. Two specific cases are for a region free from electrons, or $T^{\mu\nu}-E^{\mu\nu}=0$, or a region free of the gravitational potential or in the weak Weyl limit of the gravitational field, $\Box F_{\mu\nu}=J_{\mu\nu}-J_{\nu\mu}$ where $\Box$ is the 4-space D'Alembertian operator. The solution for this latter case is for the tensor potential $A_{\mu\nu}$,

$$F_{\mu\nu}=\frac{1}{4\pi\gamma}\left(A_{\mu\nu}-A_{\nu\mu}\right)\int\frac{de}{r} \tag{7.20}$$

if all parts of the electron are the same or uniform in charge. For the proper charge, ρ_0, we have $J^\mu=\rho_0 A^\mu$ where *de* is the differential charge.

In the limit of vector $A^\mu_\mu=0$, then ρ_0, the proper charge density, is given as $\rho_0=-\frac{\gamma^2}{12\pi}J_\mu J^\mu$ for $\gamma=\left(1-\beta^2\right)^{-\frac{1}{2}}$. In Weyl's 4D world then, matter cannot be constituted without electric charge and current. But since the density of matter is always positive the electric charge and current inside an electron must be a space-like vector, thus the square of its length is negative. To quote Eddington:

> It would seem to follow that the electron cannot be built up of elementary electrostatic charges but resolves into something more akin to magnetic charges [12].

Perhaps we can use the structure of Maxwell's equations in complex form to demonstrate that this magnetic structure is indeed the complex part of the field and ask what the source is. A fundamental question is, what gives rise to charge? What attributes of matter and field give rise to charge? It is interesting to note that the charge on a proton and electron is exactly equal and opposite even though the protons mass, m_p is ~ 1860 times greater than the mass of the electron, m_e.

Considering $F_{\mu\nu}$ and $A_{\mu\nu}$ as complex entities rather than 4-space real forms, we may be required to have complex forms of the current density. The relationship between $F_{\mu\nu}$ and $A_{\mu\nu}$ has a spatial integral over charge. If we consider $F_{\mu\nu}$ and $A_{\mu\nu}$ as complex quantities, we deduce possible implications for the charge e or differential charge de being a complex quantity. Perhaps the expression $e = e_{\text{Re}} + ie_{\text{Im}}$ is not appropriate, but a form for the charge integral is, such as: $\int de_{\text{Re}} de_{\text{Im}} / \underline{r}$ where $r = r_{\text{Re}} + ir_{\text{Im}}$ is more appropriate. Fractional charges such as for quarks, give rise to the question of the source of charge in elementary particles and its fundamental relationship to magnetic phenomena (magnetic domains) are essential considerations and may be illuminated by this or similar formalisms. Neither the source of electrics or magnetics is known, although a great deal is known about their properties.

Faraday's conclusion of the identical nature of the magnetic field of a loadstone and a moving current may need reexamination as well as the issue of Hertzian and non-Hertzian waves. A possible description of such phenomena may come from a complex geometric model [13]. As discussed, one can generalize Maxwell's equations and examine real and imaginary components which comprise a symmetry in the form of the equations. We can examine in detail what the implications of the complex electric and magnetic components have in deriving a Coulomb equation and examine the possible way, given a rotational coordinate, this formalism relates to the 5D geometries of Kaluza and Klein. The approach we take in this chapter does not involve a compact rolled up 5th dimension of the original Kaluza-Klein approach which may lead to singularity problems.

Starting with $F^{\mu\nu}$, A_μ and J^μ, Maxwell's equations can be compactly written as $\dfrac{\partial F^{\mu\nu}}{\partial x_\nu} = J^\mu$ and again, $F_{\mu\nu} = \dfrac{\partial A_\mu}{\partial x_\nu} - \dfrac{\partial A_\nu}{\partial x_\mu}$ and $F_\nu^{\mu\nu} = J^\mu$.

Suppose that an electron moves in such a way that its own field on the average just neutralizes an applied external field $F'_{\mu\nu}$ in the region occupied by the electron. The value of $F_{\mu\nu}$ averaged for all the elements of change constituting the electron is given by

$$eF_{\mu\nu} = \frac{1}{4\pi}\left(A_{\mu\nu} - A_{\nu\mu}\right)\iint \frac{de_1 de_2}{r_{12}}$$

and (7.21)

$$eF_{\mu\nu} = \frac{1}{4\pi}\left(A_{\mu\nu} - A_{\nu\mu}\right)\frac{e^2}{a}$$

where $1/a$ is the average value of $1/r_{12}$ for every pair of points in the electron and a will then be a length comparable to the radius of the sphere throughout which the charge is spread. The mass of the electron is $m = e^2 / 4\pi a$. We thus have a form of Coulomb's law; and as we have shown, the complex form of $F^{\mu\nu}$ is consistent with Coulomb's law which is incorporated into Maxwell's equations in a manner that has both a real and an imaginary form of Coulomb's law.

Self-consistency can be obtained in the model by assuming that all physical variables are complex. Thus, as before, we assumed that space, time, matter, energy, charge, etc. are on an equal footing as coordinates of a Cartesian space quantized variable model. In [14] we present a 10D space applicable to quantum theory and cosmology in terms of standard physical quantities. It is reasonable then to complexity space and time as well as the electric and magnetic fields and to determine the relationship of the equations governing standard physical phenomena. Examined in detail is any unifying properties of the model in terms of complexifying physical quantities and examining any new predictions that can be made.

Faraday discusses some possible implications of considering $A_{\mu\nu}$, rather than $F^{\mu\nu}$ as fundamental in such a way that $A_{\mu\nu}$ may act in a domain where $F^{\mu\nu}$ is not observed [13]. In a later section we present a complexification of $A_{\mu\nu}$ rather than $\underline{E}$ and $\underline{B}$ (in $F^{\mu\nu}$). Continuing with the relationship of $F^{\mu\nu}$, the vector A^{μ}, and scalar potential, ϕ and the source terms of metric space, s^{μ} let us relate our complex electromagnetic field, $F^{\mu\nu}$, to complex spacetime. We have the volume element, $d\tau = \sqrt{g}\, dxdydz$ for

$$ds^2 = g_{\mu\nu} dx^{\mu} dx^{\nu} \tag{7.22}$$

and for a particular vector component of $F_{\mu} = \sqrt{g_{\mu\nu}}\, f^{\mu}$.

Then we have

$$\nabla \cdot \underline{F} = \frac{1}{\sqrt{g}} \frac{\partial f^{\mu} \sqrt{g}}{\partial x^{\mu}}. \tag{7.23}$$

For $F = \nabla \phi$ the function f^{μ} is related to the electromagnetic potential and gravitational potential as $f^{\mu} = g^{\mu\nu} \dfrac{\partial \phi}{\partial x^{\nu}}$. As before, $\dfrac{\partial F_{\mu\nu}}{\partial x_{\nu}} = J_{\mu}$ and $\gamma_{\mu\nu} \dfrac{\partial F_{\mu\nu}}{\partial x_{\nu}} = J_{\mu}$. As before we also had $\left(F_{41}, F_{42}, F_{43}\right) = i\underline{E}$ and $\left(F_{23}, F_{31}, F_{12}\right) = \underline{B}$ then the generalized complex form of $F^{\mu\nu}$ is

$$F^{\mu\nu} = \begin{pmatrix} 0 & B_z & -B_y & -\frac{i}{c}E_x \\ -B_z & 0 & B_x & -\frac{i}{c}E_y \\ B_y & -B_x & 0 & -\frac{i}{c}E_z \\ \frac{iE_x}{c} & \frac{iE_y}{c} & \frac{iE_z}{c} & 0 \end{pmatrix} \tag{7.24}$$

which can also be written as

$$\underline{F} = \left(B, -\frac{i}{c}\underline{E} \right) \text{ or } \underline{F}^{*} = \left(-\frac{iE}{c}, B \right). \tag{7.25}$$

We can now relate the complex E and B fields of the complex spacetime coordinates.

Returning to the compact notation for the two homogeneous equations, $\nabla \times \underline{E} + \dfrac{1}{c}\dfrac{\partial \underline{B}}{\partial t} = 0$ and $\nabla \cdot \underline{B} = 0$ as

$$\frac{\partial F_{\mu\nu}}{\partial x_{\kappa}} + \frac{\partial F_{\kappa\mu}}{\partial x_{\nu}} + \frac{\partial F_{\nu\kappa}}{\partial x_{\mu}} = 0. \tag{7.26}$$

It is very clear that introducing the imaginary components into these equations as $\partial / \partial\left(ix_{\mu}\right)$ and $\partial / \partial\left(it\right)$ leaves them unchanged. Examine the inhomogeneous equations $\nabla \cdot \underline{E} = 4\pi\rho$ and $\nabla \times \underline{B} = \frac{1}{c}\frac{\partial \underline{E}}{\partial t} + \underline{J}_e$. Then

$$F_{\mu\nu} = \frac{\partial A_\nu}{\partial x^\mu} - \frac{\partial A_\mu}{\partial x^\nu}$$

or (7.27)

$$\underline{F} = \underline{\nabla} \times \underline{A} \text{ for } \underset{\sim}{A}_\mu = \left(\underline{A}_j, \phi\right),$$

as before j runs 1 to 3 and all Greek indices run 1 to 4. Then the inhomogeneous equations become in general form, $\partial F^{\mu\nu}/\partial x^\nu = s^\mu$ which sets the criterion on s for using $\partial/\partial\left(ix_{\text{Im}}\right)$; that is, $s' \to is$. To be consistent [15], we can use $A_\mu = \left(A_j, \frac{1}{c}\phi\right)$.

We then consider the group of affine connections for a linear transformation from one system Σ to another Σ' where Σ and Σ' are two frames of reference and

$$x_\mu^{'} = a_{\mu\nu} x_\nu \tag{7.28}$$

where $a_{\mu\nu} a_\mu^\kappa = \delta_{\nu\kappa}$ and $\det\left|a_{\mu\nu}\right| = 1$. In general we can form a 4 x 4 coefficient matrix for the usual diagonal condition where, $a_{11} = 1$, $a_{22} = 1, a_{33} = 1$ and $a_{44} = -1$, all the other elements are zero, i.e. the signature (+++-). We can choose arrays of $a_{\mu\nu}$'s both real and imaginary for the general case so that we obtain forms for space and time components as being complex; for example,

$$x_3^{'} = \gamma\left(x_3 + i\beta x_4\right) \tag{7.29}$$

for $x_4 = t$, $\gamma = \left(1 - \beta^2\right)^{-1/2}$ and $\beta = v/c$. Other examples involve other combinations of complex space and time which must also be consistent with unitarity. We have discussed an 8-space formalism for the usual diagonal conditions. See Table 7.1.

Let us examine the effect of a gravitational field on an electron. Then we discuss some multidimensional models in which attempts are made to relate the gravitational and electromagnetic forces. Some of these multidimensional models are real and some are complex. The structure of the metric may well be determined by the geometric constraints set up by the coupling of the gravitational and electromagnetic forces. These geometric constraints govern allowable conditions on such phenomena as types of allowable wave transmission and the manner in which remote spacetimes are connected. Nonlocality or remote spacetime connections have implications for electromagnetic phenomena such as Young's double slit experiment and Bell's theorem.

Table 7.1 Coefficient Matrix for the Affine Connection for the Transformation from Reference Frame Σ to Σ'

	x_1	x_2	x_3	x_4
x_1'	a_{11}	a_{12}	a_{12}	a_{12}
x_2'	a_{21}	a_{22}	a_{23}	a_{24}
x_3'	a_{31}	a_{32}	a_{33}	a_{34}
x_4'	a_{41}	a_{42}	a_{43}	a_{44}

In fact, these experiments are more general than just the properties of the photon, that is, both experiments above can be and have been conducted with photons and other particles; and therefore what are exhibited are general quantum mechanical properties. Remote connection and/or transmission and nonlocality are more general than just electromagnetic phenomena but certainly have their application in electrodynamics and the nonlocal properties of the spacetime metric can be tested by experiments involving classical and quantum electro-dynamic properties. See Chap. 4.

7.2 Complex *E* and *B* in Real 4-Space and the Complex Lorentz Condition

Another approach to relate the relativistic and electromagnetic theories is the approach of Wyler in his controversial work at Princeton [16]. The model of Kaluza and Klein use a 5th rotational dimension to develop a model to relate electromagnetic and gravitational phenomena. This geometry is one-to-one mapable to our complex Minkowski space. Wyler introduces a complex Lorentz group with similar motives to those of Kaluza and Klein [17,8]. Wyler's formalism appears to relate to our complex Maxwell formalism and to that of Kaluza and Klein. The fundamental formalism for the calculation of the fine-structure constant, α is most interesting but perhaps not definitive.

$$\alpha = \frac{e^2}{\hbar c 4\pi\varepsilon_0} = \frac{e^2 c\mu_0}{2h} \tag{7.30}$$

where e is elementary charge, ε_0 vacuum permittivity and μ_0 the magnetic constant or vacuum permeability. An anthropic explanation has been given as the basis for the value of the fine-structure constant by Barrow and Tipler. They suggest that stable matter and intelligent living systems would not exist if α were much different because carbon would not be produced in stellar fusion [19].

Wyler [16] introduces a complex description of spacetime by introducing complex generators of the Lorentz group. He demonstrates that the Minkowski, M^n group is conformally isomorphic to the SO(n,2) group and then introduces a Lie algebra of M^4 which is isomorphic to SO(5,2). From his five and four spaces he generates a set of coefficients that generate the value of the fine structure constant, α. It is through introducing the complex form of the Lorentz group, L(T^n) that he forms an isomorphism to SO(n,2). Wyler calculates the electromagnetic coupling constant in terms of geometric group representations. He expands the generators of the set of linear transformations, T^n, of the group L(T^n). By definition, L(T^n) is isomorphic to the Poincaré group P(M^n), where M^n is the Minkowski space with signature (+++-) or, more generally, (1, n-l). The conformal group C(M^n) is then isomorphic to the SO(n,2) group, which is of quadratic form and signature (n,2).

Wyler then chooses the complex form

$$T^n = R^n + iV^n \tag{7.31}$$

where R^n represents T_{Re}, and V^n represents T_{Im} for $y \in R^n$, or y is an element of R^n and all y's are $y > 0$. The Poincaré group, $P(\text{M}^n)$ is the semi-direct product of the Lorentz group SO(l, n-l) and the group of transformations R_n then is $g \in$ SO(n,2).

Then C(M^4)$\cong$SO(4,2) is the invariance group of Maxwell's equations. The hyperboloids of the 4-mass shell momentum operators are $p_1^2,\ldots,p_4^2 = m^2$ from the representation of the Lie group geometry of M^4 isomorphic to SO(5,2). The intersection of the D^5 (5D) hyperspace with D^4 (4D) gives a structure reduced on D^4 which is colinear to the reduction of a Casimir operator function, $f(z)$ harmonic in D^4.

The coefficients of the Poisson group D^n (n dimensional) as D^4 and D^5 give the value of $\alpha \sim 1/137.036$. Actually, it is the coefficients of the Poisson nucleus $P^n(z,\xi)$ harmonic in D^n which gives the value of α in terms of z where z is, in general, a complex function and ξ is a spinor. The value is obtained from the isomorphic groups SO(5) x SO(2) and SO(4) x SO(2) which gives $(9/8\pi 4)\,(V(D^5))^{1/4}$ = 1/137.037 where $V(D^5)$ is a Euclidean value of the D^5 domain [20].

The expression for the Poisson nucleus is given by Hau [20]. Note that the Wyler calculation is another example of the relationship between a fifth dimension and a complex "space" of Lorentz transformation. The Wyler theory appears to strongly support the fundamental nature of geometric models. If one can calculate the fine structure constant or any other force field coupling constants from first principles, this gives great impetus to the concept that geometric constraints are extremely significant and may potentially be able to explain the origin of scientific law. In particular, we may be able to at least describe the major force fields (nuclear, electromagnetic, weak, and gravitational in terms of a geometric structure and, perhaps, by this formalism demonstrate the unifying aspects of major forces of nature [14,21].

Wyler also associates the conformal group $C(M^n)\cong SO(4,2)$ with the invariant group of Maxwell equations. The 4-mass shell conditions on the hyperboloids of mass form the representation of the Lie algebras of M^4. Isomorphism to SO(5,2) and S(4,2) intersection lead to a model of the intersection of Maxwell's field and the elementary particle field, i.e. a possible unification of electromagnetic and weak interactions as another approach to the electroweak vector - axial vector model [22]. In the presence of an external gravitational field, the cosmological term is small and finite and depends on vacuum state polarization. In fact, the cosmological term is given by the sum of all vacuum diagrams. In supersymmetry the cosmological term vanishes and therefore the total zero-point energy density of the free fields vanishes [23].

We return to our complex $\underline{E}$ and $\underline{B}$ fields and suggest the relation of our formalism to the Wyler formulation. Using the invariance of line elements $s = X^2 - c^2t^2$ for $r = ct = \sqrt{X^2}$ for $X^2 = x^2 + y^2 + z^2$, to measure the distance from a test charge to an electron charge, we can write for the imaginary part of the complex Maxwell equation $\nabla\times\left(i\underline{E}_{\mathrm{Im}}\right) = \frac{1}{c}\frac{\partial\left(i\underline{B}_{\mathrm{Im}}\right)}{\partial t} + i\underline{J}_{\mathrm{Im}}$ then for $E_{\mathrm{Im}} = 0$.

$$\nabla\times\left(i\underline{E}_{\mathrm{Im}}\right) = 0 \quad \text{or} \quad \frac{1}{c}\frac{\partial\left(i\underline{B}_{\mathrm{Im}}\right)}{\partial t} = i\underline{J}_{\mathrm{Im}} \tag{7.32}$$

or

$$\frac{\partial\left(i\underline{B}_{\mathrm{Im}}\right)}{\partial r}=ic\underline{J}_{\mathrm{Im}} \quad \text{or} \quad \frac{\partial \underline{B}_{\mathrm{Im}}}{\partial r}=c\underline{J}_{\mathrm{Im}} \tag{7.33}$$

for the assumed imaginary, B_{Im} commutator relation.

The energy associated with the imaginary part of the magnetic field, $\underline{B}_{\mathrm{Im}}$ is of interest. We calculate an energy invariant by squaring and integrating the above equation as [1,4]

$$\mathcal{E}=-\int_{\tau} J_m^2 R d\tau=-\int_{\tau}\left(\frac{\partial B_z}{\partial r}\right)^2 R d\tau \leq 0 . \tag{7.34}$$

The distance function, $R(r)$ over the volume element, $d\tau$ is assumed to be point-symmetrical and vanishes for positive real energy states. The volume, $d\tau$ is constructed to include a small real domain where a point charge is located, avoiding possible divergences. The negative value of the energy integral leads us to hypothesize about the possible source of this energy, such as arising from the vacuum. Perhaps it can be related to vacuum state polarization in a Fermi-Dirac sea model, as we have presented before [24]. Another possible association is with advanced potential models such as those of de Beauregard [25,26]. A third and perhaps the most interesting association would be with the complex coordinate space [27,28].

In Weyl's non-Riemannian geometry, [10,11] he presents a model that does not apply to actual spacetime but to a graphic representation of that relational structure, which is the basis in which both electromagnetic and metric variables are interrelated [12]. This is the deep significance of the geometry and relates to work of Hanson and Newman [29] and Rauscher [27,28] on the complex Minkowski space as well as Wyler's work [16] on complex group theories, such as complex Lorentz invariance, where he attempts to reconcile Maxwell's equations and relativity theory. The examination of the hyperspheres of the de Sitter space is presented by Ellis, where he attempts to unify electromagnetic and gravitational theory [30]. Eddington has suggested that the Weyl formalism, developed around 1923, is one of the major advances in the work of Einstein. The key is that if electromagnetism and QED can be reconciled with the gravitational field, along with the electroweak theory, a unity of the four forces can be made with a simpler and perhaps more reasonable model than the current Theories of everything (TOE). The strong force must also be included.

There is a significant difference between Einstein's generalization of Galilean geometry and Weyl's generalization of Riemannian geometry. The gravitational force field renders Galilean geometry useless and therefore required a move to Riemannian geometry. In terms of Weyl's geometry, we find that the electromagnetic force, $F_{\mu\nu}$, is comparable to the surface of an electron of 4 x 10^{18} volts/cm, [12] and the size of the charge was compatible with the radius of curvature of space.

For the electromagnetic mass, $m_e = e^2 / 4\pi a$, we have

$$m_g ds = \frac{1}{8\pi} G\sqrt{g} d\tau \tag{7.35}$$

where we denote the curvature R by G for the general case of both gravitational and electromagnetic field. The ratio of the masses m_g / m_e relates to the ratio of field strengths of about 10^{-37}.

7.3 Complex Electromagnetic Forces in a Gravitational Field

We used the weak Weyl limit of the gravitational force in previous calculations of this chapter. We will outline how the complexification of $F_{\mu\nu}$ can be formulated geometrically. We demonstrate that we obtain the same results for the relationship of mass and charge. Let v^μ denote the velocity vector as $v^\mu = dX_\mu / ds$ of the electron in the field, and ρ_0 denote the proper density of charge, *e*. The current is given by $J^\mu = \rho_0 v^\mu$. The fields, $F_{\mu\nu}$ refer to the applied external force of the electron. Returning to Eddington's approach [11], we then have

$$mA^\nu A_{\mu\nu} = -F_{\mu\nu}\rho_0 A^\nu. \tag{7.36}$$

We can also write ρ_0 as *e* in the above equation.

In the limit of our gravitational field we can neglect the gravitational field as an external field or also the gravitational energy of the electron. For an electron in a gravitational field we start from the field equations with the Ricci curvature tensor, R_μ^ν and the metric tensor, g_μ^ν. For the case where no matter is present we have:

$$G_{\mu}^{\nu} = R_{\mu}^{\nu} - \frac{1}{2} g_{\mu}^{\nu} R_{\mu}^{\nu} = -\frac{8\pi}{c^4} G E_{\mu}^{\nu} \tag{7.37}$$

using the scalar curvature, $R = \dfrac{8\pi GE}{c^4} = 0$, where $F = c^4 / G$ and G is the gravitational constant. This equation simplifies to

$$R_{\mu\nu} = -8\pi E_{\mu\nu} \tag{7.38}$$

and applies to certain regions that contain electromagnetic fields but no matter and no electron charges.

For the only surviving component in the energy considerations, we have

$$F_{41} = -F_{14} = \frac{\partial \phi}{\partial r} \tag{7.39}$$

where r is the radial separation. Then $F^{41} = g^{44} F_{41}$ and $\dfrac{\partial \phi}{\partial r} \propto \dfrac{e}{r^2}$, and

$$E_1^1 = +E_2^2 = E_3^3 = -E_4^4 = \frac{1}{2}\frac{\partial \phi}{\partial r} = \frac{1}{2}\frac{e^2}{r^4}. \tag{7.40}$$

One can associate m_e, the mass of the electron, with $4\pi e$, giving $\alpha = \dfrac{2\pi e^2}{m} \sim 1.5 \times 10^{-13}$ cm and justifies identifying $4\pi\varepsilon$ with the electrical charge e for $4\pi e$ or

$$F_{41} = \frac{\partial \phi}{\partial r} = \frac{1}{4\pi}\frac{e}{r^2}. \tag{7.41}$$

We use

$$\Box F_{\mu\nu} = J_{\mu\nu} - J_{\nu\mu} \text{ for } A^{\mu} = \frac{de}{4\pi r} \tag{7.42}$$

and then

$$F_{\mu\nu} = \int \frac{de\left(A_{\mu\nu} - A_{\nu\mu}\right)}{(4\pi\gamma) r} = \frac{1}{4\pi\gamma}\left(A_{\mu\nu} - A_{\nu\mu}\right)\int \frac{de}{r} \tag{7.43}$$

because all parts of the electron obey the same relativity where

$$\frac{\partial^2 A^\mu}{\partial t^2} - \nabla^2 A^\mu = J^\mu$$

and (7.44)

$$A^\mu = \frac{1}{4\pi}\frac{ds}{dt}v^\mu \int_\tau \frac{\rho d\tau}{r}$$

for velocity, v^μ, we drop the γ since all measurements are assumed to be proper time measurements. Integrating over the electron between pairs of points on the electron surface,

$$eF_{\mu\nu} = \frac{1}{4\pi}\left(A_{\mu\nu} - A_{\nu\mu}\right)\iint \frac{de_1 de_2}{r_{12}} = \frac{1}{4\pi}\left(A_{\mu\nu} - A_{\nu\mu}\right)\frac{e^2}{a} \quad (7.45)$$

where 1/a is the average value of $1/r_{12}$. We can write Eq. (7.43) as

$$-eA^\nu F_{\mu\nu} = \frac{1}{4\pi}A^\nu\left(A_{\mu\nu} - A_{\nu\mu}\right)\frac{e^2}{a} \quad (7.46)$$

and using the equation from before, relating v^ν, $A_{\mu\nu}$, $F_{\mu\nu}$ and A^ν, $mv^\nu A_{\mu\nu} = -F_{\mu\nu}eA^\nu$, so that $m = e^2/4\pi a$ as before.

How does this relate to the deSitter spaces? In the de Sitter algebras the proper time in all inertial frames of intervals is the same or equivalent. This is the powerful absolute of the de Sitter space. The proper time interval, $d\tau$ on its geodesic world-line in the de Sitter picture is given as

$$d\tau^2 = dt^2 - e^{2t}\left(dX^2\right) \quad (7.47)$$

for $dX^2 = dx^2 + dy^2 + dz^2$ in Euclidean coordinates and t is the cosmic time. The metric form of the de Sitter universe represents the metric form consistent with the observed asymptotically flat, low density universe. The se Sitter space is constant with Einstein dynamic equations and is therefore consistent with the Hubble constant, H_R [30,31].

Ellis [32] suggests that geometry and electromagnetism can be unified by a rigorous analysis of time. The hyperspheres of de Sitter space can be represented as a 5D metric manifold which tie the geometric models of gravity and electromagnetism to the structure of matter. Time is not primary but a property of the matter of elementary particles. If $\tau = t$ is allowed in the de Sitter space, then the typical geodesies represent what appears to be electromagnetic field lines. This is the manner in which Ellis attempts to describe the electromagnetic phenomena as geometric!

Figure 7.1. Geodesic plots of de Sitter space representing the field lines of the electromagnetic field. Various conditions for signal propagation are given.

The conformal invariant is given as

$$ds^2 = \frac{1}{R^2}\left(dx^2 + dy^2 + dz^2 - dR^2\right) \tag{7.48}$$

which depends only on the ratios of distances and is thus independent of scale. Let $t = -\ell nR$, then $R = e^{-t}$ and $ds^2 = e^{2t}\left(dx^2 + dy^2 + dz^2\right) - dt^2$ which is the de Sitter metric element. Ellis' geodesies of his angle metric correspond to geodesics of the de Sitter space (Fig. 7.1). In Fig. 7.1b, time-like subluminal geodesies are represented, in Fig. 7.1c they are luminal, and in Fig. 7.1d these geodesics are space-like superluminal (see Chap. 9). The figures also contain Euclidean space planes as spheres of infinite radii.

Feinberg [33] suggested that the first step in the test of multi-dimensional geometric models is to predict some simple phenomena such as the Coulomb attraction-repulsion; and that the geodesic form in Fig. 7.3 may point a way to do this, because if we can relate this 5D geometry to the complex geometry, then we can relate this complex geometry to Coulomb interactions. The curvature of space may then be related to a rotation or angular momentum component as a Kaluza-Klein 5th dimension. We form an isomorphism of this geometry to an 8D real-complex coordinate geometry which appears to not only unify electromagnetic theory and gravitational theory but may also resolve some other apparent paradoxes [34,35].

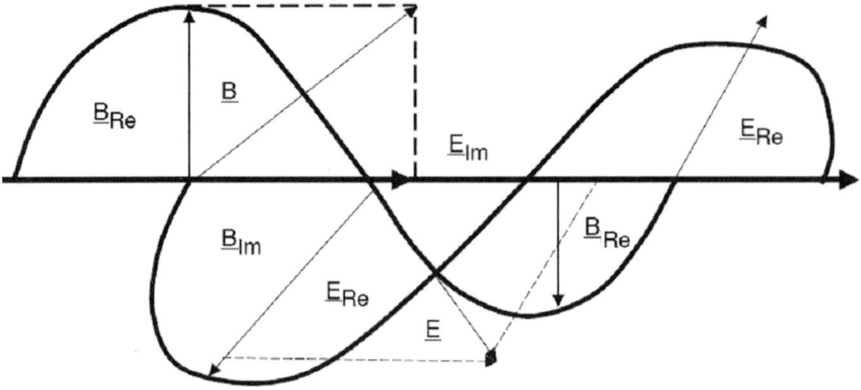

Figure 7.2. Hertzian and non-Hertzian waves. Where E and B are decoupled into two components $\underline{E}_{\mathrm{Re}}$ and $\underline{E}_{\mathrm{Im}}$ and $\underline{B}_{\mathrm{Re}}$ and $\underline{B}_{\mathrm{Im}}$.

We have observed that introducing complex E and B fields or complexifying the $F^{\mu\nu}$ field can be performed in such a way as to not distort the electric charge on the electron. We also find consistency with the 5D geometry of Kaluza and Klein, the 8D Minkowski space, and the de Sitter space where the geodesic represents the electromagnetic field lines. We can also maintain Lorentz invariance conditions for both real and complex transforms on the line element.

References

[1] Weyl, H. (1957) spacetime-Matter, New York: Dover.
[2] Wheeler, J.A. (1962) Geometrodynamics, New York: Academic Press.
[3] Einstein, A. (1955) Relativity, Princeton: Princeton Univ. Press.
[4] Inomata, I. (1976) Consciousness and complex electromagnetic fields, Electrotechnical Laboratory, MITI, Tokyo.
[5] Minkowski, H. (1915) *Ann, Phys. Lpz.* 47; 927; *Jber. Deutsche Mat. Vev.* 24; 372.
[6] Rauscher, E.A. (1982) Theoretical and field studies of extremely low frequency radiation and coherent linear phenomena, TRL Laboratory report, PSRL-726A.
[7] Frank, P. (1911) *Ann. Phys. Lpz.* 34; 599.
[8] Pauli, W. (1962) Theory of Relativity, Chicago: Permagon Press.
[9] de Broglie, L. (1955) *Mécanique Ondulatoire du Photon*, Paris: Gauthier-Villas.
[10] Weyl, H, (1918) *Math. Z.* 20, 384; *Pruss. Akad. Wiss.* 465.
[11] Weyl, H, (1920) Phys. Z. 21, 649.
[12] Eddington, A. (1922) The Mathematical theory of Relativity, London: Chelsea Publishing Co.
[13] Faraday, M. (1855) Experimental Researches in Electricity, Vols. 1-3, R. Taylor

& W. Francis (eds.) London: Univ. London Press.
[14] Rauscher, E.A. (1971) A Unifying Theory of Fundamental Process, UCAL-20808, Berkeley: LBNL Book
[15] Stratton, J.A. (1941) Electromagnetic Theory, NY: McGraw-Hill.
[16] Kaluza, T. (1921) *Sitz. Berlin Preuss. Akad. Wiss.*, 966.
[17] Klein, O. (1926) *Z. Phys.* 37, 895.
[18] Tipler, F.J. (1974) Phys. Rev. D9, 2203.
[19] Wyler, A. (1968) Arch. for Rational Mechanics & Analysis, 31, 35.
[20] Hua, L.K. (1963) Harmonic Analysis in the Classical Domain, Providence: American Mathematical Society.
[21] Gell-Mann, M., Raymond, P. & Slonsky, R. (1979) Supergravity, P. Van Nieuwenhuizen & D.Z. Freedman (eds.) Amsterdam: N. Holland.
[22] Wyler, A. (1969) *Acad. Sc.* Paris Series A, 269, 743; (1971) *Acad. Sc.* Paris Series A, 271, 186.
[23] Wess, W. & Zumino, B. (1974) Nucl. Phys. B78, 1.
[24] Andrews, T.B. (1999) Observational tests of the static universe model and the derivation of the Hubble redshift, in A.J. Bunker & W.J.M. van Breugel (eds.) The Hy-Redshift Universe: Galaxy Formation and Evolution at High Redshift, Berkeley, 21-24 June 1999, AIP Conference Proceedings.
[25] Costa de Beauregard, O. (1976) Acad. Sci. Paris, 104, 1251.
[26] Costa de Beauregard, O. (1977) Phys. Gelt. 60A, 93.
[27] Rauscher, E.A. (1973) in The Iceland Papers: Frontiers of Physics Conf. Amherst: Essentia Research Associates.
[28] Ramon, C & Rauscher, E.A. (1980) Found. Phys. 10, 661.
[29] Hanson, R.O. & Newman, E.T. (1975) Gen. Rel. & Grav. 6, 216.
[30] Ellis, H.E. (1974) Found. Phys. 4, 311.
[31] Rauscher, E.A. (1973) Early universe cosmological models, Bull. Amer. Phys. Soc. 18,1570.
[32] Rauscher, E.A. (2003) Cosmogenesis and quantum gravity, in R.L. Amoroso, B. Lehnert and J-P Vigier, Beyond the Standard Model: Searching for Unity in Physics, pp. 43-72, Oakland: The Noetic Press.
[33] Feinberg, G. (1967) Phys. Rev. 159, 1089.
[34] Rauscher, E.A.& Targ, R. (1973) Why only 4D will not explain nonlocality, J Sci. Explor. 16, 655.
[35] Ramon, C. & Rauscher, E.A. (1982) Remote connectedness in complex geometries, TRL Laboratory report, PSRL-4105.

Chapter 8

Real and Complex Amended Maxwell's Equations for Non-Abelian Gauge Groups

In this chapter we analyze, calculate and extend modification of Maxwell's equations in a complex Minkowski metric, M_4 in a $\mathbb{C}_2$ complexified space using the SU_2 gauge, SL(2,c) and other gauge groups, such as SU_n for $n > 2$ expanding the U_1 gauge theories of Weyl. Weyl identified the U_1 gauge group for the standard Maxwell's equations in its nonrelativistic form in M_4 space. We expand the form of the elegant electromagnetic equations and express them in $\mathbb{C}_4$ space in Chap. 5 for the nonrelativistic formalism and in Chap. 6 for the relativistic formalism of the equations. The advanced and retarded formalisms are also examined in Chap. 7. In the case where Maxwell's equations are solved in $\mathbb{C}_4$ space, or the complex 8-space, we can extend the theory to considerations of other gauge groups such as SL(2,c), SU_n for $n > 2$ and SU_2 expanding the approach beyond U_1 gauge conditions.

8.1 Introduction – Extended Maxwell's Equations

In addition to our work others have examined complex multidimensional geometries [1-6]. In particular we have examined the complexification of M_4 Minkowski space as an 8D complex $\mathbb{C}_4$ space [5,6]. The complex space is comprised of four real dimensions and 4 imaginary dimensions and this geometry is consistent with Lorentz invariance and analytic continuation. See Chaps. 2 and 4.

We have developed an 8D complex Minkowski space, M_4 composed of four real dimensions and four imaginary dimensions which is consistent with Lorentz invariance and analytic continuation in the complex plane [1-6]. The unique feature of this geometry is that it admits of nonlocality consistent with Bell's theorem, (EPR paradox), possibly Young's double slit experiment, the Aharonov-Bohm effect and multi-mirrored interferometric experiment [7]. See Chap. 4.

This work of amending Maxwell's equations yields additional predictions beyond the electroweak unification scheme. Some of these are:

- Modified gauge invariant conditions,
- Short range non-Abelian force terms and Abelian long range force terms in Maxwell's equations,
- Finite but small rest mass, m_γ of the photon,
- A magnetic monopole like term
- Longitudinal as well as transverse magnetic and electromagnetic field components in a complex Minkowski metric, M_4 in a $\mathbb{C}_4$ space.

Additionally, expressing Maxwell's electromagnetic equations in complex 8-space, leads to some new and interesting predictions in physics, including possible detailed explanation of some of the previously mentioned nonlocality experiments [8-13]. Complexification of Maxwell's equations requires a non-Abelian gauge group which amend the usual theory and which utilizes the usual unimodular Weyl U_1 group. We have examined the modification of gauge conditions using higher symmetry groups such as, SU_n and other groups such as the SL(2,c) double cover group of the rotational group SO(3,1) related to the Ricci curvature tensor [14]. Thus we are led to new and interesting physics involving extended metrical space constraints. In addition to the usual transverse we also have longitudinal, non-Hertzian electric and magnetic field solutions to Maxwell's equations leading to new communication systems and antennae theory, non zero solutions to $\nabla \cdot \underline{B}$, and a possible finite but small rest mass of the photon, m_γ.

Comparison of our theoretical approach is made to the work of Vigier et al. [15,16] Barrett et al. [17] and Harmuth et al. [18] on amended Maxwell's theory. We compare our predictions such as our longitudinal field to the $B^{(3)}$ term of Vigier, and our Non-Abelian gauge groups to that of Barrett and Harmuth. We interpret this work as leading to new and interesting physics, including a possible interpretation of nonlocal information transmission properties within the Dirac polarized vacuum.

8.2. Complexified Electromagnetic Fields in Minkowski Space and Nonlocality

We expand the usual 4D line element metric $ds^2 = g_{\mu\nu}dx^\nu dx^\mu$ in the following manner. We consider a complex 8D space, $\hat{M}_4$ as $\mathbb{C}_4$ so that

$Z^{\mu} = X^{\mu}_{\text{Re}} + iX^{\mu}_{\text{Im}}$ and likewise for Z^{ν} where the indices ν and μ run 1 to 4 yielding (1, 1, 1, -1). Hence, we now have a new complex 8-space metric as $ds^2 = \eta_{\nu\mu} dZ^{\nu} dZ^{*\mu}$. We have developed this space and other extended complex spaces and examined their relationship with the twister algebras and asymptotic twister space and the spinor calculus and other implications of the theory [6]. The Penrose twistor SU(2,2) or U_4 is constructed from 4D-spacetime, $U_2 \otimes \widetilde{U}_2$ where U_2 is the real part of the space and $\widetilde{U}_2$ is the imaginary part of the space, this metric appears to be a fruitful area to explore.

The twistor, $\mathbb{Z}$ can be a pair of spinors, U^A and π_A which Penrose created to represent a twistor. The condition for these representations is:

- The null infinity condition for a zero spin field is $\mathbb{Z}^{\mu}\overline{\mathbb{Z}}_{\mu} = 0$
- Conformal invariance
- Independence of the origin.

The twistor is derived from the imaginary part of the spinor field. The underlying concept of twister theory is that of conformally invariance fields occupy a fundamental role in physics and may yield some new approaches to physics. Since the twister algebra falls naturally out complex space.

Other researchers have examined complex dimensional Minkowski spaces. In [2], Newman demonstrates that M_4 space does not generate any major “weird physics” or anomalous physics predictions and is consistent with an expanded or amended special and general relativity. In fact the Kerr metric falls naturally out of this formalism as demonstrated by Newman [4] and Rauscher [5,6,19,20].

As we know twistors and spinors are related by the general Lorentz conditions in such a manner that all signals are luminal in the usual 4*n* Minkowski space but this does not preclude super or transluminal signals in spaces where $n > 4$. Stapp, for example, has interpreted Bell’s theorem experimental results in terms of transluminal signals to address the nonlocality issue of the Clauser, and Aspect experiments [8]. Newman et al. demonstrate the role of nonlocal fields in complex 8-space [2,3].

We believe that there are some very interesting properties of the complex M_4 space which include the nonlocality properties of the metric applicable in the non-Abelian algebras which are related to the quantum theory and the conformal invariance in relativity as well as new properties of Maxwell’s equations. In addition, complexification of Maxwell’s equations in $\mathbb{C}_4$ space yields interesting predictions, yet we find the usual conditions on the manifold hold [19,20]. Some of these new predictions come out of the complexification of 4-space and appear to relate to the work of Vigier, Barrett, Harmuth and others [15,17.18]. Also we find that the twistor algebra of the complex 8D $M_4 + \mathbb{C}_4$ space is mapable 1 to 1 with the twistor algebra of the Kaluza-Klein 5D electromagnetic - gravitational metric [21-23].

Some of the predictions of the complexified form of Maxwell's equations are:

- A finite but small rest mass of the photon, m_γ
- A possible magnetic monopole, $\nabla \cdot \underline{B} \neq 0$
- Transverse as well as longitudinal $B^{(3)}$ like components of $\underline{E}$ and $\underline{B}$,
- New extended gauge invariance conditions to include non-Abelian algebras
- An inherent fundamental nonlocality property on the manifold. Evans and Vigier also explore longitudinal $\underline{E}$ and $\underline{B}$ components in detail and finite rest mass of the photon, m_γ [16].

We consider both the electric and magnetic fields to be complexified as $\underline{E} = E_{\text{Re}} + iE_{\text{Im}}$ and $\underline{B} = \underline{B}_{\text{Re}} + iB_{\text{Im}}$ for $E_{\text{Re}}, E_{\text{Im}}, B_{\text{Re}}$ and B_{Im} are real quantities. Then substitution of these two equations into the complex form of Maxwell's equations above yields, upon separation of real and imaginary parts, two sets of Maxwell-like equations. The first set is

$$\begin{aligned} &\nabla \cdot \underline{E}_{\text{Re}} = 4\pi\rho_e && \nabla \times E_{\text{Re}} = -\frac{1}{c}\frac{\partial \underline{B}_{\text{Re}}}{\partial t} \\ &\nabla \cdot \underline{B}_{\text{Re}} = 0 && \nabla \times \underline{B}_{\text{Re}} = \frac{1}{c}\frac{\partial \underline{E}_{\text{Re}}}{\partial t} = \underline{J}_e ; \end{aligned} \tag{8.1}$$

the second set is

$$\begin{aligned} &\nabla \cdot \left(i\underline{B}_{\text{Im}}\right) = 4\pi i\rho_m && \nabla \times \left(i\underline{B}_{\text{Im}}\right) = \frac{1}{c}\frac{\partial\left(i\underline{E}_{\text{Im}}\right)}{\partial t} \\ &\nabla \cdot \left(i\underline{E}_{\text{Im}}\right) = 0 && \nabla \times \left(i\underline{E}\right) = \frac{1}{c}\frac{\partial\left(i\underline{B}_{\text{Im}}\right)}{\partial t} = i\underline{J}m . \end{aligned} \tag{8.2}$$

The real part of the electric and magnetic fields yield the usual Maxwell's equations and complex parts generate "mirror" equations. For example, the divergence of the real component of the magnetic field is zero, but the divergence of the imaginary part of the electric field is also zero, and so forth. The structure of the real and imaginary parts of the fields is parallel with the electric real components being substituted by the imaginary part of the magnetic fields and the real part of the magnetic field being substituted by the imaginary part of the electric field. In the second set of equations, (8.2), the i's, "go out" so that the quantities in the equations are real, hence $\nabla \cdot \underline{B}_{\mathrm{Im}} = 4\pi\rho_m$, and not zero, yielding a term that may be associated with some classes of monopole theories [20].

We express the charge density and current density as complex quantities based on the separation of Maxwell's equations above. Then, in generalized form $\rho = \rho_e = i\rho_m$ and $\underline{J} = \underline{J}_e + i\underline{J}_m$ where it may be possible to associate the imaginary complex charge with the magnetic monopole and conversely the electric current has an associated imaginary magnetic current. Using the invariance of the line element $s^2 = x^2 - c^2t^2$ for $r = ct = \sqrt{x^2}$ and for $x^2 = x^2 + y^2 + z^2$ for the distance from an electron charge, we can write the relation,

$$\frac{1}{c}\frac{\partial\left(i\underline{B}_{im}\right)}{\partial t} = i\underline{J}_m \quad \text{or} \quad \frac{1}{c}\frac{\partial \underline{B}_{im}}{\partial t} = \underline{J}_m$$

$$\nabla \times \left(i\underline{E}_{\mathrm{Im}}\right) = 0 \text{ for } \underline{E}_{\mathrm{Im}} = 0 \quad \text{or} \quad \frac{1}{c}\frac{\partial\left(i\underline{B}_{\mathrm{Im}}\right)}{\partial t} = i\underline{J}_m . \tag{8.3}$$

8.3 The General Concept of Gauge Symmetry in Current Physics

Gauge symmetry is the basic concept required in field theory to describe a field for which the equations describing the field do not change when an operation applied to all particles and fields everywhere in space is globally invariant. It is also possible to have local gauge symmetry where the operation is applied to some particular region of space. Fields with gauge symmetry are, for example, gravity, electromagnetism and QED. The gauge symmetry approach was a key development in the theory of weak, and electroweak interactions and QCD. The quantum field is restored to symmetry by its Yang-Mills gauge field. Thus the origin of the concept of broken symmetry in gauge theory which led to the development of the electroweak theory in 1967 by Weinberg and Salam. This was a key benchmark in developing a grand unified theory (GUT).

In field theory a gauge group corresponds mathematically to a fiber symmetry group, whereas gauge theory corresponds mathematically to the principle fiber bundle. The gauge group for the electromagnetic photon is a U_1 gauge group. The gauge group for the strong force, SU_3 is mediated by eight independent gluons binding the three quarks. See Table 8.1. The electroweak force, SU_2 x U_1 corresponds to the quantum gauge groups of quantum chromodynamics (QCD); the electroweak gauge force is mediated by $W^{\pm}, Z^0$ which are termed intermediate vector Bosons. Gauge Bosons couple to conserved currents.

Table 8.1 The Color Octet of Gluon Gauge

$(r\bar{b} + b\bar{r})/\sqrt{2}$	$-i(r\bar{b} - b\bar{r})/\sqrt{2}$
$(r\bar{g} + g\bar{r})/\sqrt{2}$	$-i(r\bar{g} - g\bar{r})/\sqrt{2}$
$(b\bar{g} + g\bar{b})/\sqrt{2}$	$-i(b\bar{g} - g\bar{b})/\sqrt{2}$
$(r\bar{r} - b\bar{b})/\sqrt{2}$	$(r\bar{r} + b\bar{b} - 2g\bar{g})/\sqrt{6}$

8.4 New Gauge Conditions, Complex Minkowski Space and New Implications for Physics

In a series of papers, Barrett, Harmuth and Rauscher have examined the modification of gauge conditions in modified or amended Maxwell theory. The Rauscher approach, as briefly explained in the preceding section is to write complexified Maxwell's equation in consistent form to complex Minkowski space [20].

The Barrett amended Maxwell theory utilizes non-Abelian algebras and leads to some very interesting predictions. He utilizes the non commutitative SU_2 gauge symmetry rather than the U_1 symmetry. Although the Glashow electroweak theory utilizes U_1 and SU_2, but in a different manner, but his theory does not lead to the interesting and unique predictions of the Barrett theory. Barrett, in his amended Maxwell theory, predicts that the velocity of the propagation of signals is not the velocity of light. See Chap. 12. He presents the magnetic monopole concept resulting from the amended Maxwell picture. His motive goes beyond the standard Maxwell formalism and generates new physics utilizing a non-Abelian gauge theory [17].

The SU_2 group gives us symmetry breaking to the U_1 group which can act to create a mass splitting symmetry that yield a photon of finite (but necessarily small) rest mass which may be created as self energy produced by the existence of the vacuum. This finite rest mass photon can constitute a propagation signal carrier less than the velocity of light. We can construct the generators of the SU_2 algebra in terms of the fields $\underline{E}$, $\underline{B}$, and $\underline{A}$. The usual potentials, A_μ are expressed as the important 4-vector quality, $A_\mu = (\underline{A}, \phi)$ where the index runs 1 to 4. One of the major purposes of introducing the vector and scalar potentials is to subscribe to their non-physicality because of the desire by physicists to avoid the issue of action at a distance. In fact in gauge theories, A_μ is all there is! Yet it appears that in fact these potentials yield a basis for a fundamental nonlocality and have real physical consequences!

Let us address the specific case of the SU_2 group and consider the elements of a non-Abelian algebra such as the fields with SU_2 (or even SU_n) symmetry then we have the commutation relations where $XY\text{-}YX \neq 0$ or $[X,Y] \neq 0$. This is reminiscent of the Heisenberg uncertainty principle non-Abelian gauge. Barrett explains that SU_2 fields can be transformed into U_1 fields by symmetry breaking. For the SU_2 gauge amended Maxwell theory additional terms appear in term of operations such $\underline{A} \cdot \underline{E}, \underline{A} \cdot \underline{B}$ and $\underline{A} \times \underline{B}$ and their non Abelian cases. For example $\nabla \cdot \underline{B}$ no longer equals zero but is given as $\nabla \cdot \underline{B} = -jg(\underline{A} \cdot \underline{B} - \underline{B} \cdot \underline{A}) \neq 0$ where $[A,B] \neq 0$ for the dot product of $\underline{A}$ and $\underline{B}$ and hence we have a magnetic monopole term and j is the current and g is a constant. Also Barrett gives references to the Dirac, Schwinger and 't Hooft monopole work. Further commentary on the SU_2 gauge conjecture of Harmuth [18] that under symmetry breaking, electric charge is considered but magnetic charges are not. Barrett further states that the symmetry breaking conditions chosen are to be determined by the physics of the problem. These non Abelian algebras have consistence to quantum theory.

In our analysis, using the SU_2 group there is the automatic introduction of short range forces in addition to the long range force of the U_1 group. U_1 is 1D and Abelian and SU_2 is 2D and is non-Abelian. U_1 is also a subgroup of SU_2. The U_1 group is associated with the long range $1/r^2$ force and SU_2, such as for its application to the weak force yields short range associated fields. Also SU_2 is a subgroup of the useful SL(2,c) group of non compact operations on the manifold. The SL(2,c) group is a semi-simple 4D Lie group and is a spinor group relevant to the relativistic formalism and is isomorphic to the connected Lorentz group associated with the Lorentz transformations. It is a conjugate group to the SU_2 group and contains an inverse. The double cover group of SU_2 is SL(2,c) where SL(2,c) is a complexification of SU_2. Also LS(2,c) is the double cover group of SU_3 related to the set of rotations in 3D space [24]. Topologically, SU_2 is associated with isomorphic to the 3D spherical, O_3^+ (or three dimensional rotations) and U_1 is associated with the O_2 group of rotations in two dimensions. The ratio of Abelian to non Abelian components, moving from U_1 to SU_2, gauge is 1 to 2 so that the short range components are twice as many as the long range components.

Instead of using the SU_2 gauge condition we use SL (2,c) we have a non-Abelian gauge and hence quantum theory and since this group is a spinor and is the double cover group of the Lorentz group (for spin ½) we have the conditions for a relativistic formalism. The Barrett formalism is non-relativistic. SL (2,c) is the double cover group of SU_2 but utilizing a similar approach using twister algebras yields relativistic physics.

8.5 Concluding Remarks

It appears that complex geometry can yield a new complementary unification of quantum theory, relativity and allow a domain of action for nonlocality phenomena, such as displayed in the results of the Bell's theorem tests of the EPR paradox [9,25], and in which the principles of the quantum theory hold to be universally. The properties of the nonlocal connections in complex 4-space may be mediated by non -or low dispersive loss solutions. We solved Schrödinger equation in complex Minkowski space [26-29]. See Chaps. 11 and 13. In progress is research involving other extended gauge theory models, with particular interest in the nonlocality properties on the spacetime manifold, quantum properties such as expressed in the EPR paradox and coherent states of matter.

Utilizing Coxeter graphs or Dynkin diagrams, Sirag [24] lays out a comprehensive program in terms of the A_n, D_n and E_6, E_7 and E_8 Lie algebras constructing a hyper dimensional geometry for as a classification scheme for elementary particles. Inherently, this theory utilizes complexified spaces involving twisters and Kaluza-Klein geometries. This space and the complex 8-space incorporate string theory and Grand Unification Theories (GUT) models [30,31]. We display the comparison of relativistic electromagnetic theory, quantum theory and gauge groups in Table 8.1. Gauge potentials, A^μ and gauge field strengths, A^μ are compared to U_1 and the Weyl gauge theories and to the Lie algebras of the supersymmetry groups, Su_n.

TABLE 8.2
Comparison of Quantum Theory, Relativistic Maxwell's Equations and Gauge Groups

QUANTUM THEORY	GAUGE THEORY
Physics	Mathematics
Gauge Theory	Principle Fiber Bundle
Gauge Group	Fiber Symmetry Group
Spacetime	Bose Space
Gauge Potential Field, A_μ	Connection 1-Form, U_1
Gauge Field Strength, $F_{\mu\nu}$	Curvature of Connected 2-Form (spin 2)
Gauge Particle (Boson)	Basic Elements of Lie Algebra Symmetry Groups
Matter Field	Spin or Valued Function on the Principle Bundle Basic Elements, Vector Space Acted on by a Symmetry Group

It appears that utilizing the complexification of Maxwell's equations with the extension of the gauge condition to non-Abelian algebras, yields a possible metrical unification of relativity, electromagnetism and quantum theory. This unique new approach yields a universal nonlocality [32,33]. No radical spurious predictions result from the theory, but some new predictions are made which can be experimentally examined. Also, this unique approach in terms of the twister algebras may lead to a broader understanding of macro and micro nonlocality and possible transverse electromagnetic fields observed as nonlocality in collective plasma state and other media [34]. See Chap. 11.

References

[1] Penrose, P. & Newman, E.J. (1978) Proc. Roy. Soc. A363, 445.
[2] Newman, E.T. (1973) J. Math. Phys. 14, 774.
[3] Hansen, R.O. & Newman, E.T. (1975) Gen. Rel. and Grav. 6, 216.
[4] Newman, E.T. (1976) Gen. Rel. and Grav. 7, 107.
[5] Ramon, C. & Rauscher, E.A. (1980) Found. Physics 10, 661.
[6] Rauscher, E.A. (1979) in The Iceland Papers: Frontiers of Physics, pp. 49-94, Amherst: Essentia Research Assoc; Reprinted, Ottawa: PACE.

[7] Rauscher, E.A. (1991) The dual laser nonlocal interferometry experiment, TRL, USRL-1769A, Report Proposal.
[8] Stapp, H.P. (1993) Phys. Rev. A47, 847 and Private Communication.
[9] Bell, J.S. (1964) Physics 1, 195.
[10] Clauser, J.F. & Horne, W.A. (1971) Phys. Rev. 10D, 526 and (1977) private communication with J. Clauser.
[11] Aspect, A. et al. (1982) Phys. Rev. 49, 1804 and (1977) private communication.
[12] Gisin, N. (1990) Phys. Lett. 143, 1.
[13] Tittel, W., Bredel, J., Zbinden, H. & Gisin, N. (1998) Phys. Rev. Lett. 81, 3563.
[14] Rauscher, E.A. (2002) Non-Abelian gauge groups and real amended Maxwell's equations, in R.L. Amoroso et al. (eds.) Gravitation and Cosmology: From the Hubble radius to the Planck Scale, pp. 183-188, Dordrecht: Kluwer Academic.
[15] Vigier, J.P. (1991) Foundations of Phys. 21, 125.
[16] Evans, M.W. & Vigier, J.P. (1994, 1995, 1996) The Enigmatic Photon, Vols. 1, 2 ,3, Dordrecht: Kluwer Academic.
[17] Barrett, T.W. & Grimes, D.MN. (1996) Advanced Electromagnetism, Singapore: World Scientific.
[18] Harmuth, H.F., Barrett, T.W. & Meffert, B. (2001) Modified Maxwell's Equations in Quantum Electrodynamics, Singapore: World Scientific.
[19] Rauscher, E.A. (1976) Bull. Am. Phys. Soc. 21, 1305.
[20] Rauscher, E.A. (1983) Electromagnetic Phenomena in Complex Geometries and Nonlinear Phenomena, Non-Hertzian Waves and Magnetic Monopoles, Millbrae: Tesla Books.
[21] Kaluza, T. (1921) sitz. Berlin Press, *A. Kad. Wiss,* 968.
[22] Klein, O. (1926) *Z. Phys*. 37, 805.
[23] Klein, O. (1927) *Z. Phys*. 41, 407.
[24] Sirag, S.P. (1996) A mathematical strategy for a theory of particles, in S.R. Hameroff, A.W. Kasniak & A.C. Scott (eds.) The 1st Tucson Conference, pp. 579-588, Cambridge: MIT Univ. Press.
[25] Einstein, A., Podolsky, B. & Rosen, N. (1935) Phys. Rev. 47, 777.
[26] Rauscher, E.A. (2001) Soliton Solutions to the Schrödinger Equation in Complex Minkowski Space, pp. 89-105, Proceeding of the First International Conference, IJCAS, Liege, Belgium.
[27] Rauscher, E.A. & Amoroso, R.L.(2006) The Schrödinger equation in complex Minkowski, nonlocality and anticipatory systems, Unified Theories, Budapest, Hungary, in R. L. Amoroso, I. Dienes & C. Varges (eds.) pp. 370-388, Oakland: The Noetic Press.
[28] Rauscher, EA (1981) Conceptual changes in reality models from new discoveries in physics, Proceedings of the 10th International Symposium on the Sciences, Vol. 2, pp. 1407-1431, New York: ICF Press; Wigner comments on Rauscher, p. 1479.
[29] Rauscher, E.A. (2010) Complex Minkowski Space & Nonlocality on the Metric & Quantum Processes, in progress.
[30] Binetruy, P. (2006) Supersymmetry, Oxford: Oxford Univ. Press.
[31] Aitchison, L.J.R. & Hay, A.J.G. (1989) Gauge Theories in Particle Physics, 2nd Edition, New York: Adam Hilger.

[32] Rauscher, E.A. & Van Bise, W. (1981) Relaxation of Gauge invariant conditions for VLF phenomena and their implications for magnetic and electromagnetic wave transmission, Tecnic Research Laboratory Report; and (1989) Bull. Am. Phys. Soc. 34, 82.
[33] Witten, E. (1981) Search for a realistic Kaluza-Klein Theory, Nucl. Phys. B186, 213
[34] Rauscher, E.A. (1968) J. Plasma Phys. 2, 517.

Chapter 9

Sub and Superluminal Transformations of the Complex Vector Potential

We consider in this chapter the transformations of an oscillatory form of A^μ in a complex 12D Minkowski space. The form of the transformation of A^μ or $(\underline{A},\phi)$ depends on whether such a transformation is a superluminal Lorentz transformation (SLT) or a subluminal Lorentz transformation.

9.1 Introduction

In [1] we extended our examination of the properties of complex Minkowski spaces and the mixing of real and imaginary components of space and time under the influence of a superluminal boost in the x-direction. We determined that there is a unique mixing of position and time vectors in complex Minkowski space which is not present in real 4D Minkowski space. We take real M_4 4-space as a slice through the complex Minkowski space, $\hat{M}_4 \subset \mathbb{C}_4$.

In this chapter we examine in detail the transformations of the vector and scalar potential in complex Minkowski space under an x-direction superluminal boost. We find that we have a mixing of the temporal and spatial components in the laboratory frame but only mixing of temporal components in the moving frame. In the laboratory frame, Σ an oscillatory and damped solution, both expressed in terms of space and time, whereas in the moving frame, Σ' the damped term is expressed in terms of time components only. The oscillation in terms of the spatial coordinates vanishes in going from the Σ to Σ' frame. It is also interesting to note that the vector potential normalization term A_{ox} goes to ϕ', the scalar potential term under the superluminal boost (SLT).

We also examine the relationship between the vector and scalar potential transformation under the SLT and compare this to the variation of E and B fields and their relationship to A_x and ϕ. The transformation from $v < c$ through c to $v > c$ produces a mixing or spacetime rupture which greatly modifies any existing vacuum fields. We examine the presence of tachyonic signals in [2,3], and here we demonstrate that the monopole structure may be associated with a tachyonic signal.

9.2 Complex Minkowski Spaces with Time Symmetry Considerations

In previous work, Rauscher and Ramon introduced the structure and properties of complex Minkowski spaces [1,2] and examined the mixing of real and imaginary components of space and time under the influence of superluminal boosts in the x direction and determined that the mixing is unique. We label the complex coordinates as

$$\underline{z}^{\mu} = x^{\mu}_{\text{Re}} + i x^{\mu}_{\text{Im}} \tag{9.1}$$

where $\underline{z}$ is a complex quantity and x_{Re} and x_{Im} are real quantities, where "Re" and "Im" refer to the real and imaginary parts of the complex quantity $\underline{z}$. The index μ runs 0,1,2,3 where the index 0 represents the time component and 1,2,3 represent the spatial vector components. We denote these 4-component vectors as t, x, y, z.

In complex Minkowski space these vectors are complex quantities and are given as

$$\underline{t} = t_{\text{Re}} + i t_{\text{Im}}, \quad \underline{x} = x_{\text{Re}} + i x_{\text{Im}}$$
$$\underline{y} = y_{\text{Im}} + i y_{\text{Im}}, \quad \underline{z} = z_{\text{Re}} + i z_{\text{Im}} \tag{9.2}$$

This set of vectors defines an 8D complex space [4]. A slice of this 8-space gives four real dimensions of M_4 forming a subspace in which the line elements are given by the real part of the complex quantities [5].

For a 12D space we consider time as a 3D complex quantity,

$$t = t_x \hat{x} + t_y \hat{y} + t_z \hat{z} \tag{9.3}$$

where we have the components

$$\begin{aligned} t_x &= t_{x\text{Re}} + i t_{x\text{Im}} \\ t_y &= t_{y\text{Re}} + i t_{y\text{Im}} \\ t_z &= t_{z\text{Re}} + i t_{z\text{Im}}. \end{aligned} \tag{9.4}$$

As before the subscripts Re and Im refer to the real and imaginary parts of the complex quantities.

Demers [6] introduced a symmetry principle between multidimensional time components which specify only one modulus as having physical meaning,

$$|t| = \left(t_x^2 + t_y^2 + t_z^2\right)^{\frac{1}{2}} \tag{9.5}$$

The modulus of the time vector is chosen to correspond to the usual physical time. A detailed discussion of this choice of modulus and its implications for Lorentz invariance is in [1]. In our HD model we imply that all complex temporal components are physically significant [7]. This probably makes correspondence to Cramer's Transactional Interpretation of quantum theory where he implies that all off diagonal components of a transaction are physically real [8-10].

9.3 Complex Transformations of the Vector Potential

We start with the vector potential $\underline{A}$ in the usual form

$$\underline{A} = A_x\hat{x} + A_y\hat{y} + A_z\hat{z} \tag{9.6}$$

where we choose A_x to vary as

$$\underline{A}_x = \underline{A}_{0x} \exp i\left(\omega t \pm \underline{k}_x \cdot \underline{x}\right). \tag{9.7}$$

The sign in the exponent refers to the accelerated or retarded vector potentials, respectively, and k_x is the wave number in the x-direction. The potential A_{0x} is complex and is given by

$$A_{0x} = A_{0x\mathrm{Re}} + iA_{0x\mathrm{Im}}. \tag{9.8}$$

The plus sign in the exponential in Eq. (9.7) is associated with the advanced or accelerated vector potential and the minus sign with the retarded vector potential [11].

Upon substitution of the real and imaginary parts of A_{0x} from Eq. (9.8) and for t and x from (9.2), we have

$$A_x = \left(A_{0x\mathrm{Re}} + iA_{0x\mathrm{Im}}\right)\exp\left[\omega\left(t_{x\mathrm{Re}} + it_{x\mathrm{Im}}\right) + ik_x\left(x_{\mathrm{Re}} + x_{\mathrm{Im}}\right)\right] \tag{9.9}$$

where the wave vectors are given as $k_x = w/c = k_y = k_z$ and $k_x = k = \omega / v_\phi$ where v_ϕ is the phase velocity. The frequency, ω and velocity, c are isotropic in all directions of the propagation of the potential. The phase velocity, v_ϕ is taken to be $v_\phi > c$ for a superluminal boost also called a superluminal Lorentz transformation, (SLT). We examine this case and determine for SLT's if $k_x = k_y = k_t$. That is, if the isotropy of the potential maintained. In the vacuum propagation of the vector potential, $k_x = k_y = k_z$, it is maintained but not in a material medium for the case where $k_x \neq k_y \neq k_z$, which is the case for uniaxial, biaxial or triaxial crystals, for example, in phase space for the vacuum for $v_\phi > c$.

In a vacuum, $k_x = k_y = k_z$, and in the moving frame traveling at $v > c$ for a superluminal boost (for example in the frame of a rocket for $v > c$ for deep space interstellar travel), the vector potential remains isotropic. This is not true for the observer in the laboratory frame, i.e. $v < c$, where the properties of the media are modified by the Lorentz transformation which affects the perceived properties of the object. For example, a photon appears as a photon in the "rocket" frame or superluminal frame, but it appears as a tachyon to an observer in the laboratory frame. Since the vector potential is defined by a phase velocity component and not by a group velocity parameter then we consider that the vector k is the same in dimensions of free space even under the SLT boost [2].

Then considering the form of the vector and scalar potentials under the action of a superluminal boost in the x direction and determine both a damping wave and an oscillatory wave for the advanced potential solutions only. This case will be expanded on further for the limit as the velocity approaches the velocity of light and where quantum mechanical considerations may demand that $k_x = k_y = k_z$, for example. In the limit of a transformation where nonlinear stresses exist one can define as a "rupture" in the extreme case of the subluminal, through c, to superluminal transformation.

Similar to Eq. (9.9) for the vector potential, we can define the complex scalar potential, ϕ as

$$\begin{aligned} \phi_x &= \phi_{0x} \exp\left[i\left(\omega t \pm k_x \cdot x \right) \right] \\ &= \left(\phi_{0x\mathrm{Re}} + i\phi_{0x\mathrm{Im}} \right) \exp i\omega \left(t_{x\mathrm{Re}} + it_{x\mathrm{Im}} \right) \\ &\times \exp\left[\pm ik_x \left(x_{\mathrm{Re}} + x_{\mathrm{Im}} \right) \right]. \end{aligned} \tag{9.10}$$

For Eq. (9.10) we again have assumed isotropic conditions of the vacuum in which the propagation constant is symmetric in all directions, i.e. $k_x = k_y = k_z$

$= \omega / v_\phi$ where v_ϕ, is the phase velocity propagation.

9.4 Superluminal Vector and Scalar Potential Transformation Laws

For simplicity we consider superluminal boost $v_x = \infty$ along the positive x direction [13,14]. See Chap. 2 on tachyonic signaling. See Fig. 9.1. The space and time vectors in the real 4D Minkowski space transform as follows [12]

$$x' = +t,\ y' = -iy,\ z' = iz,\ t' = x \tag{9.11}$$

for real and imaginary parts separately, where x, y, z, t are real quantities in the laboratory Σ frame, and x', y', z', t' are the real quantities in the moving Σ' frame. Now in the 12D (M^{12}) complex Minkowski space, the above transformation laws for a superluminal boost $(v_x = +\infty)$ in the positive x direction become

$$\begin{aligned} &x'_{\text{Re}} + ix'_{\text{Im}} = t_{x,\text{Re}} + it_{x,\text{Im}}, \quad y'_{\text{Re}} + iy'_{\text{Im}} = y_{\text{Im}} - iy_{\text{Re}}, \\ &z'_{\text{Re}} + iz'_{\text{Im}} = z_{\text{Im}} - iz_{\text{Re}}; \quad t'_{x,\text{Re}} + it'_{x,\text{Im}} = x_{\text{Re}} + ix_{\text{Im}}, \\ &t'_{y,\text{Re}} + it'_{y,\text{Im}} = t_{y,\text{Im}} - it_{y,\text{Re}}, \quad t'_{z,\text{Re}} + it'_{z,\text{Im}} = t_{z,\text{Im}} - it_{z,\text{Re}}. \end{aligned} \tag{9.12}$$

The transformation laws given by (9.12) preserve the magnitude of the line element but not the sign as

$$-x'^{\mu} x'^{\nu} = x^{\mu} x^{\nu} \tag{9.13}$$

where index μ and ν run over 0,1,2,3 representing 0 as time vector and 1,2,3 as spatial vectors. Therefore we have the signature (-+++).

Similar to the transformation laws for space and time vectors as given by (9.12) we can write the transformation laws for the vector and scalar potential. For a superluminal boost in positive x direction, the transformation laws for $(\underline{A}, \phi)$ are:

$$A'_x = \gamma\left(A_x - \frac{v_x^2}{c^2}\phi\right),\ A'_y = A_y,\ A'_z = A_z,\ \phi' = \gamma(\phi - v_x A_x) \tag{9.14}$$

where ϕ is the scalar potential and γ is the usual Lorentz term

$$\gamma' \equiv \frac{1}{\left(\frac{v_x^2}{c^2}-1\right)^{\frac{1}{2}}}. \tag{9.15}$$

We consider A_x', etc., transforming as a gauge. In Eq. (9.14), the vector potential A is considered to be a 4-vector real quantity, A_μ or $\underset{\sim}{A} = (A_x, A_y, A_z, \frac{i\varphi}{c})$, which preserves the length of the line element but not the sign, i.e. we have for the gauge transform

$$A_\mu A_\mu = -A_\mu' A_\mu' \tag{9.16}$$

Eq. (9.14) then simplifies to the following relationships for the velocities approaching infinity, $v_x = \infty$.

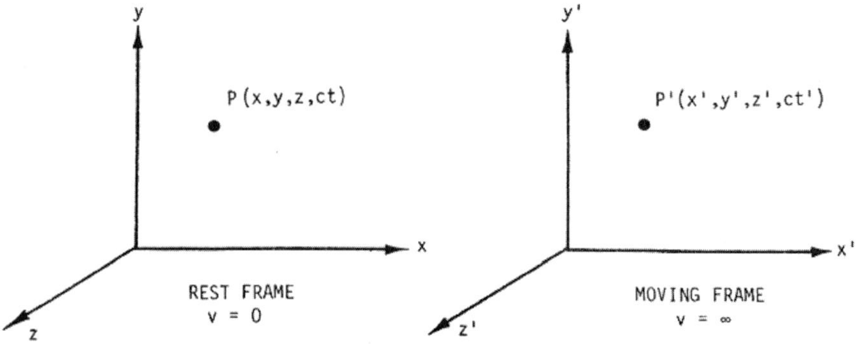

Figure 9.1. Schematic representation of a superluminal boost (Eq. 9.12) between an event P', in the moving frame and the corresponding event P in the rest frame, Σ. Relative velocity of the moving frame, Σ' is infinite, $v = \infty$. For an observer in the rest frame looking at event, P', he will see the *x*' coordinate transform to the time component, *t* in the rest frame and vice-versa.

The transformation laws for scalar and vector potentials under the superluminal boost in the positive *x* direction for $v_x = +\infty$. From the rest frame, Σ, to the moving frame, Σ', for unaccelerated vector and scalar potentials, we have

$$A_x = -\phi', \quad A_y = A_y', \quad A_z = A_z', \quad \phi = -A_x'. \tag{9.17}$$

From the moving frame, S', to the rest frame, S, for the unaccelerated vector and scalar potentials we obtain

$$A_x' = -\phi, \quad A_y' = A_y, \quad A_z' = A_z, \quad \phi' = -A_x \tag{9.18}$$

Eq. (9.18) is valid for real or complex vector and scalar potentials. Real and imaginary parts are easily separable in a complex quantity and they will transform according to Eq. (9.18) under the influence of a superluminal boost in the positive x direction. If these are the retarded (or accelerated or advanced) vector and scalar potentials, the transformation laws under the superluminal boosts will be different from the ones given by Eq. (9.18). These transformation laws are given by the combination of Eq. (9.18) and the transformation laws of the complex space and time vectors as given by Eq. (9.12).

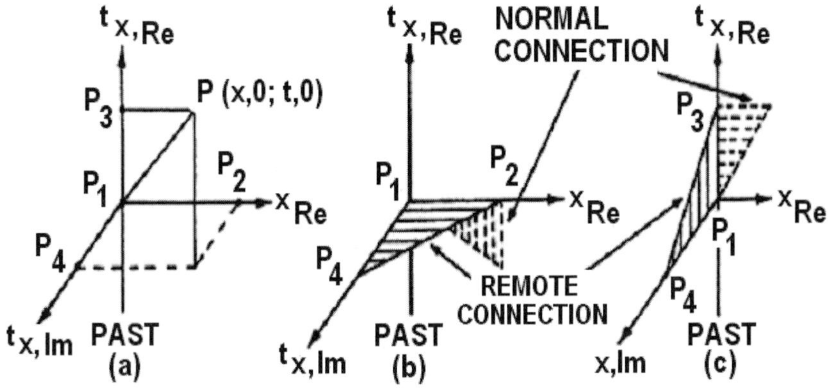

Figure 9.2. We represent the location of four points in the complex manifold. In Fig. 9.2a, point P_1 is the origin, and P is a generalized point which is spatially and temporally separated from P_1. In Fig. 9.2b, the Points P_1 and P_2 are separated in space but synchronous in time. This could be a representation of real-time nonlocal spatial separation.. In Fig. 9.2c, points P_1 and P_3 are separated temporally and spatially contiguous. This represents an anticipatory temporal connection.

These conditions are illustrated in Fig. 9.2. In Fig. 9.2a we represent a generalized point $P(x_{Re}, t_{Re}, t_{Im})$, displaced from the origin which is denoted as P_1. This point can be projected on each dimension x_{Re}, t_{Re} and t_{Im} as points P_2, P_3, and P_4 respectively. In Fig. 9.2b, we denote the case where a real-time *spatial* separation exists between points, P_1 and P_2 on the x_{Re} axis, so that

$\Delta x_{\text{Re}} \neq 0$, and there is no anticipation, so that t_{Re} = 0, and access to imaginary time t_{Im}, nonlocality can occur between the P_1 to P_4 interval, so that $\Delta t_{\text{Im}} \neq 0$. Then, our metric gives us $\Delta s^2 = 0$, where nonlocality is the contiguity between P_1 and P_2 by its access to the path to P_4. By using this complex path, the physical spatial separation between P_1 and P_2 becomes equal to zero, allowing direct nonlocal connectedness of distant spatial locations, observed as a fundamental nonlocality of remote connectedness on the spacetime manifold.

Figure 9.2c represents the case where anticipation occurs between P_1 and an apparent future anticipatory accessed event, P_3 on the t_{Re} axis. In this case, no physical spatial separation between observer and event is represented in the figure. Often such separation on the x_{Re} exists. In the case where x_{Re} = 0, then access to anticipatory information, along t_{Re} can be achieved by access to the imaginary temporal component, t_{Im}. Hence, remote, nonlocal events in 4-space or the usual Minkowski space, appear contiguous in the complex 8-space and nonlocal temporal events in the 4-space appear as anticipatory in the complex 8-space metric. Both nonlocality and anticipatory systems occur in experimental tests of Bell's Theorem and perhaps in all quantum measurement processes.

The propagation constant is considered to be isotropic in vacuum and defined as $d_x = \omega / v_\phi$, where v_ϕ, is the phase velocity and ω is the radian frequency of the propagating signal. Usually in most cases the phase velocity of propagation in vacuum is a constant $v_\phi = c$, where c is the velocity of light in vacuum. For the purpose of this volume, we will consider a tachyon traveling faster than light emitting an electromagnetic signal at frequency ω which propagates at the velocity of light. This assumption will simplify the subject matter of this book. We examine the faster than light electromagnetic signals emitted by a traveling tachyon which might lead into a Doppler effect at velocities faster than light.

Considering only the advanced potential solution from Eq. (9.11). See Chap. 2. Equation (9.11) can now be rewritten as two separate terms, so that in the Σ frame,

$$A_x = (A_{0x,\text{Re}} + iA_{0x,\text{Im}})\{\exp i[\omega t_{x,\text{Re}} - kx_{\text{Re}}] \times \exp -[\omega t_{x,\text{Im}} - kx_{\text{Im}}]\} \tag{9.19}$$

where the first exponent represents the usual type of oscillatory terms and the second exponent represents a decaying component which is not present in the usual four dimensional spacetime model. Note also that we have used the isotropy of the vector, k in Eq. (9.19) as examined in the previous section.

Examine the complex exponential of Eq. (9.19) using the transformations of Eq. (9.11) as follows so that we have for the exponents

$$\exp i[\omega x'_{\text{Re}} - kt'_{x,\text{Re}}] \times \exp - [\omega x'_{\text{Im}} - kt'_{x,\text{Im}}]. \tag{9.20}$$

We regroup terms in ω and k so that we have

$$\exp i[\omega(x'_{\text{Re}} + ix'_{\text{Im}}) - k(t'_{x,\text{Re}} - it'_{x,\text{Im}})]. \tag{9.21}$$

Now using equations from for $\underline{x}' = x'_{\text{Re}} + ix'_{\text{Im}}$ we have

$$\exp i[\omega x' - k(t'_{z,\text{Re}} - it'_{x,\text{Im}})]. \tag{9.22}$$

Note that the second part of the exponent for the k term does not reduce to t' since there is a minus before $it'_{x,\text{Im}}$. For the boost, $v_x \rightarrow \infty$ or $v > c$, we obtain for $\exp i[\omega t + kx]$ from Eq. (9.11) under this transformation going to

$$\exp i[\omega x'] \times \exp - k[t'_{x,\text{Re}} - it'_{x,\text{Im}}]\ . \tag{9.23}$$

Let us look at the example of the transformation from A'_x (in the moving frame, Σ') to its form in the restframe, Σ a mixing vector and scalar potential. In the SLT from the restframe, Σ to the moving Σ' frames; we have a change of length of the time component vector in Eq. (9.23). The vector potential term, A_{0x} transforms as

$$A'_x = \gamma\left(A_x - \frac{v_x^2}{c^2}\phi\right) \tag{9.24}$$

which is the same as Eq. (9.15), so that for the superluminal boost $v_x \rightarrow \infty$, implies that

$$\gamma \equiv \frac{1}{\sqrt{\frac{v_x^2}{c^2} - 1}} = \frac{1}{\frac{v_x}{c}\sqrt{1 - \frac{c^2}{v_x^2}}} \cong \frac{c}{v_x} \tag{9.25}$$

where the $\sqrt{1 - c^2 / v_x^2}$ term approaches unity as $v_x \rightarrow \infty$. Then we rewrite the transformed vector potential as

$$A_x^{'} = \frac{1}{\sqrt{\frac{v_x^2}{c^2} - 1}}, \qquad A_x - \frac{\frac{v_x}{c}}{\sqrt{\frac{v_x^2}{c^2} - 1}} \phi \ . \tag{9.26}$$

For $v_x \to \infty$ and from Eqs. (9.25) and (9.26),

$$A_x^{'} = \frac{cA_x}{v_x} - \frac{v_x}{c^2}\frac{c}{v_x}\phi = 0 - \frac{1}{c\phi} \equiv -\phi \tag{9.27}$$

for units in which $c = 1$. Therefore $A_x^{'} = -\phi$ for a superluminal boost, $v_x \to \infty$ and the transformation of the scalar potential, in analogy to Eq. (9.15), we have

$$\phi' = \gamma(\phi - v_x A_x) \tag{9.28}$$

and for $v_x \to \infty$, we have $\gamma \cong c / v_x$ so that in the limit of the SLT,

$$\phi' \lim_{v \to \infty} = \frac{c}{v_x}\phi - cA_x = -cA_x \tag{9.29}$$

for the units $c = 1$, then $\phi' = A_x$. Compare this equation to Eq. (9.27) and also for $A_y^{'} = A_y$ and $A_z^{'} = A_z$ we write

$$\begin{aligned} A_x &= [A_{0x,\mathrm{Re}} + iA_{0x,\mathrm{Im}}]\exp i[\omega t + kx] = \\ &[-\phi_{\mathrm{Re}}^{'} - i\phi_{\mathrm{Im}}^{'}]\exp i\omega x' \times \exp \pm k_x[t_{x,\mathrm{Re}} - it_{x,\mathrm{Im}}^{'}] \end{aligned} \tag{9.30}$$

where $x' = x_{\mathrm{Re}}^{'} + ix_{\mathrm{Im}}^{'}$ and using the result of Eqs. (9.27) and (9.29) for the non-exponent part and the exponential term which is given in Eq. (9.22), Eq. (9.30) gives us the vector and scalar form in the moving, Σ' frame.

If we consider only the accelerated potential, then we consider only the plus sign in Eq. (9.30). By use of the definition of complex quantities, Eq. (9.31) can be rewritten in a compact, simplified form:

$$A_x = -\phi_{0x}^{'} \exp(i\omega x') \cdot \exp(ik_x t_x^{'}) \ . \tag{9.31}$$

Using Eq. (9.31) we can describe the x component of the complex vector potential in moving frame, Σ' after a superluminal boost in the positive x direction. The same vector potential in the rest frame is defined. The transformation of the A_y and A_z components of the complex vector potential under a superluminal boost in the positive x direction can similarly be written as

$$\begin{aligned} A_y &= A'_{0y} \exp[-\omega(t'_{y,\mathrm{Re}} + it'_{y,\mathrm{Im}})] \cdot \exp[-ky(z'_{\mathrm{Re}} + iy'_{\mathrm{Im}})] \\ &= A'_{0z} \exp[-\omega(t'_{z,\mathrm{Re}} + it'_{z,\mathrm{Im}})] \cdot \exp[-ky(z'_{\mathrm{Re}} + iz'_{\mathrm{Im}})]. \end{aligned} \tag{9.32}$$

The scalar potential is defined by a complex quantity, so that

$$\phi' = \phi'_{\mathrm{Re}} + i\phi'_{\mathrm{Im}} \tag{9.33}$$

which we use for the non-exponential term of Eq. (9.32) which then becomes

$$A_x = -\varphi' \exp i\omega x' \cdot \exp k[t_{x\mathrm{Re}} - it_{x'\mathrm{Im}}]. \tag{9.34}$$

We compare the vector potential forms of A_x in Eq. (9.29) in the Σ or laboratory frame, and A_x of Eq. (9.34) in the Σ' frame or moving frame. (See Table 9.1)

TABLE 9.1 Comparison of the Exponential Part of the Vector Potential A_x in the Σ and Σ' Frames of Reference

	OSCILLATORY	DAMPED
Σ Frame	$A_{0x} \propto \exp i[\omega t_{x,\mathrm{Re}} - kx_{\mathrm{Re}}]$	$\exp-[\omega t_{x,\mathrm{Im}} - kx_{\mathrm{Im}}]$
Σ' Frame	$\phi' \propto \exp i[\omega x']$	$\exp k[t'_{x,\mathrm{Re}} - it'_{x,\mathrm{Im}}]$

In the oscillatory solution of the Σ' frame for ϕ', we find no dependence on the wave number factor k and hence we have apparent media independence, recalling $x' = x_{\mathrm{Re}} + ix_{\mathrm{Im}}$, whereas in the Σ frame for A_{ox}, we have dependence on ω and k.

For the damped solution, we have ω and k dependence in the Σ frame for A_{ox}, which is a pure real exponential and hence not oscillatory. In the Σ' frame, ϕ' sometimes has a damped solution dependent on k which has a real and imaginary component. The exponential factor can be written as

$$t'_{x\mathrm{Re}} - it'_{x\mathrm{Im}} = x_{\mathrm{Re}} - ix_{\mathrm{Im}} . \tag{9.35}$$

Time dilation and vector length are modified in the complex 12D space. We find that a superluminal, unidimensional x-dimensional boost in complex Minkowski space not only modifies space and time (as well as mass) by the γ factor, it also modifies $\underset{\sim}{A} = (\underline{A}, \varphi)$ and we find a mixing of $\underline{A}$ and ϕ for $\underline{A} = A_j$ where j runs 1 to 3 for space-like quantities and ϕ transforms as a temporal quantity for subluminal transformations.

References

[1] Ramon, C. & Rauscher, E.A. (1980) Found. Physics, 10, 661.
[2] Ramon, C. & Rauscher, E.A. (1982) Remote connectedness in complex geometries and its application to the nature of the human mind, pp. 1423-1442, New York: ICF Press.
[3] Rauscher, E.A. (1993)
[4] Rauscher, E.A. (1993) Electromagnetic Phenomena in Complex Geometries and Nonlinear Phenomena, Non-Hertzian Waves and Magnetic Monopoles, Millbrae: Tesla Books.
[5] Newman, E.T. (1973) J. Math. Phys. 14, 774.
[6] Demers, P. (1975) Canada J. Phys. 14, 774.
[7] Amoroso, R.L. & Rauscher, E.A. (2009) The Holographic Anthropic Multiverse: Formalizing the Complex Geometry of Reality, Singapore: World Scientific.
[8] Cramer, J.G. (1980) Phys. Rev. D22, 362.
[9] Cramer, J.G. (2006) Found. Phys. Let. 19, 63.
[10] Cramer, J.G. (1986) Rev. Mod. Phys. 58, 647.
[11] de Beauregard, P. Costa (1979) Nuovo Cimento, 51B, 429.
[12] Evan, G.T. & Sen, D.K. (1973) J. Math. Phys. 14, 1668.
[13] Leibowitz, E. (1974) J. Math. Phys. 15, 306.
[14] Rauscher, E.A. (1978) Bull. Amer. Phys. Soc. 23, 84.

Chapter 10

The Schrödinger Equation in Complex Minkowski Space

We develop a formalism for the Schrödinger equation in an eight dimensional (8D) complex Minkowski space and discuss its relation to the Dirac equation, properties of nonlocality, remote connectedness, Young's double-slit experiment, Bell's Theorem, the EPR paradox and anticipatory parameters of spacetime; and also identify an imaginary temporal component as a small nonlinear term and find soliton or solitary wave solutions. These coherent solutions can carry information over long distances, are consistent with Lorentz invariance and appear to provide a fundamental methodology for describing the issue of quantum measurement and a new context for the basis of quantum theory. In the Copenhagen view models of reality are not desirable. However our new approach may enable the redefinition of concepts of reality from a new nonlocal and anticipatory quantum theory. Certainly the most desirable consequence of scientific discovery is the ability to refine our concepts of reality.

10.1 Remote Connectedness and Coherent Collective Phenomena

The interpretation of the extremely successful quantum theory, beyond the Copenhagen Theory, carries within it the vital need for the interpretation of what it means to make a measurement, primarily in the microdomain. The rapid and major development of the structure, content and interpretation of quantum theory in the 1920s and 1930s, as exemplified by the Heisenberg Uncertainty Principle and Schrödinger Cat Paradox and EPR Paradox [1], led to conceptual paradoxes beyond the practical application of quantum theory. The Schrödinger Cat Paradox arises over the issue of the collapse of the wave function. For two equally probable states arising from a microscopic process, only observation can determine which state exists. Heisenberg and Bohr demonstrated that the act of observation necessarily leaves the system in a new state through what Wheeler terms "participation" [2] (that is quantum theory can only predict the probability of the outcome of a specific experiment). The Copenhagen interpretation of quantum mechanics was an attempt to dismiss the observer's participation, but by this dismissal, we can no longer build models of reality.

The test of the universality of quantum theory's experimental validity demands that nonlocality is a fundamental property of the quantum domain. The issue of nonlocality as a fundamental property of spacetime has been thoroughly proven by experimental verification. If quantum theory is universally valid, nonlocality is necessarily true Bohm termed the nonlocal correlations [3].

Einstein's dissatisfaction with the lack of determinism of quantum theory, and its probabilistic nature, led him to write the Einstein, Podolsky, and Rosen, EPR, paper. He had hoped to find a flaw in the quantum theory that would allow a way around the Heisenberg Uncertainty Principle and the probabilistic nature of the quantum theory [1] He was not the only physicist to be discontented with the, "spooky action of a distance." Bell reformulated the EPR Paradox into a rigorous inequality that could be experimentally tested. In more recent years, the formalism of the EPR Paradox terms of Bell's theorem [4], and its extensive tests which demonstrate that quantum theory holds in all known quantum experiments which necessarily demands the properties of nonlocality on the spacetime manifold.

What are some of the possible implications from the quantum description, if we choose to pursue the development of models of reality and perhaps relax the pure objectivity constraint in physical theory? This issue is well exemplified by the Bell's theorem formulation of the Einstein, Podolsky, Rosen Paradox [1]. An indication that non-locality is a principle in Nature is contained in Bell's theorem, which asserts that no deterministic local "hidden variable" theories can give all the predictions of quantum theory [5]. However, most physicists believe that Nature is non-deterministic and that there are no hidden variables. The prevailing view is that Bell's theorem merely confirms these ideas, rather than that it is an indication of a fundamental statement of nonlocality. However, in recent years this view has changed.

Stapp demonstrates that determinism and hidden variables occupy no essential role in the proof of Bell's theorem, which Stapp has reformulated [6]. Stapp asserts that no theory which predicts the outcome of individual observations which conform to the predictions of quantum theory can be local. A less restrictive interpretation of Bell's theorem is that either locality or realism fail [7]. Realism is a philosophical view in which external reality is assumed to exist and have definite properties fundamentally independent of an observer [7,8] Stapp presents reasonable and comprehensive models of reality in which nonlocality, as implied by Bell's theorem, is inconsistent with "objective reality," in which observable attributes can become definite, independent of the observer, the so-called "collapse of the wave function".

In Young's double slit experiment, photons from a source can go through one of two slits or openings of a slit interference arrangement. Through which slit did the photon go that blackens a photographic plate at the other end of the apparatus. The answer is not yet defined because of the Heisenberg Uncertainty Principle. One can observe interference fringes when both slits are open, but at the cost of not knowing through which slit the photon went. Or, one can know through which slit the photon went when one slit is closed, but at the cost of not having any interference fringes. Again, the choice appears to be that of the observer [9]. This experiment also brings the role of the observer into consideration and may also involve nonlocality and anticipation [10]. Certainly, one of the most desirable consequences of scientific discovery is the ability to discover and refine our concepts of reality. See Chap. 4.

10.2 Complex 8-Space and the Formation of Nonlocality

We have introduced a complex multi-dimensional geometry of the four real dimensions of space, X_{Re} of x_{Re}, y_{Re}, z_{Re}, and t_{Re} and four imaginary dimensions X_{Im} of ix_{Im}, iy_{Im}, iz_{Im} and it_{Im}, such that we can describe nonlocal macroscopic connections of events that do not violate causality [11]. There are several motivations for introducing such a model; one of which relates to a possible macroscopic formulation of a Bell's theorem-like nonlocal correlation function that may have macroscopic implications, leading to a new interpretation of the Bell's theorem experimental results and to a more fundamental interpretation of the quantum measurement issue. The complex Minkowski Space, M_4 is constructed so as to maintain causality and analytic continuation in the complex manifold [11-13]. The four real dimensional space can be considered a slice though the hyperdimensional complex 8-space [13].

Events that appear remote in 4-space, M_4, are contiguous in the complex 8-space, $M_4 \subset \mathbb{C}_4$. We have demonstrated a fundamental relationship between the complex 8-space and the topology of the Penrose twister algebra [8,14,15]. In this model, spacetime events can become contiguous in the complex 8-space, demonstrating that the remoteness of the observer and observed can become contiguous in the complex 8-space in which causality conditions are preserved and the acquisition of apparent remote information is allowed.

We have solved the Schrödinger equation in the complex 8D space and, with the inclusion of a relatively small, but significant, nonlinear term,

$g^2(t_{\text{Im}})$ we find soliton and solitary wave solutions. The non-linear term, which depends on the imaginary time component, overcomes dispersion giving the non-dispersive soliton waves. The coherence over remote space and time of these wave solutions relates to macroscopic-related phenomena as it does for Bell's theorem, Young's double-slit experiment and other nonlocal anticipatory phenomena. The non-linear form of the Schrödinger equation may be related formally to the non-linear gravitational phenomena [15] and also has implications for the quantum measurement problem [16]. Resolution of the observer-participant problem may be at hand as demonstrated by a new interpretation of the Schrödinger equation. In this formation, remote spacetime events are contiguous so that the observer has direct acquisition to remote observable information, in such a manner as to preserve causality.

10.3 Space-Like Remote Connectedness, Bell's Theorem and its Experimental Test

A most significant theorem about the nature of physical systems is Bell's formulation [4] of the Einstein, Podolsky and Rosen (EPR) "completeness" formulation of quantum mechanics [1]. The EPR paper was written in response to Bohr's proposal that the non-commuting operators (Heisenberg uncertainty principle) comprise a *complete* theory called the Copenhagen interpretation of quantum mechanics. Einstein, Podolsky and Rosen define a complete theory as one in which every element of the theory corresponds to an element of "reality". Bohm introduced additional quantum non-observable variables or "hidden variables," as we presented in the last section, in order to make the EPR quantum interpretation consistent with *causality* and *locality* [17]. In 1964, Bell "formulated" the EPR statement and showed mathematically that locality is incompatible with the statistical predictions of quantum mechanics. The locality or separability assumption states that the result of a measurement on one system is *unaffected* by operations on a *distant* system with which it may have previously interacted or had become entangled.

Bell discusses a specific experiment, Stern-Gerlach measurements of two spin one-half particles in the singlet spin state moving freely in opposite directions. The spins are called s_1 and s_2; we make our component spin measurements remote from each other at position P_1 and P_2, such that the Stern-Gerlach magnet at P_1 does not affect one at P_2 and vice versa. Since we can predict, in advance, the result of measuring any chosen component of s_2 at P_2 by previously measuring the same component of s_1 and P_1, this implies that the result of the second measurement must actually be predetermined by the result of the first (remote from P_2) measurement. In Bell's proof, he introduces a more complete specification of the parameters of a system by introducing parameters which in essence are hidden variables. Bell's proof is most eloquent and clear. He calculates the conditions on the correlation function for measurements at P_1 and P_2, as an inequality.

Bell's precise statement made it possible for Clauser and Horne to test the predicted statistical distribution of quantum processes and demonstrate a laboratory instance of quantum connectedness and nonlocality [18,19]. Indeed, in Clauser's calcium two photon cascade system, two photodetectors remote from each other are each preceded by independent, randomly-oriented polarizers. The statistical predictions of quantum mechanics is borne out in the measurements made at the two photomultiplier tubes (PMT). In Bell's words, "there must be a mechanism whereby the setting of one measuring device can influence the reading of another instrument, however remote." Moreover, the signal involved must propagate instantaneously so that a theory could not be Lorentz invariant. Lorentz invariance, in the usual sense, implies $v \leq c$. Feinberg discusses the relationship between Lorentz invariance and superluminal signals which he found not to be incompatible. It is not completely clear that superluminal signals must be invoked to derive Bell's theorem [20] but we have demonstrated that indeed Bell's theorem may imply either $v \leq c$ in a complex 8-space [21] or an additional type of , $t = 0$, holographic-like instantaneity that is a Witten Vertex type Riemann Ising Model rotation of 12D space.

In fact additional research should reveal that a Kantian antinomy or duality exists here that will be revealed between Newton's instantaneous and Einstein's, $v = c$ gravity wave propagation models by additional research. Recall that the issue of a putative quantum gravity is still an open question. We are in the process of redefining the graviton. It appears that there is no M_4 spin 2 field exchange quanta; but that the quadrupole structure is more complex requiring additional structural-phenomenology even beyond that hinted at by our intermediate $\hat{M}_4 \subset \mathbb{C}_4$ complex 8-space. Some researchers have invoked the term wavicle to invoke some sort of entity beyond the either or wave-particle duality scenario. In any case this would not be sufficient to describe a graviton from the HAM cosmological perspective where there is no phenomenological exchange as such. The duality requires a form of topological switching of boundary conditions that are an ontological-phenomenological duality; where the phenomenological component is virtual in the sense that there is no measurable spin 2 quanta in 4-space.

The conclusion from Bell's theorem, then, is that any hidden variable theory that reproduces all statistical predictions of quantum mechanics must be nonlocal (implying remote connectedness). Of course, thus far, all these formulations involve microproperties only, but some recent formulations seem to imply possible macroscopic consequences of Bell's theorem as well.

It is believed that the key lies in the formulation of the correlation function which represents the interconnectedness of previously correlated events. Stapp has demonstrated that hidden variable theory is not necessary to the formalism of Bell's theory [22]. Stapp has recently expanded the pragmatic view of Bell's theorem and discusses the role of the macroscopic detection apparatus as well as the possible role of superluminal signals. He explores both cases for superluminal propagation or subluminal connection issuing from the points in common to the backward lightcones coming from the two regions.

We can write a general correlation function, for example, for an angle, θ between polarization vectors in two polarizers as $C(\theta) = \frac{1}{2} - \frac{1}{2}\cos 2\theta = \cos^2\theta$ for Clauser's experiment, or for odd integers we can write $nC(\theta) - C(n\theta) - (n-1) \leq 0$, which is Bell's inequality – specifically for n = 3; $3C(\theta) - C(3\theta) - 2 \leq 0$. We can write in general $C(\theta) = \frac{1}{2} + g\cos 2\theta$ where g is determined by the particular experiment under consideration. The magnitude correlation function constant, *g* relates to the type of nonlocal correlation experiment. For $g = \frac{1}{2}$, we have the Bell's theorem photon-photon correlation.

An exciting and extremely important test of Bell's inequality was designed and implemented by Clauser et al. in the early 1970s at the UC Berkeley, which author (EAR) had the privilege to observe [7,18], as well as the work of Aspect, et al. at the University of Orsay [23]. These extremely well designed and implemented experiments demonstrate a fundamental and unique remote causal connections and nonlocality on the spacetime manifold. Photon correlations have been observed over meter distances in the Aspect experiment. More recently, Gisin et al. has tested Bell's inequality over kilometer distances [24,25]. Rauscher and Targ apply the complex 8-space and its description of nonlocality, such as exemplified in the Bell's theorem tests, to the nonlocal aspects of consciousness [26,27]. Precognition and retrocognition comprise an anticipatory system. Clauser expressed his impression of these nonlocality experiments to the above authors. He stated that quantum experiments have been carried out with photons, electrons, atoms, and even 60-carbon-atom Buckyballs. He said that, "it may be impossible to keep anything in a box anymore." Bell emphasizes, "no theory of reality compatible with quantum theory can require spatially separate events to be independent." That is to say, the measurement of the polarization of one photon determines the polarization of the other photon at their respective measurement sites. Stapp also stated to one of us (EAR) that these quantum connections could be the, "most profound discovery in all of science" [26].

Bohm has conducted research on the concept of the undivided nonlocal whole, and Bohm and Hiley [3], having extensive discussions with one of us

(EAR). Also Wheeler's fundamental explanations on the concept of nonlocal interactions and the foundations of the quantum theory in publications and discussion with author (EAR) are fundamental to anticipatory systems [2]. Wheeler's design of his delayed choice experiment demonstrates that, according to quantum theory, the choice to measure one or another pair of complimentary variables at a given time can apparently affect the physical state of events for considerable periods of time *before* such a decision is made. Such complimentary variables are typically momentum and distance, or in Wheeler's experiment refer to the dual wave and particle nature of light, as observed in a two slit interference apparatus. See Chap. 4.

Wigner attempted to formulate a nonlinear quantum theory and stated support of the complex Minkowski 8-space which has macroscopic nonlocal consequences [28]. The fundamental issue he addressed is when are where does the measurement observation occur for a stochastic causal system. Earlier, von Neumann had suggested a sequence of observations, or von Neumann chain. Wigner also addresses the issue of multiple observers of a quantum generated event [28].

10.4 Complex 8-Space and Nonlocality

Within the context of a fundamental observation and theoretical formalism of nonlocality and anticipation, such a theory must be consistent with the main body of the principles of physics. The major universal principles are used to determine the structure and nature of physical laws and act as constraints on physical phenomena. These are *Poincaré invariance* and its corollary, *Lorentz invariance* (which expresses the spacetime independence of scientific laws in different frames of reference), *analyticity* (which is a general statement of causality conditions in the complex space), and *unitarity* (which can be related to the conservation of physical quantities such as energy and momentum). See Chap. 3.

These principles apply to microscopic as well as macroscopic phenomena. The quantum description of elementary particles has led to the formulation of the analyticity principle in the complex momentum plane [29]. Complex geometries occupy a vital role in many areas of physics and engineering. Analyticity relates to the manner in which events are correlated with each other in the spacetime metric (that is, causality). When we apply this critical principle to the complex eight-dimensional space we can reconcile nonlocality and anticipatory systems with physics, without violating causality. It has been mathematically demonstrated that the equations of Newton, Maxwell, Einstein, and Schrödinger are consistent with the eight-dimensional complex space described here [12-14,20,30-33]. In addition, nondispersive solitary wave solutions are obtained for the complex 8-space Schrödinger equation [21].

The least number of dimensions that has the property of nonlocality and that is consistent with Poincaré invariance or Lorentz invariance is eight dimensions. In this space, each physical *spatial* distance has an imaginary temporal counterpart, such that there is a zero *spatial* separation in the higher dimensional space. Likewise for every real physically temporal separation, there is a counterpart imaginary *spatial* separation that subtracts to zero on the metric, allowing access to future information and bringing it into the present, which acts as an anticipatory system.

We have also demonstrated the properties of nonlocality with the formalism of Maxwell's equations in complex 8-space [29-31]. In the next section, we present a brief description of the complex Minkowski 8-space and its properties and implications. Then we present in section 10.5 the solution to the Schrödinger equation in complex 8-space and nonlinear recursive solutions which are consistent with and explanatory of Bell's nonlocality and the general principles of nonlocality and anticipatory phenomena in the quantum domain.

Both special and general relativistic forms of the complex 8-space have been formulated and examined in applications [11,13,15]. We present a brief description of the formalism which we utilize to solve the Schrödinger equation. We express the solution of the Schrödinger equation in complex 8-space. In the usual 4D Minkowski space, where Einstein considered time as the fourth dimension of three space, this formulated as a 4D lightcone diagram displayed in two dimensions, in which the ordinate is the time coordinate and the abscissa is the space coordinate, representing the three dimensions of space as $X = x,y,z$. the sides of the forward and backward lightcone form signal connections at the velocity of light, c, and the apex of the cone represents "now" spacetime. Inside the forward, future time, and backward, past time, lightcone event connections are represented by signaling for $v < c$ called time-like signaling. The space-like signaling outside of the lightcone represents greater than light speed, or space-like signaling, or $v >$ c.

Bell's nonlocality test implies spacetime signaling and hence, even though experimentally well-verified, some physicists find nonlocality unsatisfactory. However, as we know, the truth is in what Nature shows us, not in our particular biased beliefs. The complex 8-space formalism not only yields a mathematical description of nonlocality, but the complexified Schrödinger formalism gives a detailed picture of the quantum nonlocality that is consistent with the statistical nature of the quantum theory, but is also consistent with the formalism of relativity. Apparent superluminal signaling can occur for the connection of correlated past time events that remain correlated for present measurement and are related by luminal velocity of light signaling in the complex 8-space. Also, this formalism allows anticipatory measurements such as in the Aspect, Gisin experiments and Wheeler's delayed choice experimental proposal.

The conditions for causality in the usual 4-space, distance ds^2 is invariant and given as $ds^2 = g_{ab}dx^a dx^b$ where the indices a and b run 1 to 4. We use the metrical signature (+,+,+,-) for the three spatial and one temporal component in the metric $g_{ab.}$ This metric is expressed as a 16-element 4 x 4 matrix which represents a measure of the form and shape of space. This is the metric defined on the lightcone, connecting time-like events. A second four imaginary dimensional space lightcone can be constructed, which intersects with the usual 4D Minkowski space, can be constructed. These two lightcones coincide in there "now" spacetime realities. The complexified 8-space metric is denoted as M_4 because it represents the complexification of four spacetime dimensions. The complex space is expressed in terms of the complex 8-space variable Z^μ, where $Z^\mu = X^\mu_{\mathrm{Re}} + iX^\mu_{\mathrm{Im}}$, and $Z^{*\nu}$ is the complex conjugate of Z^μ so that $Z^\nu = X^\nu_{\mathrm{Re}} - iX^\nu_{\mathrm{Im}}$. We now form the complex 8-space differential line element $dS^2 = \eta_{\mu\nu} dZ^\mu dZ^{*\nu}$ where the indices run 1 to 8 and $\eta_{\mu\nu}$ is the complex metric of 8-space. The generalized complex metric in the previous equation is analogous to the usual Einsteinian 4-space metric. In our formalism, we proceed by extending the usual 4D Minkowski space into a 4D complex spacetime. This new manifold (or spacetime structure) is analytically expressed in the complexified 8-space.

As stated before we represent X_{Re} by x_{Re}, y_{Re}, z_{Re} and t_{Re}, i.e. the dimensions of our usual four space. Likewise, X_{Im} represent the four additional imaginary dimensions of x_{Im}, y_{Im}, z_{Im} and t_{Im}. Hence, we represent the dimensions of our complex space as Z^μ or x_{Re}, y_{Re}, z_{Re}, t_{Re}, x_{Im}, y_{Im}, z_{Im}, and t_{Im}. These are all real quantities. It is the i before the x_{Im}, etc. that complexifies the space. We write the expression showing the separation of the real and imaginary parts of the differential form of the metric: $dZ^\mu dZ^{*\mu} = \left(dX^\mu_{\mathrm{Re}}\right)^2 + \left(dX^\mu_{\mathrm{Im}}\right)^2$. We can write in general for real and imaginary space and time components in the special relativistic formalism.

$$ds^2 = \left(dx^2{}_{\mathrm{Re}} + dx^2{}_{\mathrm{Im}}\right) + \left(dy^2{}_{\mathrm{Re}} + dy^2{}_{\mathrm{Im}}\right) + \left(dz^2{}_{\mathrm{Re}} + dz^2{}_{\mathrm{Im}}\right) - c^2\left(dt^2{}_{\mathrm{Re}} + dt^2{}_{\mathrm{Im}}\right). \quad (10.1)$$

We now use lower case x and t for the three dimensions of space and of time. We represent the three real spatial components dx_{Re}, dy_{Re}, dz_{Re} as dx_{Re}

and the three imaginary spatial components dx_{Im}, dy_{Im}, dz_{Im} as dx_{Im} and similarly for the real time component $dt_{\text{Re}} = dt$, the ordinary time and imaginary time component dt_{Im} remains dt_{Im}. We then introduce complex space -time coordinates as a space-like part x_{Im} and time-like part t_{Im} as imaginary parts of x and t. Now we have the invariant line elements as,

$$s^2 = |x'|^2 - c^2|t'|^2 = |x'|^2 - |t'|^2 \tag{10.2}$$

again where we take the units $c^2 = c = 1$ which is made for convenience

$$x' = x_{\text{Re}} + ix_{\text{Im}} \tag{10.3a}$$

and

$$t' = t_{\text{Re}} + it_{\text{Im}} \tag{10.3b}$$

as our complex dimensional components so that [11,26,27]

$$x'^2 = |x'|^2 = x_{\text{Re}}^2 + x_{\text{Im}}^2 \tag{10.4a}$$

and

$$t^2 = |t|^2 = t_{\text{Re}}^2 + t_{\text{Im}}^2\,. \tag{10.4b}$$

Recalling that the square of a complex number is given as,

$$|\,x'\,|^2 = x'\,x'^{*} = (x_{Re} + ix_{Im})\,(x_{Re} - ix_{Im}) \tag{10.5}$$

where the modulus of a complex number squared is $|x'|^2 = x_{\text{Re}}^2 + x_{\text{Im}}^2$ so that x_{Re} and x_{Im} are real numbers. This is a very important point, as we can only measure events described in terms of the mathematics of real numbers. Therefore, we have the 8-space line element where spatial and temporal distances are taken from the origin.

$$s^2 = x_{\text{Re}}^2 - c^2 t_{\text{Re}}^2 + x_{\text{Im}}^2 - c^2 t_{\text{Im}}^2 \tag{10.6a}$$

$$s^2 = x_{\text{Re}}^2 - t_{\text{Re}}^2 + x_{\text{Im}}^2 - t_{\text{Im}}^2\,. \tag{10.6b}$$

Causality is defined by remaining on the right cone, in real spacetime as,

$$s^2 = x_{\text{Re}}^2 - c^2 t_{\text{Re}}^2 = x_{\text{Re}}^2 - t_{\text{Re}}^2 \tag{10.7}$$

using the units of $c = 1$. Then the generalized causality in complex spacetime is defined by

$$s^2 = x_{\text{Re}}^2 - t_{\text{Re}}^2 + x_{\text{Im}}^2 - t_{\text{Im}}^2 \tag{10.8}$$

where the coordinates in complex 8-space can be represented by $x_{\text{Re}}, t_{\text{Re}}, x_{\text{Im}}, t_{\text{Im}}$ on two generalized lightcones 8D space [11,12,31].
We calculate the interval separation between two events or occurrences, Z_1 and Z_2 with real separation $\Delta x_{\text{Re}} = x_{\text{Re},2} - x_{\text{Re},1}$ and imaginary separation $\Delta x_{\text{Im}} = x_{\text{Im},2} - x_{\text{Im},1}$. Then the distance along the line element is $\Delta s^2 = \Delta\left(x_{\text{Re}}^2 + x_{\text{Im}}^2 - t_{\text{Re}}^2 - t_{\text{Im}}^2\right)$ and it must be true that the line interval is a real separation. The spatial and temporal distances that are generalized are not taken only from the origin, but from any two points in space and time. Then,

$$\begin{aligned}\Delta s^2 = \left(x_{\text{Re},2} - x_{\text{Re},1}\right)^2 + \left(x_{\text{Im},2} - x_{\text{Im},1}\right)^2 - \\ \left(t_{\text{Re},2} - t_{\text{Re},1}\right)^2 - \left(t_{\text{Im},2} - t_{\text{Im},2}\right)^2.\end{aligned} \tag{10.9a}$$

Or we can write Eq. (10.9a) as:

$$\begin{aligned}\Delta s^2 = \left(x_{\text{Re},2} - x_{\text{Re},1}\right)^2 + \left(x_{\text{Im},2} - x_{\text{Im},1}\right)^2 - \\ \left(t_{\text{Re},2} - t_{\text{Re}.1}\right)^2 - \left(t_{\text{Im},2} - t_{\text{Im},1}\right)^2\end{aligned} \tag{10.9b}$$

In Eq. (10.9b), the upper left diagonal term $\left(x_{\text{Re},2} - x_{\text{Re},1}\right)^2$ is be offset or "cancelled" by the lower right diagonal term $-\left(t_{\text{Im},2} - t_{\text{Im},1}\right)^2$, and the lower left diagonal term $-\left(t_{\text{Re},1} - t_{\text{Re},1}\right)^2$ is off set by the upper right diagonal term $\left(x_{\text{Im},2} - x_{\text{Im},1}\right)^2$. Because of the relative signs of the real and imaginary space and time components, and in order to achieve the causality connectedness condition between the two events, or $\Delta s^2 = 0$, we must "mix" space and time. That is, we use the imaginary time component to effect a zero space separation. We identify $\left(x_{\text{Re},1}, t_{\text{Re},1}\right)$ with a subject receiver remotely perceiving information from an even target $\left(x_{\text{Re},2}, t_{\text{Re},1}\right)$.

The nonlocality of Bell's theorem and its experimental test involves a real physical separation $\Delta x_{\text{Re}} = x_{\text{Re},2} - x_{\text{Re},1} \neq 0$ and can either involve a current time observation such that $\Delta t_{\text{Re}} = t_{\text{Re},2} - t_{\text{Re},1} = 0$ or a anticipatory time interval $\Delta t_{\text{Re}} = t_{\text{Re},2} - t_{\text{Re},1} > 0$. The case where there is no anticipatory time element $\Delta t_{\text{Re}} = 0$. The simplest causal connection then is one in which

$\Delta x_{\text{Im}} = 0$, and we have,

$$\Delta s^2 = 0 = (x_{Re,2} - x_{Re,1})^2 - (t_{Im,2} - t_{Im,1})^2. \qquad (10.10)$$

These conditions are illustrated in fig. 10.1 In fig. 10.1a we represent a generalized point $P(x_{\text{Re}}, t_{\text{Re}}, t_{\text{Im}})$, displaced from the origin which is denoted as P_1. This point can be projected on each dimension x_{Re}, t_{Re} and t_{Im} as points P_2, P_3, and P_4 respectively. In Figure 1b, we denote the case where a real-time *spatial* separation exists between points, P_1 and P_2 on the x_{Re} axis, so that $\Delta x_{\text{Re}} \neq 0$, and there is no anticipation, so that $t_{\text{Re}} = 0$, and access to imaginary time t_{Im}, nonlocality can occur between the P_1 to P_4 interval, so that $\Delta t_{\text{Im}} \neq 0$. Then, our metric gives us $\Delta s^2 = 0$, where nonlocality is the contiguity between P_1 and P_2 by its access to the path to P_4. By using this complex path, the physical spatial separation between P_1 and P_2 becomes equal to zero, allowing direct nonlocal connectedness of distant spatial locations, observed as a fundamental nonlocality of remote connectedness on the spacetime manifold.

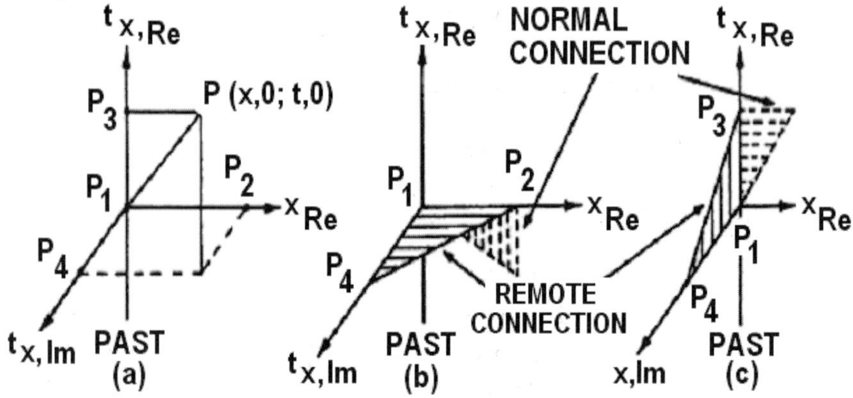

Figure 10.1. We represent the location of four points in the complex manifold. In figure 1a, point P_1 is the origin, and P is a generalized point which is spatially and temporally separated from P_1. In figure 1b, the Points P_1 and P_2 are separated in space but synchronous in time. This could be a representation of real-time nonlocal spatial separation.. In figure 1c, points P_1 and P_3 are separated temporally and spatially contiguous. This represents an anticipatory temporal connection.

Figure 10.1c represents the case where anticipation occurs between P_1 and an apparent future anticipatory accessed event, P_3 on the t_{Re} axis. In this case, no physical spatial separation between observer and event is represented in the figure. Often such separation on the x_{Re} exists. In the case where x_{Re} = 0, then access to anticipatory information, along t_{Re} can be achieved by access to the imaginary spatial component, x_{Im}. Hence, remote, nonlocal events in 4-space or the usual Minkowski space, appear contiguous in the complex 8-space and nonlocal temporal events in the four space appear as anticipatory in the complex 8-space metric. Both nonlocality and anticipatory systems occur in experimental tests of Bell's Theorem and perhaps in all quantum measurement processes.

10.5 Solitary Wave and Coherent Non-dispersive Solutions in Complex Geometries

The properties and some of the implications of complex Minkowski spaces hold fundamental significance. We have presented the formalism for complex geometries in the previous section and also for superluminal x direction boosts in these geometries and the possible implications for remote connectedness, and anticipatory systems [11]. Also the symmetry relations of the vector and scalar electromagnetic potential and other properties of Maxwell's equations, the x-directional superluminal boost, have been formulated [18]. The relationship of this approach to the Schrödinger equation in this work is of interest.

In this section we determine solutions to the Schrödinger equation formulated in a complex Minkowski space and demonstrate the relationship of the solutions to inter-connectedness and the nonlocality principle. The solutions are solitary or soliton waves which exhibit little or no dispersion over long distances. We present several implications of this formalism, for the test of Bell's Theorem, anticipatory processes and an explanation for some coherent, nonlinear, non-dispersive phenomena, such as nonlinear plasma phenomena [34,35].

We examine the relationship between our multi-dimensional remote connectedness geometry and possible coherent, non-dispersive solutions to the Schrödinger equation. These non-dissipative or non-dispersive solutions are termed soliton solutions, or solitary wave solutions, and are well known in macroscopic hydrodynamic phenomena. There has been some recent interest in the use of the soliton or instanton model to describe the gluon quark structure for "infinitely" bound quarks, in part, to explain the lack of experimentally observed free quarks.

The solution to linear wave equations are dispersive in space and time, that is, their amplitude diminishes and width at half maximum becomes larger as a function of time. The term soliton is commonly used to define a wave which retains its amplitude and "half width" over space and can interact and remain intact with other solitons. The term instanton, or evanescent wave, is used to describe a structure which experiences both spatial and temporal displacement. The term instanton seems to imply a short-lived structure but actually instantons can retain their spatial and temporal configuration indefinitely and interact with other instantons in a particle-like manner as do solitons. These unique solutions can explain the existence of long spatial and temporal phenomena such as Bell's remote connectedness phenomenon, Young's double slit experiment, plasma coherent collective states and other coherent phenomena.

Starting from the Schrödinger equation in complex spacetime, as seen previously [8,11], complex geometries have properties consistent with the above mentioned phenomena. We proceed from the time-dependent Schrödinger equation in a vacuum with no potential term, $V\psi$. Which is considered later [21]. In real spacetime, we have

$$\frac{\hbar\nabla^2\psi}{2m} = \frac{1}{i}\frac{\partial\psi}{\partial t} \quad . \tag{10.11}$$

Monochromatic plane wave solutions for one dimension of space, or x-direction, such as

$$\psi = e\frac{i(kx - \omega t)}{\hbar} \tag{10.12a}$$

or

$$\psi^* = e\frac{-i(kx - \omega t)}{\hbar} \tag{10.12b}$$

which comprise the usual solutions. We can also write Eq. (10.12a) as

$$\psi = e^{i\alpha} \quad for \quad \alpha = \frac{kx - \omega t}{\hbar} \tag{10.13}$$

and we can write Eq. (10.13) as

$$\psi = e^{i\alpha} = \cos\alpha + i\sin\alpha \tag{10.14a}$$

and also

$$\psi = e^{i\alpha} = \sinh i\alpha + \cos i\alpha \, . \tag{10.14b}$$

Equation (10.11) is the usual linear form of the Schrödinger equation in which the superposition principle holds and the quantum measurement issue arises.

We proceed to formulate the Schrödinger equation in complex spacetime. The form of complex derivative utilized here is given in [8,11]. Only 1D forms of the derivative are considered in the del operator, ∇. We consider x-directional spatial dependence only for the real component of x as x_{Re}

$$\hbar\frac{\nabla^2}{2m}\psi \rightarrow \frac{\hbar}{2m}\frac{\partial^2}{\partial x_{\mathrm{Re}}^2}\psi\,. \tag{10.15}$$

Using the imaginary components of space and time x_{Im} and t_{Im}, we have

$$\hbar\frac{\nabla_{\mathrm{Im}}^2}{2m}\psi \rightarrow \frac{\hbar}{2m}\frac{\partial^2}{\partial x_{\mathrm{Im}}^2}\psi\,. \tag{10.16}$$

Note that the sign change occurs for the spatial second derivative for $ix \rightarrow x_{\mathrm{Im}}$. The imaginary time derivative yields

$$\frac{\partial}{\partial it^*} \rightarrow \frac{1}{i}\frac{\partial}{\partial t_{\mathrm{Im}}} \tag{10.17}$$

which is an imaginary term derivative.

The imaginary form of the Schrödinger equation becomes

$$\frac{\hbar}{2m}\nabla_{\mathrm{Im}}^2\psi = \frac{\partial\psi}{\partial t_{\mathrm{Im}}}\,. \tag{10.18}$$

Because the Schrödinger equation is second order in space and first order in time and no imaginary term occurs in Eq. (10.18), the harmonic solutions in Eqs. (10.13, 10.14a, 10.14b) are not solutions to the imaginary components of the Schrödinger equation. Since the Dirac equation is first order in space and time, and the Klein-Gordon equation and classical wave equation are second order in space and time, quite a different picture emerges.

Starting from a real solution, which is a plane exponential growth function

$$\psi = e^{\alpha} \quad for \quad \alpha \equiv \frac{kx + \omega t}{\hbar} \tag{10.19}$$

we then have from Eq. (10.18),

$$\frac{\partial \psi}{\partial x_{\text{Im}}} = \frac{k}{\hbar}\psi \quad or \quad \frac{\partial^2 \psi}{\partial x_{\text{Im}}^2} = \frac{k^2}{\hbar^2} \tag{10.20}$$

and

$$\frac{\partial \psi}{\partial t_{\text{Im}}} = \frac{\omega}{\hbar}. \tag{10.21}$$

and Eq. (10.19) satisfies Eq. (10.18). Note that $\alpha \equiv (kx - \omega t)/\hbar$ does not satisfy Eq. (10.18) because of the minus sign which then occurs in Eq. (10.21). All quantities $k^2, \hbar^2, \omega^2$ are real as is x_{Im} and t_{Im}.

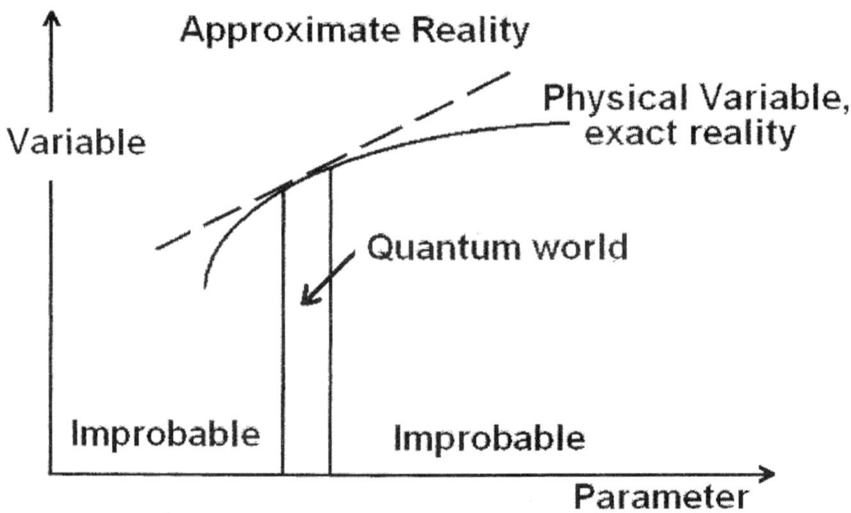

Figure 10.2. We approximate the quantum domain as a linear variable dependent on a parameter. The full "space" of exact reality is nonlinear.

The form of the solution in Eq. (10.19) for α positive definite, for all quantities greater than zero, yields an undamped growth function, that is we find that solutions in an imaginary spacetime geometry yield growth equations. Equation (10.19) is of a linear form. We also have another solution in Eq. (10.25a), but Eq. (10.25b) is not a solution:

$$\psi = e^{-\alpha} \quad for \quad \alpha \equiv \left(\frac{kx - \omega t}{\hbar}\right) \tag{10.22a}$$

and

$$\psi = e^{-\alpha} \quad for \quad \alpha \equiv \left(\frac{kx + \omega t}{\hbar} \right) \tag{10.22b}$$

where in kx, x is x_{Im} and standing wave solutions cannot occur. Before we examined the solution of the Schrödinger equation in complex spacetime for $x' = x_{Re} + ix_{Im}$ and $t' = t_{Re} + it$. Let us briefly discuss the introduction of a nonlinear term with a small coupling constant.

10.5.1 *Nonlinear Schrödinger Equation With Complex Temporal Perturbation*

We introduce a 'potential' like term which is coupled by a small coupling constant, g^2, and is associated with an attractive force. If the coupling term is small, then solutions can be determined in terms of a perturbation expansion. A $g^2 > 0$ implies an attractive force when it is regarded as a second quantized Fermi field. This field satisfies the Dirac equation and introduces an additional term in the Lagrangian. In reference [11] we detail this formalism, in which causality conditions in terms of analytic continuation in the energy plane gives motivation for identifying the nonlinear coupling term with the imaginary temporal coordinate, as $t^* = it_{Im}$.

By analogy to this form of the Dirac equation, we can write

$$\frac{\hbar \nabla^2_{\text{Im}} \psi}{2m} + g^2 (\psi^+ \psi) \psi = 0 \tag{10.23}$$

for the time-dependent equation where ψ^+ is the Hermitian conjugate of ψ. For the real time-dependent equation, we have

$$\frac{\hbar \nabla^2_{\text{Im}} \psi}{2m} + g^2 (\psi^+ \psi) \psi = \frac{1}{i} \frac{\partial \psi}{\partial t_{\text{Im}}}. \tag{10.24}$$

For the Schrödinger and Dirac equation, we can find solutions which we can identify in a field theory, in which each point is identifiable with a kinetic, potential and amplitude function. Linearity can be approximated for $g^2 \sim 0$, for g^2 expressed in terms of it_{Im}. In the following subsection we examine the complexification of the Schrödinger equation.

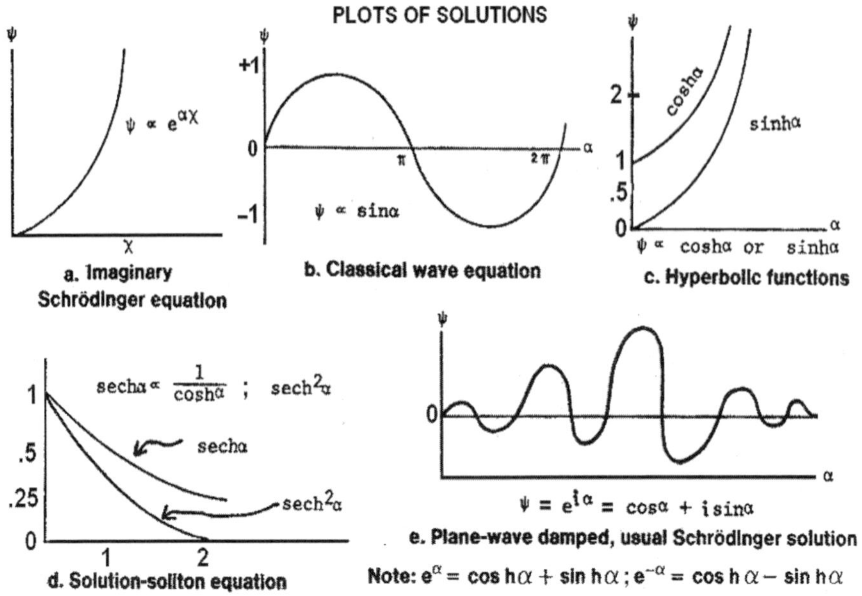

Figure 10.3. Five plots of various solutions related to the Schrödinger equation.

10.5.2 *The Schrödinger Equation in Complex Space and Time*

Returning to our definition of complex space and time,

$$x' = x_{\text{Re}} + ix_{\text{Im}}, \quad t' = t_{\text{Re}} + it_{\text{Im}} \tag{10.25}$$

where x_{Re} and t_{Re} are the real parts of space and time and x_{Im} and t_{Im} are the imaginary parts of space and time and are themselves real quantities. In the most general case we have functional dependencies x_{Im} (x,t) and t_{Im} (x,t) where x and t are x_{Re} and t_{Re}. With the quantum superposition principle, we can combine real and imaginary parts. For the x-directional form of Eq. (10.11), we have

$$\frac{\hbar}{2m}\frac{\partial^2\psi_1}{\partial x_{\text{Re}}^2} = \frac{1}{i}\frac{\partial\psi_1}{\partial t_{\text{Re}}}. \tag{10.26}$$

For the imaginary part, we have from Eq. (10.18)

$$\frac{\hbar}{2m}\frac{\partial^2\psi_1}{\partial x_{\rm Im}^2}=\frac{\partial\psi_2}{\partial t_{\rm Im}}. \tag{10.27}$$

By linear superposition, we can combine the above equation, as

$$\frac{\hbar}{2m}\left(\frac{\partial^2}{\partial x_{\rm Re}^2}+\frac{\partial^2}{\partial x_{\rm Im}^2}\right)\psi=\left(\frac{1}{i}\frac{\partial}{\partial t_{\rm Re}}+=\frac{\partial}{\partial t_{\rm Im}}\right)\psi. \tag{10.28}$$

Note that we make an assumption that the mass in Eq. (10.26) is the same as in Eq. (10.27). We discuss this assumption and tachyonic implications in [11]. We now form solutions $\psi(x_{\rm Re},x_{\rm Im},t_{\rm Re},t_{\rm Im})$ in terms of linear combinations of $\psi_1(x_{\rm Re},t_{\rm Re})$ and $\psi_2(x_{\rm Im},t_{\rm Im})$.

Equation (10.27) is defined on a 4D space $(x_{\rm Re},x_{\rm Im},t_{\rm Re},t_{\rm Im})$. In the first approximation, we will choose $\partial^2\psi/\partial x_{\rm Im}^2=0$ so that we have

$$\frac{\hbar^2}{2m}\frac{\partial^2}{\partial x_{\rm Re}^2}\psi=\left(\frac{1}{i}\frac{\partial}{\partial t_{\rm Re}}+\frac{\partial}{\partial t_{\rm Im}}\right)\psi. \tag{10.29}$$

Motivation for this approximation can be found in our discussion of remote connectedness properties, diagrammed in Figs. 10.1c and 10.1b of the previous section.

Let us rewrite Eq. (10.26) as

$$\frac{\hbar^2}{2m}\frac{\partial^2}{\partial x_{\rm Re}^2}\psi-\frac{\partial}{\partial t_{\rm Im}}\psi=\frac{1}{i}\frac{\partial}{\partial t_{\rm Re}}\psi \tag{10.30}$$

where ψ is a function of $(x_{\rm Re},t_{\rm Re},t_{\rm Im})$. From examination of the forms of Eq. (10.24) and (10.29), we can identify the g^2 term with the imaginary time derivative $\partial/\partial t_{\rm Im}$. This result is similar to the more comprehensive field theoretic argument for the Dirac equation. The associated metric space for $(x_{\rm Re},t_{\rm Re},t_{\rm Im})$ defines a remote connectedness geometry. We then have

$$\frac{\hbar}{2m}\frac{\partial^2\psi}{\partial x_{\rm Re}^2}+G^2\psi=\frac{1}{i}\frac{\partial\psi}{\partial t_{\rm Im}} \tag{10.31}$$

where $G^2 = g^2(\psi^+\psi)$ is identified with the $\partial / \partial t_{\text{Im}}$ term. We proceed from the assumption that $(x_{\text{Re}}, t_{\text{Re}}, t_{\text{Im}})$ are independent variables of each other.

We can define three cases for the right side of Eq. (10.31), that is, the real time-dependent case, (a) zero, time dependent cases, (b) $\frac{1}{i} E_n \psi$, and (c) $\frac{1}{i}\frac{\partial \phi}{\partial t}$. In determining the coupling constant G^2, we define solutions $\phi(x_{\text{Re}}, t_{\text{Re}}, t_{\text{Im}})$ for the third case. We have, in general,

$$\frac{\partial^2 \phi}{\partial x^2} + G^2(\tau)\frac{\partial \phi}{\partial \tau} = \frac{1}{i}\frac{\partial \phi}{\partial t}. \tag{10.32}$$

We define the quantity $\xi = kx_{\text{Re}} - \omega t_{\text{Re}} - \alpha t_{\text{Im}}$. For case (a) above we have solutions

$$\phi = \phi_0 + A \sec h^2 a\xi \tag{10.33}$$

where

$$G^2(t_{\text{Im}}) = \frac{a\hbar^2 k^2}{2\alpha m} \tanh a\xi \tag{10.34}$$

where k is the wave number or k_{Im}. The constant, a can be expressed in terms of $\hbar$ and m where $m' = im = m^* = m_{Im}$ which is the tachyonic mass, which we formulate in complex 8-space. For case (c), we find a similar solution for ϕ for

$$G^2(t_{\text{Im}}) = \frac{\frac{a\hbar^2 k^2}{2m}}{(\alpha + \frac{1}{2}\omega)} \tanh a\xi \,. \tag{10.35}$$

Solutions and the form of G^2 (t_{Im}) is more complicated for case (b). Note the analogy to the solutions for the Korteweg-deVries equation [21] for

$$u(x,t) = A \sec h^2 k \quad \text{for} \quad K = x - ct / L \tag{10.36}$$

where L is a characteristic length dimension of a soliton wave which is expressed in terms of the amplitude A and the hydrodynamic media depth h or $L = \sqrt{h^3 / 3A}$ [35].

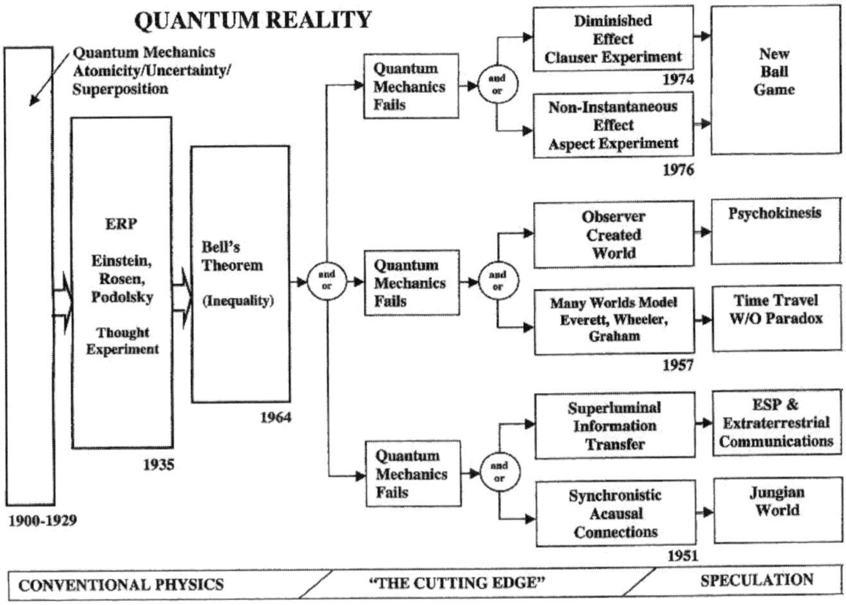

Figure 10.4. Historical development of quantum theory.

The form of $G^2(t_{\text{Im}})$ is nonlinear and is compatible with the soliton solutions. The non-dispersive nature of the solutions may be associated with a complex space "signal" which defines the connection of remote parts of the multi-dimensional geometric space [11]. Several types of solutions are displayed in Fig. 10.3. See Fig. 10.4 for the implications of the Quantum Theory and Bell's theorem.

10.5.3 *Discussion and Application of Coherent State Solutions*

The soliton solution is a unique solution in that it is non-dispersive. All other solutions to the Schrödinger equation are dispersive to various degrees. Each state solution has a particular amplitude at a specific point in space and instant in time. One can calculate the probability of this existence of a specific amplitude as a function of x and t. A unique feature of the soliton is that it retains its amplitude in space and time and therefore we have a reasonable certainty in our measure of it for each space and time.

In practice, there are no completely non-dispersive waves but soliton solutions are defined in terms of coherent, non-dispersive states that retain their identity and amplitude over many iterations. Hence the soliton acts like a particle, in that soliton solution collisions do not disrupt the wave form or amplitude in elastic processes [36].

In hydrodynamics, the interpretation of the soliton or solitary wave is not completely clear [37]. One possible interpretation of this particular type of solution to the wave equation in this particular complex geometry, including the small coupling nonlinear term, is that the geometry selects the particular wave function. Note that this possible interpretation may have deep implications for the quantum measurement issue or the "collapse of the wave function". In the usual nuclear energy levels, a particular state may be composed of a sum of various states of angular momentum and spin which sum to the total I and l values. The amplitude of these states vary, with one predominant term [38]. In the current case, the soliton non-dispersive wave could represent the predominant, fixed amplitude solution with other smaller dispersive terms.

We have examined coherent collective states in plasmas with high temperature fusion media and electron gases in metal conductors. It is felt that these and other types of collective, coherent, dynamical phenomena can be explained by the soliton formalism. Other such phenomena which may also involve an intermediate temperature plasma is the illustrative so-called "ball lightning" [32].

10.6 Conclusion

We have formulated a complex multi-dimensional Minkowski space and associated twistor algebra which has nonlocal and anticipatory properties. One unique property of this geometry is its remote connectedness. We have formulated the Schrödinger equation in this multi-dimensional geometry. We identify the imaginary temporal component term as a small nonlinear term and determine soliton or solitary wave solutions. These non-dispersive, coherent waves are appropriate to define signals, in the space, which exhibit remote connectedness properties. Phenomena which involve remote correlation of events, such as Bell's Theorem, Young's double slit experiment, and super-coherence phenomena, demand nonlocality. The twistor algebra can be constructed to be mapable 1:1 with the spinor calculus and allows us to develop a unique formalism of Bell's inequality.

We also speculate that the nonlinear quantum model with coherent non-dispersive solutions to the Schrödinger equations, which is an expression of the remote nonlocality property of the space, may lend insight into the quantum measurement problem. A mechanism may be formulated which defines a connection between the observer and the observed. The properties of certain systems appear to demand a nonlinear, nonlocal anticipatory description.

References

[1] Einstein, A., Podolsky, B., & Rosen, N. (1935) Can a quantum mechanical description of physical reality be considered complete? Phys Rev, 47, pp. 777-780.
[2] Misner, C.W., Thorne, K.S. & Wheeler, J.A. (1973) Gravitation, San Francisco: W.H. Freedman; and private communications with EAR 1976-1979.
[3] Bohm, D. & Hiley, B. (1993) The Undivided Universe, London: Routledge; and private communications with EAR 1976-1979.
[4] Bell, J.S. (1964) On the Einstein, Podolsky, Rosen paradox, Phys. 1, pp. 195-200.
[5] Bell, J.S. (1966) On the problem of hidden variables in quantum theory, Review of Modern Physics, 38, 447.
[6] Stapp, H.P. (1971) Phys. Rev. 3D, 1303; private comm.. with EAR 1964-1993.
[7] Clauser, J.F. & Shimony, A. (1978) Rep. Prog. Phys. 41, 1881; and private communications with EAR 1971-1992.
[8] Rauscher, E.A. (1979) Some physical models potentially applicable to nonlocal measurement, in The Iceland Papers: Frontiers of Physics Conference, pp. 50-93, Amherst: Essentia Research Associates.
[9] Amoroso, R.L., Vigier, J-P, Kafatos, M. & Hunter, G. (2002) Comparison of near and far-field double-slit interferometry for dispersion of the photon wave packet, in R.L. Amoroso, G. Hunter, M. Kafatos & J-P Vigier (eds.) Gravitation and Cosmology: From the Hubble Radius to the Planck Scale, Dordrecht: Kluwer.
[10] Antippa, A.F. & Dubois, D.M. (2004) Anticipation, orbital stability and energy conservation in the discrete harmonic oscillator, in D.M. Dubois (ed.) AIP Conference Proceedings 718, CASYS03, Liege, Belgium, pp. 3-44, Melville: AIP.
[11] Ramon, C. & Rauscher, E.A. (1980) Superluminal transformations in complex Minkowski spaces, LBL Report 9752 ; Foundations of Physics (1980) 10, 661.
[12] Hansen, R.O. & Newman, E.T. (1975) A complex Minkowski approach to twistors, General Relativity and Gravitation, 6, 361-385.
[13] Newman, E.T. (1976) H-space and its properties, General Relativity and Gravitation, 7, 107-111.
[14] Newman, E.T., Hansen, R.O., Penrose, R. & Ton, K.P. (1978) The metric and curvature properties of H-space, Proc. Royal Society of London, A363, 445-468.
[15] Haramein, N. & Rauscher, E.A. (2008) Complex Minkowski space formalism of the Penrose twistor and the spinor calculus, in D. Dubois (ed.) Proceedings of CASYS07, Liege, Belgium, in press; Rauscher, E.A. (1971) A unifying theory of fundamental processes, LBNL/UCB Press, UCRL 20808, June, and refs. therein.
[16] Rauscher, E.A. (1972) A set of generalized Heisenberg relations and a possible new form of quantization, Letters il Nuovo Cimento, 4, 757.
[17] Bohm, D. (1952) Physical Review, 85, 166; private communication with EAR.
[18] Clauser, J.F. & Horne, W.A. (1974) Physical Review, 10D, 526.
[19] Freedman, S. & Clauser, J.F. (1972) Experimental test of local hidden variable theories, Physical Review Letters, 28, 934-941.

[20] Feinberg, G. (1967) Possibility of faster-than-light particles, Phys Rev., 159, 1089.
[21] Rauscher, E.A. (2007) A detailed formalism of Bell's theorem in complex 8-space, in progress.
[22] Stapp, H.P. (1975) Theory of reality, LBL Report 3837.
[23] Aspect, A., Grangier, P. & Roger, G. (1992) Experimental tests of Bell's inequalities using time-varying analyzers, Phys. Rev. L. 49, 1804-1907; and private comm. with EAR
[24] Gisin, N., Tittel, W. Brendel, J. & Zbinden, H. (1998) Violation of Bell's inequalities by photons more than 10 km apart, Phys. Rev. Let. 81, 3563-3566.
[25] Gisin, N., Tittel, W. Brendel, J. & Zbinden, H. (1998) Quantum correlation over more than 10 km, Optics and Photonics News, 9, 41.
[26] Rauscher, E.A. & Targ, R. (2001) The speed of thought: Investigations of a complex spacetime metric to describe psychic phenomena, J. Sci. Expl., 15, 331.
[27] Rauscher, E.A. & Targ, R. (2006) Investigation of a complex spacetime metric to describe precognition of the future, in D.P. Sheehan (ed.) Frontiers of Time: Retrocausation – Experiment and Theory, Melville: AIP Conference Proceedings.
[28] Wigner, E. (1967) Symmetries and Reflections: Scientific Essays, Bloomington: Indiana Univ. Press; and private comm. with EAR.
[29] Chew, G. (1964) The Analytic S-Matrix, Benjamin: Frontiers of Physics; and private comm. with EAR.
[30] Rauscher, E.A. (2002) Non-Abelian gauge groups for real and complex amended Maxwell's equations, in R.L. Amoroso, G. Hunter, M. Kafatos & J-P Vigier (eds.) Gravitation and Cosmology: From the Hubble Radius to the Planck Scale, pp. 183-188, Dordrecht: Kluwer Academic; Also (2992) Bulletin Am. Phys. Society, 47, 167.
[31] Rauscher, E.A. & Amoroso, R.L. (2006) The physical implications of multidimensional geometries and measurement, in D. Dubois (ed.) Int. J. Computing Anticipatory Systems, vol. 19, Liege: University of Liege, CHAOS, Institute of Mathematics;
[32] Rauscher, E.A. (1983) Electromagnetic Phenomena in Complex Geometries and Nonlinear Phenomena and Non-Hertzian Waves, Milbrae: Tesla Books; 2nd Edition (2008) Oakland: The Noetic Press.
[33] Amoroso, R.L. and Rauscher, E.A. the Holographic Anthropic Multiverse Formalizing the Complex Geometry of Reality", World Scientific, (2009).
[34] Sewell. G.L. (2002) Quantum Mechanics and its Emergent Macrophysics, Princeton: Princeton Univ. Press.
[35] Landau, L.J. (1987) Experimental tests of general quantum theories, Lett. In Math. Physics, 14, 33-40.
[36] Rauscher, E.A. (1968) Electron interactions and quantum plasmas, J. Plas.Phys., 2, 517.
[37] Lonngren, K.& Scott, A. (1978) Solitons in Action, New York: Academic Press.
[38] Osborne, A.R. & Burke, T. (1980) Science, 208, 451.

Chapter 11

Spinors, Twistors, Quaternions, and Complex Space

Utilizing the spinor approach, electromagnetic and gravitational metrics are mapable to the twistor algebra, which corresponds to the complexified Minkowski space. Quaternion transformations relate to spin and rotation corresponding to the twistor analysis.

11.1 Introduction

In this chapter we will present a formalism that uniquely relates electromagnetic and gravitational fields. Through this formalism and the relationship of the spinor calculus and the twistor algebra we can demonstrate the fundamental conditions of such a system which accommodates macroscopic astrophysical phenomena as well as microscopic quantum phenomena. The generalized Minkowski formalism has large scale astrophysical as well as quantum level consequences.

The generalized hyperdimensional Minkowski manifold has nonlocal as well as anticipatory properties. Also briefly we discuss the 720° symmetry of the so-termed Dirac string trick within the context of the relativistic form of the Dirac formalism. Twistors and spinors are examined and are applicable to the quaternion formalism. The quaternion formalism can be related to the hyperdimensional complexified Minkowski space, Lie groups, SU_n, as well as Riemannian topologies and the Dirac equation. See Chap. 12.

In Sec. 11.2 we present the formalism for the role of the spinor calculus which is utilized to relate the expression for the metric tensor to gravitational and electromagnetic field components through the relationship of the twistor algebra and spinor calculus. The Minkowski space formalism consistent with this approach uniquely relates to the twistors in Section 11.3. In this section, we demonstrate the manner in which the approaches presented in this paper relate to the current supersymmetry and GUT models as well as string theory. See Chap.13. We further elaborate on the symmetry principles of the complexification of Minkowski space, twistors and their properties. A fundamental relationship between complex Minkowski space, twistor algebra and quaternions are developed in Sec. 11.4. Of interest are the non-Abelian nature of quaternions, the SU_n groups, quantum theory and Penrose topology.

11.2 The Spinor Formalism and their Relationship to Twistors

The approach to unification of the electromagnetic and gravitational fields was developed by Kaluza [1] and Klein [2] in the 1920s and their work was seriously considered by Einstein in the 1930's. This 5D geometry utilizes the spinor calculus to account for the coupling of the electromagnetic field to the gravitational field, in which the spinor is treated as a rolled up dimension rather than as the four extended dimensions of the gravitational field. The concept of small Planck scale rotational "extra dimensions" (XD) is postulated in current 10D and 11D supersymmetry models. They are considered to be ultramicroscopic because they are not seen. However following Randall our cosmology utilizes large scale XD in our model. Kaluza-Klein Theory is treated as a subset of this supersymmetry, including the grand unification theory (GUT) and theory of everything (TOE).

The Kaluza-Klein Theory requires a periodicity of the 5D spinor fields to unify electromagnetism and gravity based on the homeomorphism between the Lorentz group and the unimodular transformation of Maxwell's equations and the weak Weyl limit of the gravitational field. A discussion of the Kaluza-Klein model, Rauscher [3,4], Newman [5] and Hansen and Newman complex 8-space is given in [6]. In the approach of the latter three references, the spinor calculus is demonstrated as mapable one-to-one with the twistor algebra of the complex 8-space and, hence, the Penrose twistor [3].

The coupling of the electromagnetic field with the gravitational field in the Kaluza-Klein may also yield a connection through the photon description of the twistor algebra. The photon is the quanta of the electromagnetic field and the interaction mediation between leptons, such as the electron. The 5D spinor calculus has been developed within a 5D relativistic formalism [1-3]. The spinor calculus developed in the 5D spinor formalism accounts for the coupling of the electromagnetic field to the gravitational metric.

This approach is manifestly 5-covariant in a special 5D space. The specific spin frames of reference of the 5D Kaluza-Klein geometry reduces to the spinor formalism of curved spacetime. The theory of spinors in 4D space is based upon the transformation L' and the group of unimodular transformation U_1 in $SL(2,C)$. This formalism is related to 2-toroidal space

$U_1 \times U_1$. Unimodular action of the symplectic automorphism group $SL(2,R)$ of the Heisenberg 2-step nilpotent Lie group, N has the discrete subgroups $SL(2,Z)$ of $SL(2,R)$. The 2D compact unit sphere, S_2 (Riemannian sphere) and the 3D spherical component unit sphere can map as $S_3 \to R^4$.

It has been established that the 5D 4-component spinor calculus is related to the 4D spinor formalism in order to account for the coupling of the electromagnetic field as a periodic 5D spinor field to the curved space of the gravitational Riemannian metric [7]. We can utilize projective geometry to relate 5D spinor calculus to the 4D twistor space.

An isomorphism between vectors v^{β} and spinors $v^{AA'}$ satisfies the condition

$$\overline{\pi}^{AA'} = \pi^{AA'} \tag{11.1}$$

so that the spinor equivalent to a vector v^{β} is

$$\pi^{AA'} = \tau_{\beta}^{AA'} v^{\beta} \tag{11.2}$$

where $\tau_{\beta}^{AA'}$ is a tensor.

Therefore,

$$v^{\beta} = \tau_{AA'}^{\beta}\, \pi^{AA'} \tag{11.3}$$

where v^{β} is real for $\overline{\pi}^{AA'} = \pi^{AA'}$. The covering map $SL(2,C)$ goes to $O(1,3)$ by using the vector-spinor correspondence.

We present some of the properties and structure of this significant advancement in developing a unified force theory for the electromagnetic and gravitational fields which can be related to the twistor algebra. In addition to the general coordinate transformations of the four coordinates x^{μ}, the preferred coordinate system permits the group relation,

$$x'^5 = x^5 + f\left(x^1, x^2, x^3, x^4\right). \tag{11.4}$$

Using this condition and the 4D cylindrical metric or $ds^2 = \gamma_{ik} dx^i dx^k$ yields the form

$$ds^2 = \left(dx^5 + \gamma_{\mu 5} dx^{\mu}\right)^2 + g_{\mu\nu} dx^{\mu} dx^{\nu} \tag{11.5}$$

where the second term is the usual 4-space metric. The quantity $\gamma_{\mu 5}$ in the above equation, transforms like a gauge [7]

$$\gamma'_{\mu 5} = \gamma_{\mu 5} - \frac{\partial f}{\partial x^{\mu}} \tag{11.6}$$

where the function f is introduced as an arbitrary function. Returning to our 4D metrical form in its 5-compact form and 4D and 5D form gives,

$$\gamma_{\mu\nu} = g_{\mu\nu} + \gamma_{\mu 5}\gamma_{\nu 5}\,. \tag{11.7}$$

Proceeding from the metrical form in a "cylindrical" space, $ds^2 = \gamma_{ik} dx^i dx^k$ where indices i, k run 1 to 5, we introduce the condition of cylindricity which can be described in a coordinate system in which the γ_{ik} are independent of x^5 or

$$\frac{\partial \gamma_{ik}}{\partial x^5} = 0\,. \tag{11.8}$$

Kaluza and Klein assumed $\gamma_{55} = 1$ or the positive sign $\gamma_{55} > 0$ for the condition of the 5th dimension to ensure that the 5th dimension is metrically space-like. In geometric terms, one can interpret x^5 as an angle variable, so that all values of x^5 differ by an integral multiple of 2π corresponding to the same point of the 5D space, if the values of the x^{μ} are the same. Greek indices μ, ν run from 1 to 4, and Latin indices i, k run from 1 to 5 and for this specific case, each point of the 5D space passes exactly one geodesic curve which returns to the same point. In this case, there always exists a perpendicular coordinate system in which $\gamma_{55} = 1$ and

$$\frac{\partial \gamma_{5\mu}}{\partial x^5} = 0\,. \tag{11.9}$$

It follows from those properties that $g_{\mu\nu}$ and γ_{ik} can be made analogous so that $g_{\mu\nu} = \gamma_{ik}$ then

$$\gamma^{55} = 1 + \gamma^{\mu\nu}\gamma_{\mu 5}\gamma_{\nu 5} \tag{11.10a}$$

(also see Eq. (11.7)) and

$$\gamma^{\mu 5} = -g^{\mu\nu}\gamma_{\mu 5}\,. \tag{11.10b}$$

The gauge-like form alone is analogous to the gauge group, which suggests the identification of $\gamma_{\mu 5}$ with the electromagnetic potential, ϕ_μ. We can write an expression for an antisymmetric tensor,

$$\frac{\partial\,\gamma_{\mu 5}}{\partial x^\mu} - \frac{\partial\,\gamma_{\mu 5}}{\partial x^\nu} = f_{\mu\nu} \tag{11.11}$$

which is an invariant with respect to the "gauge transformation" (Chap. 8).

We now use the independence of γ_{ik} of x_5 or $\partial\,\gamma_{ik}/\partial x^5 = 0$. The geodesics of the metric in five-space can be interpreted by the expression

$$\frac{dx^5}{ds} + \gamma_{\mu 5}\frac{dx^\nu}{ds} = C \tag{11.12}$$

where C is a constant and s is a distance parameter. If we consider the generalized 5D curvature tensor, and using the form for $f_{\mu\nu}$ we can express it in terms of $F_{\mu\nu}$, the electromagnetic field strength,

$$f_{\mu\nu} = \sqrt{\frac{16\pi G}{c^4}}F_{\mu\nu} \tag{11.13}$$

where $\sqrt{G/c^4} = 1\big/\sqrt{F}$ where F is the quantized force introduced by Rauscher [3,8-10] which relates to the driving force for the perceived expansion of the universe. This is the Rauscher force term that appears in the stress energy term in Einstein's field equations [11]. Then we can write,

$$\gamma_{\mu 5} = \sqrt{\frac{16\pi G}{c4}}\phi_\mu\,. \tag{11.14}$$

The integration constant, above, can be identified as proportional to the ratio e/m of charge to mass of a particle traveling geodesics in the Kaluza-Klein space [1-3].

Under the specific conditions of the conformal mappings in the complex Minkowski space, one can represent twistors in terms of spinors. The spinor is said to "represent" the twistor. The twistor is described as a complex two-plane in the complex Minkowski space (see Section 11.3 and [3] for references on twistor theory and the spinor calculus. Twistors and spinors can be easily related by the general Lorentz conditions in such a manner as to retain the condition that all signals are luminal in real 4-space. The conformal invariance of the tensor field, which can be Hermitian, can be defined in terms of twistors and these fields can be identified with particles [11,12].

It is through the representation of spinors as twistors in complex Minkowski space that we can relate the complex eight-space model to the Kaluza-Klein geometries and to the grand unification or GUT theory. See Chap. 13. In the 5D Kaluza-Klein geometries, the XD is considered to be a spatial rotational dimension in terms of $\gamma_{\mu 5}$. The Hanson-Newman [6], and Rauscher [4,5] complex Minkowski space has introduced with it an angular momentum or helix or spiral dimension called a twistor which is expressed in terms of spinors [7].

The spinor formalism was used by Dirac to define the Schrödinger equation in a relativistic invariant form so that the complex scalar time dependent field of Schrödinger is in terms of a two component spinor field. Using this formalism Dirac obtained a 2-valued solution which predicted the observed electron and positron pair. The spinor field or spinor variable, utilized in the Kaluza-Klein geometry, directly relates to the spin degrees of freedom that are observed by the Zeeman Effect in atomic spectra. The spin degrees of freedom appear to be fundamental to quantum theory and to relativity and are a good starting point to treat spin in a fundamental manner. The Lorentz 4-space representation of relativity can be reduced to the direct product of two 2D complex representations. The spinor variable is the most fundamental representation of a relativistically invariant theory and spin degrees of freedom may be formulated relativistically and, hence, also in a possible "quantum gravity" picture which applies to the Dirac equation. This approach may be applicable to the Penrose twistor (Chap. 12).

This approach appears to fit well with the spinor approach in the Dirac formalism in the quantum domain, that is, that the Lorentz conditions applied by Einstein in relativity may be the origin of the spinor and, hence, be the fundamental theory that yields the spinor formalism and the role of spin. Other implications of the relationship between the Penrose twistor formalism and the complex Minkowski space, which includes anticipatory systems related to causality and spatial and temporal nonlocality, are given in references [12-18].

11.3 The Penrose Twistor, Harmonic Sequencing and Particle Spin

Interest in the twistor program has been in the form of quantizing gravity in order to unify the physics of the micro- and macro-cosmos in 1971 and 2005. Such a procedure has been taken by Penrose et al. and is based on the concept of a more general theory that has limits in the quantum theory and the relativistic theory [15]. In addition, there have been approaches to the underlying structure of spacetime in the quantum [11] and structural regime [8]. A structured and/or quantized spacetime [1] may allow a formalism that unequally relates the electromagnetic fields with the gravitational metric [9]. Feynman [13] and Penrose graphs [11,12] may overcome the divergences of such an approach. In order to translate the equations of motion and Lagrangians from spinors to twistors, one can use the eigenfunctions of the Casimir operators of the Lie algebra of $U(2,2)$.

For the simplest case of a zero rest mass field (photon-like) for $n/2$ spin for $n \neq 0$, we can write

$$\nabla_{AA'}\varphi^{A...N} = 0 \tag{11.15}$$

for $A,....,N$ written in terms of N indices, and for $N = 1$, we have the Dirac equation for massless particles. For a spin zero field, we have the Klein-Gordon equation

$$\nabla^{AA'}\nabla_{AA'}\varphi = 0 \tag{11.16}$$

and in Eq. (11.15) for $n = 2$, we have the source-free Maxwell equation $\Box F^{\mu\nu} = 0$ for spin 1 or U_1 fields, and for $n = 4$, we have the spin free Einstein field equations, $R_{\mu\nu} = 0$. The indices μ and ν run 0 to 3. For a system with charge, then $\Box F^{\mu\nu} = J_{\mu\nu} - J_{\nu\mu}$, or this can be written as $\frac{F_{\mu\nu}}{\partial x_\nu} = J_\mu$ and then we can write

$$\gamma_{\mu\nu}\frac{\partial F_{\mu\nu}}{\partial x_\nu} = J_\mu . \tag{11.17}$$

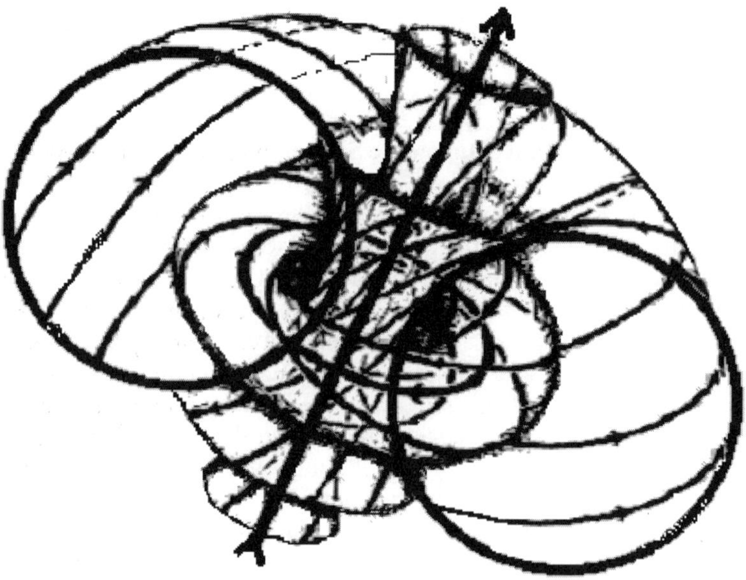

Figure 11.1. A Penrose developed a well-known special diagram for the twistor from what he calls the Robinson congruence. This Robinson congruence is a twistor representing the propagation of a photon along the arrow at the top of the diagram or a time-slice (t = 0) of a Robinson congruence. Redrawn from [11].

In this section, we outline a program to relate the twistor topology to the spinor space and specifically to the Dirac spinors. Both Fermi-Dirac and Bose-Einstein statistics are considered. The relationship between twistor theory and the Dirac "string trick" model is further discussed in our chapter on the complexification of the Dirac equation (Chap. 12). The Penrose spin approach is designed to facilitate the calculation of angular momentum states for SL(2). The spinor formalism, in the Dirac equation, established spinors within quantum theory. The twistor formalisms are related to the structure of spacetime and the relation of the spinors and twistors is also of interest because it identifies a relationship between quantum mechanics and relativity [11,12,18,19].

Twistor theory has been related to conformal field theory and string theory [20]. Also, twistor theory has been related to quaternions and complex quaternionic manifolds [21,22]. The projective twistor space, PT, corresponds to two copies of the associated complex projective space of CP^3 or $CP^3 \times CP^3$. It is through the conformal geometry of surfaces in S^4, utilizing the fact that CP^3 is an S^2 bundle over S^4, that quaternions can be related to twistors [23].

We can demonstrate a useful relationship between the complex eight-space and the Penrose twistor topology; the twistor is derived from the imaginary part of the spinor field. The Kerr Theorem results naturally from this approach in which twisting is shear free in the limit of asymptotic flat space. The twistor is described as a 2-plane in complex Minkowski space, M_4. Twistors define the conformal invariance of the tensor field, which can be identified with spin or spinless particles. For particles with a specific intrinsic spin, s, we have $Z^\alpha \overline{Z}_\alpha = 2s$, and for zero spin, such as the photon, $Z^\alpha \overline{Z}_\alpha = 0$ where $\overline{Z}_\alpha$ is the Hermitian conjugate of Z^α, and Z^α and Z_α can be regarded as canonical variables such as $\underline{x}$, $\underline{p}$ in the quantum theory phase space analysis. The twist free conditions, $Z^\alpha \overline{Z}_\alpha$, hold precisely when Z^α is a null twistor. The upper case Latin indices are used for spinors, and the Greek indices for twistors. The spinor field of a twistor is conformally invariant and independent of the choice of origin [24]. For the spinor, the indexes A and A' take on values 1, 2 [11,12]). We briefly follow along the lines of Hanson and Newman in the formalism relating the complex Minkowski space to the twistor algebra.

Penrose states regarding the Robinson congruence:

I had, earlier, worked out the geometry of a general Robinson congruence: in each time-slice t = constant of M, the projections of the null directions into the slice are the tangents to a twisting family of linked circles (stereographically projected Clifford parallels on S^4 – a picture with which I was well familiar), and the configuration moves with the speed of light in the (negative) direction of the one straight line among the circles.

I decided that the time had come to count the number of dimensions of the space R of Robinson congruences. I was surprised to find, by examining the freedom involved that the number of real dimensions was only six (so of only three complex dimensions) whereas the special Robinson congruences, being determined by single rays, had five real dimensions. The general Robinson congruences must twist either right-handedly or left-handedly, so R had two disconnected components R^+ and R^-, these having a common 5D boundary S representing the special Robinson congruences. The complex 3-space of Robinson congruences was indeed divided into two halves R^+ and R^- by S.

I had found my space! The points of S indeed had a very direct and

> satisfyingly relevant physical interpretation as "rays", i.e. as the classical paths of massless particles. And the "complexification" of these rays led, as I had decided that I required, to the adding merely of one extra real dimension to S, yielding the complex 3-manifold PT = $S \cup R^{-} \cup R^{+}$ [11].

Twistors and spinors are related by the general Lorentz conditions in such a manner as to retain the fact that all signals are luminal in the real four-space, which does not preclude superluminal signals in an $n > 4\mathrm{D}$ space. The twistor Z^{α} can be expressed in terms of a pair of spinors, ω^{A} and π_{A}, which are said to represent the twistor. We write

$$\mathrm{Z}^{\alpha} = \left(\omega^{A}, \pi_{A'}\right) \tag{11.18}$$

where $\omega^{A} = i\, r^{AA'} \pi_{A'}$

Every twistor Z^{α} is associated with a point in complex Minkowski space, which yields an associated spinor, ω^{A}, $\pi_{A'}$. The spinor is associated with a tensor which can be Hermitian or not. The spinor can be written equivalently as a bivector forming antisymmetry. In terms of spinors ω^{A} and $\pi_{A'}$, they are said to represent the twistor Z^{α} as $\mathrm{Z}^{\alpha} = \left(\omega^{A}, \pi_{A'}\right)$ (see Eq. (11.18)). In terms of components of the twistor space in Hermitian form, φ for $\varphi_{AA'} = \varphi_{A'A}$, we have,

$$\varphi\left(\mathrm{Z}^{\alpha}\mathrm{Z}^{\beta}\right) = \overline{\mathrm{Z}^{0}\mathrm{Z}^{2}} + \overline{\mathrm{Z}^{1}\mathrm{Z}^{3}} + \overline{\mathrm{Z}^{2}\mathrm{Z}^{0}} + \overline{\mathrm{Z}^{3}\mathrm{Z}^{1}} \tag{11.19}$$

where the α index runs 0 to 3. The components of Z^{α} are $\mathrm{Z}^{0}, \mathrm{Z}^{1}, \mathrm{Z}^{2}, \mathrm{Z}^{3}$ and are identifiable with a pair of spinors, ω^{A} and $\pi_{A'}$, so that

$$\omega' = \mathrm{Z}^{1},\ \pi_{0'} = \mathrm{Z}^{2},\ \pi_{1'} = \mathrm{Z}^{3} \tag{11.20}$$

so that we have

$$\mathrm{Z}\overline{\mathrm{Z}}_{\mu} = \mu^{0}\overline{\pi}_{0} + \mu'\overline{\pi}_{1} + \pi_{0}\,\overline{\mu}^{\prime 0'} + \pi_{1'}\overline{\mu}^{1'}. \tag{11.21}$$

Note that the spinor ω^{A} is the more general case of μ^{A}. This approach ensures that the transformations on the spin space preserve the linear transformations on twistor space, which preserves the Hermitian form, φ.

The underlying concept of twistor theory is that of conformal invariance or the invariance of certain fields under different scalings of the metric $g_{\mu\nu}$. Related to the Kerr theorem, for asymptotic shear-free null flat space, the analytic functions in the complex space of twistors may be considered a twisting of shear-free geodesics. In certain specific cases, shear inclusive geodesics can be accommodated.

Twistors are formally connected to the topology of certain surfaces in complex Minkowski space, M^4. This space, the complex space, C^4, is the cover space of R^4, the 4D Riemannian space. On the Riemann surface, one can interpret spinors as roots of the conformal tangent plane of a Riemann surface into R^3. This approach is significant because it ensures the diffeomorphism of the manifold. Complexification is formulated as $Z^\mu = X^\mu_{\mathrm{Re}} + X^\mu_{\mathrm{Im}}$, which constitutes the complexification of the Minkowski space, M^4. The usual form Minkowski space is a submanifold of complex Minkowski space. Twistors are spacetime structures in Minkowski space, which is based upon the representation of twistors in terms of a pair of spinors as we have shown [4,14]. Twistors provide a unique formulation of complexification. The interpretation of twistors in terms of asymptotic continuation accommodate curved spacetime. One feature of this approach to quantum theory in twistor space leads to a quantum gravity theory [14].

This spinor representation of a twistor makes it possible to interpret a twistor as a 2-plane in complex Minkowski space, M^4. Then we can relate ω^A and $\pi_{B'}$ so that $\xi^{AA'}$ is a solution as

$$\omega^A = i\xi^{AB'}\pi_{B'} \tag{11.22}$$

for the position vector $\xi^{AB'}$ in the complex Minkowski space. We can also consider the relationship of $Z^{AA'}$ and $\pi^{A'}$ to a complex position vector as

$$Z^{AA'} = \xi^{AA'} + \omega^A\pi^{A'} \tag{11.23}$$

where ω^A is a variable spinor. Just as in the conformal group on Minkowski space, spin space forms a two-valued representation of the Lorentz group. Note that SU_2 is the four value covering group of $\mathrm{C}(1,2)$, the conformal group of Minkowski space. The element of a 4D space can be carried over to the complex 8-space.

For spin, n the Dirac spinor space is a covering group of SO_n where this cohomology theory will allow us to admit spin structure and can be related to the SU_2 Lie group. Now let us consider the spin conditions associated with the Dirac equation and further formulate the manner in which the Dirac "string trick" relates to the electron path having chirality. For a spin, $s = \frac{1}{2}$ particle, the spin vector $u(p)$ is written as $\begin{pmatrix} 1 \\ 0 \end{pmatrix}$ for spin up and $\begin{pmatrix} 0 \\ 1 \end{pmatrix}$ for spin down for momentum, p. For a particle with mass we have for $c \neq 1$,

$$\left(-i\hbar c\alpha_\mu \frac{\partial}{\partial x_\mu} + \beta mc^2 \right)\psi = 0 \tag{11.24}$$

for the time independent equation, and we can divide Eq. (11.24) by $i\hbar c$ and have,

$$\left(\gamma_\mu \frac{\partial}{\partial x_\mu} + \frac{mc}{\hbar} \right)\psi = 0 \tag{11.25}$$

where $k_\circ = mc/\hbar$ and $\gamma_\mu = i\hbar c\alpha_\mu$ where indices μ run 0 to 3. The time dependent Dirac equation is given as,

$$\left(-i\hbar c\alpha_\mu \frac{\partial}{\partial x_\mu} + \beta mc^2 \right)\psi + \frac{i}{\hbar}\frac{\partial\psi}{\partial t} = 0. \tag{11.26}$$

The solution to the Dirac equation is in terms of spin $u(p)$ as

$$\psi = u(p)\exp\left[\frac{i}{\hbar}(p \cdot \underline{x} - Et) \right] \tag{11.27}$$

the Dirac spin matrices $\gamma_\mu = i\hbar c\alpha_\mu$. The spinor calculus is related to twistor algebra, which relates a 2-space to an associated complex 8-space [25].

An example of the usefulness of spinors is in the Dirac equation. For example, we have the Dirac spin matrices, $\gamma_\mu = \begin{pmatrix} 0 & \sigma_\mu \\ \sigma_\mu & 0 \end{pmatrix} = -i\,\beta\alpha_\kappa$ where terms such as $\gamma_\mu(1-\gamma_5)$ come into the electroweak vector-axial vector formalism. The three Dirac spinors (also called Pauli spin matrices) are given as

$$\sigma_x = \begin{vmatrix} 0 & 1 \\ 1 & 0 \end{vmatrix}, \sigma_y = \begin{vmatrix} 0 & -i \\ i & 0 \end{vmatrix} \quad \text{and } \sigma_z = \begin{vmatrix} 1 & 0 \\ 0 & -1 \end{vmatrix} \tag{11.28}$$

where indices 1,2,3 stand for *x, y, z* and $\gamma_5 \equiv i\gamma_0\gamma_1\gamma_2\gamma_3 = i\gamma^0\gamma^1\gamma^2\gamma^3$ for $\gamma_0 = \beta$ is given as,

$$\gamma_0 = \begin{pmatrix} 1 & 0 & 0 & 0 \\ 0 & 1 & 0 & 0 \\ 0 & 0 & -1 & 0 \\ 0 & 0 & 0 & -1 \end{pmatrix} \tag{11.29}$$

for trace, $\text{tr}\beta = 0$, that is, Eq. (11.29) can be written as,

$$\gamma_0 = \beta = \begin{pmatrix} I_2 & 0 \\ 0 & -I_2 \end{pmatrix} \tag{11.30}$$

where we have the 2×2 spin matrix as $I_2 = \begin{vmatrix} 1 & 0 \\ 0 & 1 \end{vmatrix}$ where $\text{tr}I_2 = 2$. Note that the Dirac spinors are the standard generators of the Lie algebra of SU_2.

The commutation relations of the Dirac spin matrices is given as

$$\{\gamma^\mu, \gamma^\nu\}_+ = \gamma^\mu\gamma^\nu + \gamma^\mu\gamma^\nu = ig^{\mu\nu} \underset{\sim}{I} \tag{11.31}$$

and $\det|\gamma_{\mu\nu}| = \det|g_{\mu\nu}|$ where $g_{\mu\nu}$ is the metric tensor. The Dirac spin matrices come into use in the electroweak vector-axial vector model as $\gamma_\mu(1-\gamma_5)$ for γ_5 as,

$$\gamma_5 = i\,\gamma_0\gamma_1\gamma_2\gamma_3 = i\,\gamma^0\gamma^1\gamma^2\gamma^3 \tag{11.32}$$

where indices run 0 to 3.

We can also write,

$$\gamma_{\mu\nu}(x^5, x^\mu) = \sum_{n=-\infty}^{\infty} \gamma_{\mu\nu}^{(n)}(x^\nu)\; e^{inx^5} \tag{11.33}$$

which expresses some of the properties of a 5D space having $\gamma_0, \gamma_1, \gamma_2, \gamma_3$

and γ_5. Note that γ_5 is associated with a 5D metric tensor. This 5D space passes exactly one geodesic curve which returns to the same point with a continuous direction. Note that this is a similar formalism to that of the Dirac string trick 720° path which appears to demand a hyperdimensional $n > 4$ space in analogy to the Mobius strip from dimension 2D $\rightarrow$ 3D and the Klein bottle from 3D $\rightarrow$ 4D.

A connection can also be made to the electromagnetic potential; and the metric of the Kaluza-Klein geometry. We can express $\gamma_{\mu 5}$ in terms of a potential ϕ_μ so that

$$\gamma_{\mu 5} = \sqrt{2\kappa}\phi_\mu \tag{11.34}$$

where $\kappa \equiv {8\pi}/{F}$ and where $F = {c^4}/{G}$ or the quantized cosmological force [8-10] (also see Eq. (11.34)). Then we have a 5-space vector as,

$$\gamma_{\nu 5} = \begin{pmatrix} 0 \\ 0 \\ 0 \\ 0 \\ 1 \end{pmatrix}. \tag{11.35}$$

Through this approach, we can relate covariance and gauge invariance [14]. Using Poisson's equation,

$$\nabla \varphi_\mu = \frac{1}{2}\kappa c^4 \mu_0 \tag{11.36}$$

where again $\kappa \equiv {8\pi}/{F}$ as above. The electromagnetic field, $F_{\mu\nu}$, can be expressed as,

$$F_{\mu\nu} = \frac{\partial \varphi_\mu}{\partial x^\nu} - \frac{\partial \varphi_\nu}{\partial x^\mu} \tag{11.37}$$

which yields an interesting relation of the gravitational metric to the electromagnetic field. Also the Lagrangian is given as $L = \dfrac{1}{2}F^{\mu\nu}F_{\mu\nu}$ so that

$\mathsf{L} = L\sqrt{-g}$ for the metric g. Note $L = \int \sqrt{g}\, d\tau$, where $d\tau$ represents a 4-space. Now we return to our discussion of twistor algebra and relate it to the spinor calculus. The Penrose twistor space also yields a 5D formalism as is also formulated by the Kaluza-Klein theory.

Both projective and non-projective twistors are considered as images in a complex Riemannian manifold in its strong conformal field condition. Duality, analytic continuation, unitary and other symmetry principles can be incorporated by using appropriate (Bose-Einstein or Fermi-Dirac) spin statistics in analogy to the Hartree-Fock spaces or Fock space (Chap. 3). Particles can be considered as states as the Fock space elements or the "end" of each disconnected portion of the boundary of the manifold. The quanta are associated with a quantum field of particles that carry both momentum and energy. The total energy Hamiltonian can be defined in terms of a number of simple phonon states which can be expressed in terms of a_n^+ creation and a_n destruction operator states. Since all creation operators commute, these states are completely symmetric and satisfy Bose-Einstein statistics. Such phonon states, having a definite number of phonons, are called Fock states, which is the vector sum of the momentum of each of the photons in the state. The ground state $|0\rangle$ can be considered the photon vacuum state or Fock state where the photon is taken as a phonon state. The creation and destruction operators commute as $\{a_n, a_n^+\} = \delta_{nn'}$ for the delta function, $\delta_{nn'}$ [26].

In this picture, we can consider an n-function as a "twistor wave" function for a state of n-particles. Penrose [11] considers a set of n-massless particles as a first order approximation. We form a series on a complex manifold as elements of the space C_n as

$$f_0,\ f_1(z^\alpha),\ f_2(z^\alpha, y^\alpha),\ f_3(z^\alpha, y^\alpha, x^\alpha), \ldots \qquad (11.38)$$

which are, respectively, the 0th function, 1st function, 2nd function, and 3rd function, etc. of the twistor space, which are also elements of C_n. We can also consider $f_0,\ f_1,\ f_2,\ f_3, \ldots.$ as the functions of several nested twistors in which f_0 is the central term of the wave of the twistor space. The f_n could represent nested tori that can act as a recursive sequence.

Penrose [11,12] suggests that, to a first approximation, f_1 corresponds to the amplitude of a massless, spin 1 particle, f_2 to a lepton spin ½ particle, and f_3 to hadron particle states, and f_4 to higher energy and exotic hadron particle states. Mass results from the breaking of conformal invariances for

f_n for $n = 2$ or greater; similar to the $S-$matrix approach [27]. The harmonic functions f_n form a harmonic sequence, where f_n for $n = 2$ form the Fermion states, and f_n for $n = 3$ form the Hadron twistor states. Essentially, in the twistor space, we have a center state f_0 around which f_1, f_2, ... occur. Each of these sequences of waves forms a torus, hence, f_1 and f_2 form a double nested tori set consistent with both spin 1 and spin ½ particle states where all n states are elements of the twistor, z, as $n \in z$.

In the specific case of a massless particle with spin for f_1, the 2-surface in complex Minkowski space corresponding to the twistor represents the center of mass of the system so that the surface does not intersect the real Minkowski space. This reflects the system's intrinsic spin. We see an analogy to the 3-torus Calabi-Yau M-Theory [28]. Calabi-Yau manifolds (a form of Kahler manifold) preserve the correct supersymmetry for the theory to reproduce the features of the standard model. This form of M-Theory, which features a 3-cycle toroidal symmetry is one of the better M-Theories with 10^{1000} or ($10^{\text{googolplex}}$ as sometimes called)[1] possible candidates for the string vacuum. In fact utilizing the continuous-state hypothesis we have been able to derive a unique candidate for the string vacuum [29]. The higher order f_n may describe higher order string modes or oscillations of $Z^\alpha \overline{Z}_\alpha = 0$ or f_0. This occurs for the case using f_1, f_2, and f_3 and, hence, all known particle states.

We can consider the topology of three Penrose projective twistor states which are PT, PT^+, and PT^-. The PT^+, and PT^- are meant to represent the domain of PT where we denote these two states in which $(-1,1)$ are elements of t where ε is small. We denote two line elements which are denoted in terms of twistors as a surface on the sphere S^3 as $PT^\pm$ which corresponds to $\underset{-t}{Z^\alpha} \overline{\underset{-t}{Z_\alpha}} = 0$ and $\underset{t}{Z^\alpha} \overline{\underset{t}{Z_\alpha}} = 0$ for $t = 1 - \varepsilon$ for PT^+, and PT^- gives $t = 1 - \varepsilon = \varepsilon - 1$. These two branches correspond to a transformation matrix,

[1] Googolplex: a googolplex cannot be written out since a googol of '0's will not fit into the observable universe.

$$\begin{pmatrix} 1 & 0 & t & 0 \\ 0 & 1 & 0 & t \\ t & 0 & 1 & 0 \\ 0 & t & 0 & 1 \end{pmatrix}. \tag{11.39}$$

This gives us a translation formulation for vectors into the states of spinors in terms of t, in terms of the spinors

$$\begin{pmatrix} \omega^{0}_{t\,1} \\ \omega^{1}_{t} \\ \pi'_{t\,0} \\ \pi'_{t\,1} \end{pmatrix} = \begin{pmatrix} 1 & 0 & t & 0 \\ 0 & 1 & 0 & t \\ t & 0 & 1 & 0 \\ 0 & t & 0 & 1 \end{pmatrix} \begin{pmatrix} \omega_1^{0} \\ \omega^{1} \\ \pi'_{0} \\ \pi'_{1} \end{pmatrix} \tag{11.40}$$

which is Z_t^{α} and $t \sim \pm 1$ since ε is small. Then in terms of twistors,

$$\hat{\omega}^{A} = \omega^{A} + \varepsilon \xi^{AB} \frac{\partial f}{\partial \omega^{B}} \tag{11.41}$$

for $\hat{\pi}_{A'} = \pi_{A'}$ where ω and π are orthogonal spinors. The term $\varepsilon \xi^{AB} \dfrac{\partial f}{\partial \omega^{B}}$ is small compared to ω^{A} and π_{A} since ε is small. The unit spinors or vectors are $\hat{\omega}^{A}$ and $\hat{\pi}_{A'}$ for both $A, B = 1,2$.

The projective twistor space, PT, corresponds to two copies of CP^{3}, which has an associated complex projective space. The PT space is the space which yields the torus topology of the Riemann surface of genus, $g = 1$. The genus-1 topology contains one "hole" or singularity, genus-2, two holes, etc. The two-hole system is a continuous manifold which can represent two connected tori or a double torus producing an equatorial planar membrane. This topology is related to the high-energy plasma dynamics found around black hole ergospheres and their equatorial accretion disks. It is, as well, observed in stars, and gas and dust circulation within galactic disks and halos. Observation of double tori topology at the cosmological level may, as well, be evidence of a structured polarized vacuum interacting with the high energy plasma dynamics at these scales. Haramein and Rauscher utilize torus topology to describe astrophysical objects such as supernovas and astrophysical plasmas [30].

11.4 Penrose Twistor Fields, Particles and Nested Tori

We explore some unique features of the torus or Kahler manifold topology (Recall that one of the popular forms of Calabi-Yau space is a form of dual 3-torus) We consider the relationship between the $T = U_1 \times U_1$ group and the S^2 group. An example of the n- dimensional manifold, which is not a product of n -1D manifolds, is given by the sphere, S^n. When one deals with two or more real or complex variables, there is usually a manifold, M, on which these functions are definable. The surface of a sphere of unit radius 3D Euclidean space, S^2, can be triangulated on the boundary of a tetrahedron. For the torus, T, its triangulation, K, consists of seven 0-simplexes and fourteen 2-simplexes. The contractible 1D sub-polyhedron of K contains all vertices of K. The two generators commute so that the torus group is generated by the two commuting generators $\simeq Z \oplus Z$ (see Section 11.5).

The manifold T^n is the nD torus. If $n = 2$, then $T^2 = S^1 \times S^1$ defines a torus. The torus is a subset of R^3, where R is the topology on the real numbers. The sets X and Y are called the topological space. If X is a set as a discrete topology, then Y can be a collection of all subsets of X, i.e., the set 2^x. Any finite or infinite subcollection $\{Z_\alpha\}$ of the X_α has the property that $\bigcup Z_{\alpha i} \in Y$, or the union of $Z_{\alpha i}$ are elements of Y. The torus is a subset of R^3, and $T^2 = S^1 \times S^1$ is the Cartesian product of two subsets of R^2 so that it is at least a subset of $R^2 \times R^2 = R^4$. The torus, which is in R^3, is not flat, but the torus $S^1 \times S^1$ in R^4 can be considered flat. The topology of the dual tori are the same, which has to do with the precise definition of flatness and curvature [31].

The definition of curvature depends on the specification of a Riemannian metric [32]. Once we specify the Riemannian metric, then we can define our flatness of T^2. This entails the specification of the metric $g_{\mu\nu}$ or $\eta_{\mu\nu}$ which allows us to specify the restrictions that the points in R^3 lie on the torus. Then, with respect to the metric, $\eta_{\mu\nu}$ we have a curved space torus. For $T^2 = S^1 \times S^1$, which defines two points (x, y) and (x', y') in T^2, the difference is expressed as $\left[(x - x')^2 + (y - y')^2\right]^{1/2}$ for the usual $g_{\mu\nu}$. See Chap. 2. For this metric, T^2 is flat and does not lie on R^3. The reason for this condition is that for a 2D compact connected surface to lie in R^3, it must have at least one non-zero curvature, where R_3 is the topology of real numbers [30,31].

In defining a vector space on a sphere, S^2, or torus T, we consider a simple observation of a 2D surface in R^3. For example, a disk $x^2 + y^2 \leq a^2$ for $z = 0$ has a top side and a bottom side, or a sphere S^2 has an inside and an outside, as does the torus T^2. These 2-sided surfaces are defined as orientable since we can use their two-sided properties to define directions or orientations of vectors projected from their surfaces in R^3. Hence, we have two normals at each point, an inward, or outward pointing normal vector, $\hat{n}$. We are guaranteed, in general, a diffeomorphic manifold for a torus in curved space, but not in general, for a spherical topology. Therefore, for any non-Euclidian space, diffeomorphism holds for the torus topology. Hence the Penrose topology is diffeomorphic.

11.5. Quaternions, Groups, and Allowable Spatial Structures

The complexified rotational dimensionality of quaternions may be the most appropriate approach to the description of twistor space in the context of a fundamental rotational force embedded in the structure of spacetime itself. We explore some of their interesting and related properties in this section.

11.5.1 *The Quaternion Formalism and Simple Topological Spaces*

The quaternion group is isomorphic to the group with elements $1, -1, -i, j, k, -k$, and $i^2 = j^2 = k^2 = -1$ and $ij = k,\ jk = i,\ ki = j$. These properties operate similar to complex numbers where $i = \sqrt{-1}$ and $i = -1$. In the case of the quaternions, i, j, k can represent orthogonal dimensions in three-space. The isomorphism condition states that the group elements of two groups can have a one-to-one correspondence, which is preserved under combinations of elements. Then one can construct a group table as a square array; this is only necessary for higher order groups. Quaternion groups have SU_2 or SU_3 subgroups and can be related to $O_3 +$.

Symmetric groups such as the quaternion group, which is a two-dimensional unimodular unitary group, are simply reducible groups. Following Hamilton, we identify Euclidean 4-space with the space of quaternions so that $H = \{\rho + xi + yi + zk\}$ where $\rho, x_o, y, z \in R^4$ are elements of the Riemannian space R^4. The Euclidean 3-space is the subset of imaginary quaternion, $H_{im} = \{ xi + yi + zk \}$ where $x, y, z \in R^3$ (see Section 11.3).

11.5.2 *Quaternions and Quantum Theory*

The key is that the Dirac string trick represents the properties of the symmetric group which is SU_2. The SU_2 is isomorphic with the unit length of the quaternion in 5D space. Quaternions, constructed by Hamilton, can represent rotations in 3-space, which can be performed without matrices. They also obey non-Abelian algebra. Furthermore, correspondence of quaternions can be made to vectors and tensors. Quaternions are a viable algebra for understanding rotations in 3D and 4D space. Due to symmetry considerations in the Dirac electron theory, a 720° twist is required for the electron to return to the exact same quaternion state, where a 360° rotation will not and must be doubled.

Quaternions are a complex number system with properties similar to the Rauscher [4] and Newman [5] complex 8-space. In the usual notation, we start from any complex number, $a + ib$ where a and b are real, where $a \times 1 = a$ and ib is imaginary. The quaternion is written as $t + ia + jb + kc$ where $t, a, b,$ and c are real and they are multiples of a real unit 1 and imaginary units $i, j,$ and k. The following conditions,

$$jk = -kj = i \tag{11.42a}$$

$$ki = -ik = j \tag{11.42b}$$

$$ij = -ji = k \tag{11.42c}$$

and

$$i^2 = j^2 = k^2 = -1 \tag{11.42d}$$

and

$$ijk = -1 \tag{11.42e}$$

also

$$i^2 = j^2 = k^2 = ijk = -1 \tag{11.42f}$$

which yields a set of recursive relationships.

Quaternions also have multiplicative properties similar to the complex Minkowski 8-space. Let $w = t + ia + jb + kc$, then the conjugate of w is $\overline{w}$ and is given as $\overline{w} = t - ia - jb - kc$, and the modulus is given as, $w\overline{w}$ or,

$$w\overline{w} = t^2 + a^2 + b^2 + c^2 . \tag{11.43}$$

In fact, quaternions contain all the properties of complex numbers except for commutivity and thus comprise a non-Abelian algebra such as in the quantum theory. Note that we have used a slightly different notation from Hamilton; that is, we write $ia,\ jb$, etc., instead of $ai,\ bj$, etc. Quaternions are comprehensively explored by Kauffman [31] and Rowlands [32].

If $t = 0$, then we have a pure imaginary quaternion or

$$u = ia + jb + kc \tag{11.44a}$$

and then

$$u^2 = -\left(a^2 + b^2 + c^2\right) \tag{11.44b}$$

and are of a unit length

$$a^2 + b^2 + c^2 = 1 \tag{11.45}$$

so that $u^2 = -1$. Also for two pure imaginary quaternions

$$uv = -u \cdot v + u \times v \tag{11.46}$$

as the dot and cross product of vector-like quantities in 3-space. The addition of the scalar component connotes a coordinate in the fourth dimension and hence we see the analogy of quaternions to the 4D Minkowski space, where t is time, and a corresponds to x, b to y, and c to z. What is unique then about the quaternionic "space" is that we have, for example, the permutation relations from Eqs. (11.42a) to (11.42f), and thus quaternions form a non-Euclidean set with the properties for pure quaternions uv in Eq. (11.46). We can form a set of pure quaternions on a 2D sphere of -1 in each of the three quaternion directions $i,\ j,\ k$. Note that the complex Minkowski space is formed by one imaginary component i, multiplied by x, y, and z. Now consider A and B real numbers and u is a unit length of a pure quaternion, then $u^2 = -1$ and the powers of $A + Bu$ occupy the same form as powers of complex numbers. That is, u is indistinguishable from any other $\sqrt{-1} = i$.

Let us now relate the quaternions to a complex number $Z = A + uB$ which we can write as $Z = \cos\theta + R\sin\theta$ or, in general,

$$Z^n = R^n \cos(n\theta) + R^n \sin(n\theta)u\,. \tag{11.47}$$

We can proceed with mapping of the nth roots of the quaternions. Consider a space of $n+1$ dimensions in which we represent $n+1$ space in the form of $A + Bu$, where A is a scalar and B is a real number. Now u is a limit vector in an N – space represented as R^N which is a Euclidian N – space. The vector-like quantity u belongs to the unit sphere, S^{N-1} about the origin, R^N and is taken to have squares equal to minus one, or $u^2 = -1$ for all vectors S^{N-1}. In general, uv is not defined in a HD geometry such as the 8D Minkowski space of Rauscher [4] and Newman [5]. We can, however, create power maps of the form $Z^n + K$ where K is a vector in R^{N+1} and $Z = A + Bu$ for $u^2 = -1$ for all u in S^{N-1}. With this approach, we can form classes of hypercomplex iterative processes with incursion in any arbitrary dimensional space [33].

One of the basic principles of the quaternion twist holds for the Dirac string trick for 720° degree rotation. A half cycle of twist, or 360 degrees, is expressed in terms of quaternions as $ijk = -1$. To return to +1, another twist through 360° must occur. Spin must involve a preferred geometry in space [33]. The geometry of a preferred direction can be constructed by the magnitude of total electron transfer. The Penrose spin approach is utilized to calculate angular momentum and $SL(2)$.

In terms of complex analysis involving quaternions, a single 180 degree turn is an instance of $i = \sqrt{-1}$ where $i^2 = -1$ and represents a 360 degree right- or left-handed turn. The case for $i^3 = -i$ is a non-trivial rotation and $i^4 = 1$ returns the rotation of the electron and observer to their original states, through the 720 degree rotation – hence, the interpretation of the quaternionic formalism of one square root of -1 for every direction in three-dimensional space. The electron moves on the bounded space to have contiguous surfaces at the equatorial plane. In order for the electron to pass through a 720 degree rotation and return the spin and chirality to its original state, the electron path must be different than that of a sphere.

In quantum theory, the symmetry group is the SU_2 group rather than the 3D space rotation group such as O_3^+. The SU_2 group is isomorphic with the quaternions of unit length in 4D space. In [33], the group theoretic approach that relates spinors, twistors, and quaternions is detailed. A spinor is a vector in two complex variables. Antisymmetric conditions lead to the second twist involving the quaternions to create the cycle of the electron to its original state. The antisymmetric conditions utilizing spin calculations can be conducted with Clebsch-Gordan coefficients, $3j$ and $6j$ symbols and other components of angular momentum [34,35]. Through these means, one can calculate the correct spin interactions involving multi-particle quaternion states. Suffice it to say that the iterative properties, formulated here, have a variety of applications such as scalable inclusive relations from the quantum domain to astrophysical and cosmological systems [10].

11.6 Conclusion

We have demonstrated a unique relationship of the spinor calculus, twistor algebra, the quaternionic formalism and the complex 8-Space. This topology appears to be ubiquitous in Nature. The twistor formalism appears to also occupy a role in unification models through the E_8 group utilized in supersymmetry models.

References

[1] Kaluza, T. (1921) *Sitz. Berlin Preuss, Adad. Wiss* 966.
[2] Klein, O. (1926) *Z. Phys*. 37, 895.
[3] Rauscher, E.A. (1979) The physics of nonlocality in complex eight-space, in B.J. Josephson (ed.) The Iceland Papers, Frontiers of Physics, p 93, Amherst: Essentia Research Associates; and (1996) Ottawa: P.A.C.E.
[4] Ramon, C. & Rauscher, E.A. (1980) Superluminal transformations in complex Minkowski spaces, Foundations of Physics 10, 661.
[5] Newman, E.T. (1973) J. Math. Phys. 14, 774.
[6] Hansen, R.O. & Newman, E.T. (1975) A complex Minkowski space approach to twistors, Gen. Rel. and Gravitation 6, 261-385.
[7] Leibowitz, E. (1974) Spinor calculus in five dimensional relativity, J. Math. Phys. 15, 306 and references therein
[8] Rauscher, E.A. (1968) A Unifying Theory of Fundamental Processes, Lawrence Berkeley National Laboratory Report (UCRL-20808 June 1971) and Bull. Am. Phys. Soc. 13, 1643
[9] Rauscher, E.A. (1972) Closed cosmological solutions to Einstein's field equations, Lett. Nuovo Cimento 3, 661.
[10] Rauscher, E.A. (1968) A unified theory of fundamental processes, University of California, Los Angeles, Bull. Am. Phys. Soc., 13, 1643.
[11] Penrose, R. (1986) Pretzel twistor spaces, Twistor Newsletter, 21, 7-11; (2004) Road to Reality, London: Vintage Books.
[12] Penrose R. & Mac Collem, M.A.H. (1973) Phys. Rev., 6C, 241.
[13] Rauscher, E.A. (1968) Electron interactions and quantum plasma physics, J. Plasma Phys. 2, 517.
[14] Rauscher, E.A. (1983) Electromagnetic Phenomena in Complex Geometries and Nonlinear Phenomena, Non-Hertzian Waves and Magnetic Monopoles, Tesla Book Co., Millbroe, CA; and (2008) Oakland: The Noetic Press.
[15] Rauscher, E.A. & Amoroso, R. L. (2005) The Schrödinger equation in complex Minkowski, nonlocal and anticipatory systems, in R. L. Amoroso, I. Dienes & C. Varges (eds.) Proceedings of the 1st Unified Theories, Symposium, Budapest, Hungary, Oakland: The Noetic Press.

[16] Amoroso, R.L. Vigier, J-P, Kafotos, M. & Hunter, G. (2002) Comparison of near and far field double-slit interferometry for the dispersion of the photon wave packet, in Gravitation and Cosmology from the Hubble Radius to the Plank Scale, R. L. Amoroso, G. Hunter, M. Kafotos, & J-P Vigier (eds.) Dordrecht: Kluwer Academic.
[17] Bell, J.S. (1964) Physics (N. Y.) 1, 195.
[18] Clauser, J.F. & Horne, W.A. (1974) Phys. Rev. D 10, 526.
[19] Isham, J., Penrose, R. & Sciomci, D.W. (eds.) (1975) Quantum Gravity, Oxford: Clarendon Press.
[20] Baston, R. & Eastwood, M.B. (1985) The Penrose transformation for complex homogenous spaces, Twistor Newsletter 20, 34.
[21] Le Brun, C. (1985) Quaternion manifolds and the future tube, Twistor Newsletter 20, 59.
[22] Eastwood, M.B. (1985) Complex quaternionic Kahler manifolds, Twistor Newsletter 20, 63.
[23] Burstall, F.E. (2004) Construction of fuzzy S^4, Phys. Rev., D70, 126004.
[24] Nash, C. & Sen, S. (2000) Topology and Geometry for Physicists, New York: Academic Press.
[25] Hughston, L.P. & Ward, R.S. (1979) Advances in Twistor Theory, Pitman.
[26] Rauscher, E.A. (1973) Early universe cosmological models, Bull. Am. Phys. Soc. 18, 1570.
[27] Chew, G. (1964) The Analytic S-matrix, New York: Benjamin & Co.
[28] Zwieback, B. (2004) String Theory, Cambridge: Cambridge University Press.
[29] Amoroso, R.L. & Rauscher, E.A. (2009) The Holographic Anthropic Multiverse: Formalizing the Complex Geometry of Reality, Singapore: World Scientific.
[30] Haramein, N. & Rauscher, E.A. (2005) Collective coherent oscillation plasma modes in surrounding media of black holes and vacuum structure – Quantum processes with considerations of spacetime torque and Coreolis forces, in R.L. Amoroso, B. Lehnert, and J-P Vigier (eds.) Beyond the Standard Model, Oakland: The Noetic Press.
[31] Kauffman, L.H. (1991) Knots and Physics, Singapore: World Scientific,.
[32] Rowlands, P. (2007) Zero to Infinity, Singapore: World Scientific.
[33] Ward, R.S. & Wells, R.O. (1991) Twistor Geometry and Field Theory, Cambridge: Cambridge Univ. Press.
[34] Burnside, W. (1955) Theory of Groups of Finite Order, New York: Dover; and Cambridge: Cambridge University Press.
[35] Rauscher, E.A., Rasmussen, J.D. & Harada, K. (1967) Coupled channel alpha decay rate theory applied to Po^{212m}, Nuclear Physics, A94, 33.

Chapter 12

Relativistic Dirac Quantum Theory in Complex Minkowski Space and Tachyonic Signaling

In this chapter we develop complex solutions to the Dirac equation and discuss various implications and applications.

12.1 Introduction

The Dirac electron theory is unique in that it is relativistic invariant and that it predicts two states of matter having opposite charges [1-4]. In addition is the concept of a full vacuum; one in which positive energy states, such as those of a gamma ray, can activate electron-positron pair production in which an electron is kicked out of the Fermi-Dirac sea to a positive energy state, leaving an electron hole position in the negative energy sea. The two sign solution lead to the postulate of antimatter which has been well identified. In the 1920's when Dirac developed his mathematical description of the relativistic electron, the obtainment of an antielectron or positron solution, in addition to the electron solution, did not lead immediately to hypothesis of antimatter and appeared to be an anomalous solution. In 1932 Carl Anderson discovered the positron in cloud chamber photographs leading to a good example of prediction and confirmation. With the advent of the prediction of the antiproton and its identification at the Berkeley Bevatron bubble chamber by Emilo Segre and Owen Chamberlain in 1958, the pairing of matter and antimatter lead the conundrum of why we observe more matter then antimatter in the Universe. Matter and antimatter when they collide produce massive amounts of energy, $E = mc^2$ producing high energy gamma rays through the annihilation process.

Further development of the theory led to the concept of a full vacuum termed the Fermi-Dirac sea. A gamma ray can impact a heavy nuclei producing an electron-positron pair. In the Fermi-Dirac sea model of the vacuum there are the normal positive energy states $E > 0$ and zero energy states $E = 0$ as the surface of the Sea and negative vertical energy electron states, $E < 0$. The energetic photon kicks out an electron into the positive energy states, leaving a hole in the Fermi-Dirac sea. This hole is the positron.

The Fermi-Dirac sea model has numerous applications from Feynman diagram techniques to modeling semiconductor substrates [5-7]. The presence of the full vacuum picture has been useful in describing many states of matter including particularly more exotic state of matter such as plasmas. In plasmas, the energy of the ionized plasma gas, activates the electron-positron pair production by polarizing or biasing the vacuum. Using Feynman graphical techniques, one can definitely demonstrate the actual effects of the Fermi-Dirac energy sea on such plasma dielectric constant, conductivity and other properties in medium to high temperature plasmas. The fit of these plasma parameters are to the formalism including the full vacuum picture, than just the classical or semi-classical approach [7, 8]. In this chapter, we solve the Dirac equation in the Complex Minkowski 8-Space and examine conditions in which the imaginary components of the complex 8 space contribute to small nonlinear terms in the Dirac equation. We also examine the spinor calculus and the Dirac string trick in their interpretations in M_4 and C_4 space. Historical interpretation of some of the major theories in the foundation of physics are examined.

12.2 The Basic Structure of Physics Theories and Their Interrelation

In the attempt to develop a unified theory the thorny issue of quantum mechanics and relativity arises as to the manner in which to find a quantum gravity formalism. The reconciliation of two distinctly structured theories, having different domains of applicability has been a conundrum to physicists for over seventy years. The basic structure of gravity, described by general relativity is a nonlinear tensor force and the basic formalism of the quantum theory is that of linear superposition. We examine this latter issue in Chap. 10 on the consideration of additional terms that introduce small nonlinear terms in the Schrödinger equation, which are formulated in terms of the complex 8-space.

Essentially the reconciliation of the formalism of gravity and quantum mechanics is essential to develop a unification of the forces and processes of nature as a "theory of everything" (TOE), see Chap. 13. Historically the development of these two uniquely different theories has their roots in the classical Hamilton-Jacobi theory. A major link between quantum and relativistic theories is the Dirac equation [9]. In Fig. 12.1 is represented the development of physics from the past and the top of the Fig. to current time at the bottom of the Fig.. The concept of canonically conjugate or paired variables obeying an Abelian algebra was developed in the Hamilton-Jacobi classical mechanics as the (p,q) phase space variables, where p is momentum and q is a spatial dimension, x [9]. This structure is fundamental to the non-Abelian algebras of the quantum theory, exemplified by the Heisenberg uncertainty principle, $\Delta \times \Delta p \geq \hbar$. Bohr's complementary principle is fundamental to the dual paired variables (p, x) of the quantum theory. The paired variables (E, t) for energy and the temporal dimension can also be considered for $\Delta E \Delta t \geq \hbar$. The relationship between the classical Hamilton-Jacobi theory and the quantum picture is Bohr's correspondence principle. The structure of general relativity and quantum mechanics is fundamentally very distinct. The standard quantum picture involves linear superposition where as general relativity formulates non-inertial frames or gravity which is intrinsically nonlinear. Galileo's law of fallowing bodies exemplifies the nonlinearity of gravity in a very cogent manner, that is the distance of fall, s to the time, t^2 is given as $s = ½(gt^2)$ where g is the acceleration of Earth's gravity.

The Hamiltonian equations are based on energy conservation $H = T + V$ where T is the kinetic energy and V is the potential energy. For the Lagrangian, $L = T - V$. Then $H = \frac{p^2}{2m} + \lambda q^2$ so that the equation of motion is written as $\ddot{q} + \lambda q = 0$ where p is the momentum and q is the spatial variable for the canonically conjugate variables of phase space (p,q). then the Hamiltonian expressions apply,

$$\dot{p} = \frac{\partial p}{\partial t} = -\frac{\partial H}{\partial q} \tag{12.1a}$$

$$\dot{q} = \frac{\partial q}{\partial t} = -\frac{\partial H}{\partial p} \tag{12.1b}$$

See [9-11].

The concept of energy conservation lies at the center of most major physics formulations. The conservation of total energy, E_T is expressed as the sum of the kinetic and potential energies respectively as $E_T = E + V$ in classical mechanics. The conservations principle, as the first law of thermodynamics, has applicability in many diverse fields of knowledge such as information theory. The Schrödinger equation is a basic expression of the quantum theory and is expressed in terms of the total energy Hamiltonian as the sum of the kinetic energy, E and potential energy, V written as $H\psi + V\psi = \psi$ where ψ is the eigenfunction and E is the eigenenergy.

Energy or energy-momentum conservation is essential to both the quantum theory and general relativity. We examine, in detail the derivation of the Schrödinger equation from its classical origins [9]. We represent the origin on developments of the non-relativistic quantum theory on the right of Fig. 12.1.

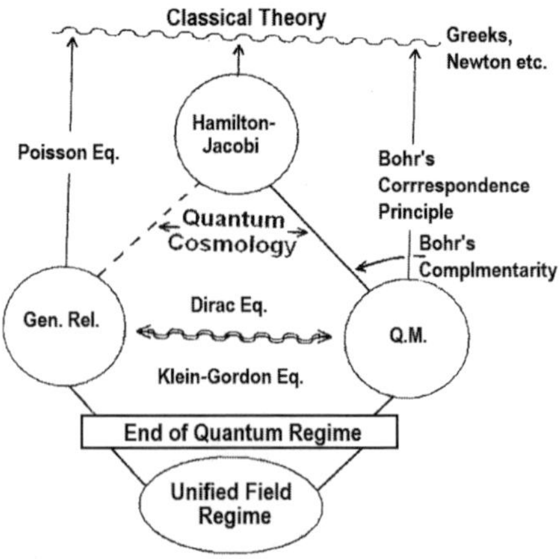

Fig. 12.1. A schematic representation of the history and structure of the fundamental equations of physics. Earlier time to present is represented from top to bottom of the Fig.. Gravity may not be quantized if the quantum regime ends in a way similar to the boundary between classical mechanics and quantum mechanics.

Essentially the development of a quantum gravity theory, which forms a synergy of these two pillars of physics is fundamental to developing a unified theory; see Chap. 13. Supersymmetry theories, GUT and TOE theories incorporating superstring theories are approaches and attempts to unify the four force fields of the strong force, electromagnetic force, weak force (electroweak force) and gravity. We have considered the efforts of adding small nonlinear terms arising from the complexified 8D Minkowski space into the quantum picture. In Chap. 10, we examined the consequences of this approach for the structure and solutions to the Schrödinger equation. In this chapter, we examine a similar approach to complexification the Dirac equation.

In the left vertical history in Fig. 12.1 we represent the evolution of the structure of the general relativistic field equations from the Poisson equation. In the structure of these basic theories, conditions are required such as covariance which is basic to relativity or a relativistic quantum theory. The condition of covariance means that the equations that describe the system are constant so far as the quantities on both sides of the equation transform in the same manner covariantly. For example, the expressions of both sides of the equations must be scalars, vectors or tensors. We consider the origin of Einstein's field equations from the classical mechanics, represented on the left of Fig. 12.1.

As suggested in Fig. 12.1, if there is a limit to the quantum regime in the same way quantum theory makes correspondence to the limit of classical mechanics; there may be no quantum gravity in the manner currently sought because the quantum regime may end before the regime of the unified field. Indeed, Feynman said:

> ...maybe we should not try to quantize gravity. Is it possible that gravity is not quantized and all the rest of the world is?...Now the postulate defining quantum mechanical behavior is that there is an amplitude for different processes. It cannot be that a particle which is described by an amplitude, such as an electron, has an interaction which is not described by an amplitude but by a probability...it seems that it should be impossible to destroy the quantum nature of fields. In spite of these arguments, we should like to keep an open mind. It is still possible that quantum theory does not absolutely guarantee that gravity has to be quantized. From Feynman, 1962, Lectures on Gravitation.

Gauge Theory is an approximation, which could mean there is no spin 2 graviton detectible in Minkowski space, no Higgs' mechanism, no super-partners or sparticles and why no magnetic monopole has been detected. What is looked for instead are complex HD topological parameters (topological charges in Calabi-Yau cavities) where brane boundary conditions handle these properties in a new way as Feynman suggests.

Basic to the classical formalism of electromagnetism and relativistic physics is the Poisson equation of the form $\nabla^2\phi = 4\pi\rho$ where the divergence ∇^2 of the potential field ϕ is proportional to the energy (or energy mass) density, ρ in the space considered. The Laplace equation, $\nabla^2\phi = 0$ is written for a density free space.

From classical mechanics, we can describe the gravitational field by Poisson's equation

$$\nabla^2\phi = 4\pi\rho G \tag{12.2}$$

Where ϕ is the gravitational potential and ρ is the matter density for ϕ and ρ are scalars. We can generalize this equation in the linearized theory to

$$\Box\psi^{\mu\nu} = -\frac{16\pi G}{c}(\tau^{\mu\nu} + T^{\mu\nu}) \tag{12.3}$$

Where $\psi^{\mu\nu}$ describes the gravitational field, $\tau^{\mu\nu}$ corresponds to non-gravitational sources and the $T^{\mu\nu}$ term expresses the fact that the gravitational field can act as its own source. The $-16\pi G / c^4$ term assures that the classical limit obeys the Poisson equation. Also $F = c^4/G$ is the universal force [11]. In deriving Einstein's field equations, we first examine the non-relativistic limit of the linearized field equations. Assuming static conditions then $T^{00} = mc^2$ the only component of the energy-momentum tensor. Neglecting $t^{\mu\nu}$ for now, we have

$$\Box\psi^{\mu\nu} \rightarrow \nabla^2\psi^{00} = -\frac{16\pi G}{c^4}T^{00} = {}^2\!/_F\, T^{00} = -16\pi Gm/c^2 \qquad (12.4)$$

where T^{00} is a scalar and m is the mass having mass density ρ. To convert back into Poisson's equation, we must have $\psi^{00} = -4\phi/c^2$ where ϕ is the Newton's potential and ψ^{00} is a scalar. The 8π and 16π term correspond to the relativistic and non-relativistic terms, for $E = \frac{1}{2}mv^2$ or $E = mc^2$.

We can describe the gravitational field by Poisson's equation of classical mechanics $\nabla^2\phi = 4\pi\rho G$ where ϕ is the gravitational potential and ρ is the matter density. A more general form of this equation is expressed including the continuity equation (for energy, mass and charge conservation) as $\nabla^2\phi = 4\pi G(\rho + 3p/c^2)$ where *p* is momentum.

The generalized form for the above equation is given in Eq. (12.4) where $\psi^{\mu\nu}$ describes the gravitational field and $-2\pi/F$ insures the proper dimensionality. We now consider the solutions to Eq. (12.3) to demonstrate that the Poisson equation leads to Einstein's field equation solutions, we proceed as follows again utilizing Poisson's equation and the continuity equation. For

$$\nabla^2\psi = 4\pi G(\rho + 3p/c^2) \qquad (12.5)$$

$$\frac{d\underline{v}}{dt} = -\nabla\psi \qquad (12.6a)$$

then we have

$$\frac{\partial\rho}{\partial t} = -\nabla\underline{v}(\rho + p/c^{2)} \qquad (12.6b)$$

for $\underline{v} = \dot{S}\gamma_0$ where *S* is arbitrary term within a constant multiplication factor which depends on the time chosen so that $S(t_0)=1$. If we define $S(t) = \frac{R(t)}{c}|C|^{1/2}$ then $S(t_0) = \frac{|C|^{1/2}}{c}$ where *R*(*t*) is the curvature of space and *C* is a constant. Using the equations for $\nabla^2\psi, \frac{d\underline{v}}{dt}$ and $\frac{\partial\rho}{\partial t}$ or Eqs. (12.5,

12.6a and 12.6b) or acceleration, then we have $\dot{S}^2 = \frac{8\pi G}{3}\rho S^2 - C$ and the relationship for *S*(*t*) then $\dot{R}^2 = \frac{8\pi G}{3}\rho R^2 - kc^2$ where $k = 0, \pm 1$ which is one of the solutions to Einstein's field equations.

It is clear that it is essential to examine the structure of the basic equations of physics that describe the micro and macro domains. Their origins from the classical Hamilton-Jacobi theory and classical concepts in general give us clues as to the manner in which to reconcile these theories and develop an approach to a unified theory. The Dirac equation stands unique in that it is relativistically Lorentz invariant. See Table 12.1 for force field type range and possible velocity of propagation.

Table 12.1 lists some types of physical phenomena, relevant forces involved and their velocity domain, $v = 0$. $v = c$ and possibly $v = \infty$ or $v > c$ in complex 8-space as well as their theoretical speculative range. Six branches of physics are given with their forces and range. In The three domains of signal propagation as related to five branches of physics. These modes of signal propagation are manifest in other branches of physics also. We compare this to the signal propagation velocity associated with local and nonlocal phenomena.

Table 12.1

DOMAINS FOR $v = 0$, $v = c$ AND $v = \infty$ AND BRANCHES OF PHYSICS

Branch of Physics	Domain	Type of Force	Theoretical Range
• Hamilton-Jacobi Mechanics	$v = 0$	Mechanical	∞
• Electromagnetism	$v = c$	Electromagnetic	∞
• General Relativity Cosmology	$v = c$	Gravitational	∞
• Superconductivity Macro-Quantum	$v = \infty$	?	∞ ?
• Young's Double Slit Quantum Mechanics	$v = \infty$	Electromagnetic	Finite
• Bell's Nonlocality	$v = \infty$	Electromagnetic and Atomic	Infinite ?

12.3 The Basis and Structure of the Dirac Equation

The Dirac equation obeys the proper relativistic invariant conditions so it comprises a quantum theory that obeys relativistic constraints on the lightcone. A geometry defines a space which is an idealization of the physical 4D space of objects and momentum and locations. The lightcone with its hyperbolic topology is a covariant representation of spacetime regions. A Lie group is a topological group. For the relativistic form of the Dirac equation we use $E=(p_x^2c^2+m^2c^4)$ where E is the relativistic energy and p_x is the momentum in the x direction. We start from $E = mc^2$ and $p_x = mv_x$ so that m $= E/c^2$. We have $p_x/m = v_x = p_x/(E/c^2) = \frac{p_x c^2}{E}$. For the relativistic form of the energy

$$E=\frac{mc^2}{\sqrt{1-\beta^2}}=\frac{mc^2}{\sqrt{1-v_x^2/c^2}}. \tag{12.7}$$

Then eliminating v_x between $v_x = \frac{p_c c^2}{E}$ and Eq. (12.7) in the form of

$$\frac{mc^2}{E}=\sqrt{1-v_x{}^2/c^2} \tag{12.8}$$

and then taking the inverse relation,

$$\left(\frac{E}{mc^2}\right)^2=1-\frac{v_x{}^2}{c^2}, \tag{12.9}$$

so that v_x is given by

$$v_x=\sqrt{(-\frac{E^2}{m^2c^4}+1)c^2}=c\sqrt{1-\frac{E^2}{m^2c^2}}. \tag{12.10}$$

Then eliminating v_x from Eq. (12.10) and $v_x = p_x{}^2c^2/E$ so we have

$$p_x c=E\sqrt{\frac{m^2c^4-E^2}{m^2c^4}} \tag{12.11}$$

Then $p^2c^2=m^2c^4-E^2$ so that $E=\sqrt{-p_x{}^2c^2+m^2c^4}$ so we have the usual relativistic energy equation

$$E = \sqrt{p^2c^2 + m^2c^4} \tag{12.12}$$

For the three components of the momentum,

$$E = \pm c\sqrt{p_x{}^2 + p_y{}^2 + p_z{}^2 + m^2c^2} \tag{12.13}$$

To derive the Dirac equation based on the two operators $p_{xop} \rightarrow \frac{\hbar}{i}\frac{\partial}{\partial x}$ and $E_{op} \rightarrow -\frac{\hbar}{i}\frac{\partial}{\partial t}$ so that

$$-\frac{\hbar}{i}\frac{\partial}{\partial t}\psi = \pm c\sqrt{-\hbar^2 \frac{\partial^2}{\partial x^2} + \frac{\partial^2}{\partial y^2} + \frac{\partial^2}{\partial z^2} + n^2c^4\psi} \tag{12.14}$$

where ψ is the wave function solution. The Hamiltonian is written as

$$H = \left(m^2c^4 + p^2c^2\right)^{1/2} \tag{12.15}$$

having two solutions which are given in terms of the energy equation

$$E = \pm\sqrt{m^2c^4 + p^2c^2} \tag{12.16}$$

which is the basic energy equation for the relativistic Dirac equation. Also other Hamiltonian forms can be written for a charged particle in an electromagnetic field as,

$$H = [m^2c^2 + \left(cp - eA^2\right)^{1/2} + e\phi \tag{12.17}$$

where A is the vector potential and ϕ is the scalar potential. See chaps. 5,6. Because we are dealing with a first order equation in space and time dependence, we have a square root giving two solutions, one is for the usual electron and the second is for a positive electron or positron. Dirac stuck to his two charge solution prediction which was later verified and led to the whole concept and discovery of antimatter [13].

12.4 The Relativistic Dirac Equation

Proceeding from the Schrödinger equation, we express the Hamiltonian in spherical coordinates as

$$H = \frac{1}{2m}\left(p_r^2 + \frac{L^2}{r^2}\right) + V(r)\,, \tag{12.18}$$

where p_r is the radial momentum $(m\dot{r})$ and L the angular momentum vector.

As is well known, the three components of angular momentum, derived from each other by cyclic permutation, are $L_z = xp_y - yp_x$, $L_x = yp_z - zp_y$, $L_y = zp_x - xp_z$ and $\underline{L} = \underline{r} \times \underline{\rho}$ where the total angular momentum, $L^2 = L_x^2 + L_y^2 + L_z^2$ has commutation rules $\underline{L} \times \underline{L} = i\hbar \underline{L}$ [14-16]. The SO(3) rotation generators l_1, l_2 and l_3 satisfy $l_1 l_2 - l_2 l_1 = l_3$, $l_2 l_3 - l_3 l_2 = l_1$, $l_3 l_1 - l_1 l_3 = l_2$; related quantum mechanically to angular momentum components L_1, L_2, L_3 with $L_x = i\hbar l_1, L_y = i\hbar l_2$ and $L_z = i\hbar l_3$ about Cartesian axes giving commutation rules $L_x L_y - L_y L_x = i\hbar L_z$, $L_y L_z - L_z L_y = i\hbar L_x$ and $L_z L_x - L_x L_z = i\hbar L_y$. Angular momentum refers to intrinsic spin about a massive particles center of mass and its magnetic moment obeys SO(3) Lie algebra which is non-Abelian acting on two component spinor wave functions $\{\psi_0(x), \psi_1(x)\} \equiv \Psi_A$; but by the uncertainty relation, $\Delta x \Delta \rho \cong \hbar$ only one set of these operators may commute at a time. Non-relativistic Fermi spin $1/2\hbar$, or simply spin ½, particles with spin angular momentum operator $s = 1/2\hbar\sigma$ can be expressed as the three anticommuting Pauli 2 x 2 spin matrices Eq. (12.19) satisfying $\sigma_x \sigma_y = -\sigma_y \sigma_x = i\sigma_z$ as derived empirically from the Stern-Gerlach experiments [13-18]

$$L_x = \frac{\hbar}{2}\sigma_x = \begin{pmatrix} 0 & 1 \\ 1 & 0 \end{pmatrix},\ L_y = \frac{\hbar}{2}\sigma_y = \begin{pmatrix} 0 & -i \\ i & 0 \end{pmatrix},\ L_z = \frac{\hbar}{2}\sigma_z = \begin{pmatrix} 1 & 0 \\ 0 & -1 \end{pmatrix} \quad (12.19)$$

where the total spin operator is given as "total spin" operator, $J^2 = L_x^2 + L_y^2 + L_z^2$ commutes with all three components of L in 3D.

Spinor space and spin spaces, such as hypercharge are developed in independent topological spaces. Spinors and spin space can be complexified and occupy a hyperspace continuum. For example, the special unitary Lie groups, which are topological groups having infinitesimal elements of the Lie algebras, are utilized to represent the symmetry operations in particle physics and in infinitesimal Lorentz transformations. For example, the generators of the special unitary SU_2 group is composed of the three isospin operators, I as I_+, I_- and I_z having commutation relations $[I_+, I_-] = iI_z$. The generators of SU_3 are the three components of I, isospin, and hypercharge Y, and for other quantities which involve Y and electric charge Q. Thus, there are 8 independent generators for the traceless 3 x 3 matrices of SU_3. The O_3^+ group of rotations is homomorphic to the SU_3 group. Just as in the conformal group on Minkowski space, spin space forms a two-valued representation of the Lorentz group. Note that SU_2 is the four value covering group of C(1,2), the conformal group of Minkowski space. The element of a four dimensional space can be carried over to the complex 8-space.

For spin, n the Dirac spinor space is a covering group of SO_n where this cohomology theory will allow us to admit spin structure and can be related to the SU_2 Lie group. Now let us consider the spin conditions associated with the Dirac equation and further formulate the manner in which the Dirac "string trick" relates to the electron path having chirality [13,16,18]. See Chap. 11.

Relativistic spin $1/2\hbar$ particles are described by Dirac's formalism for the wave equation which has been expressed by a number of notations such as

$$E\psi + c(\alpha \cdot p)\psi + mc^2 \beta\psi = 0$$

or (12.20)

$$i\hbar\frac{\partial\psi}{\partial t} - i\hbar c\alpha \cdot \nabla\psi + mc^2\beta\psi = 0$$

for $c \neq 1$ and for the time dependent equation, which is first order in space and time with fermion particle mass, m

$$\left(-i\hbar c\alpha_\mu \frac{\partial}{\partial x_\mu} + \beta mc^2\right)\psi + \frac{i}{\hbar}\frac{\partial\psi}{\partial t} = 0. \tag{12.21}$$

We express the 4 x 4 Dirac α_μ and β matrices as, $\gamma_0 = \beta$, which are Hermitian and are expressed in terms of the 2 x 2 Pauli matrices, α for example

$$\gamma_0 = \beta = \begin{bmatrix} 1 & 0 & 0 & 0 \\ 0 & 1 & 0 & 0 \\ 0 & 0 & -1 & 0 \\ 0 & 0 & 0 & -1 \end{bmatrix} \quad \alpha_x = \begin{bmatrix} 0 & \sigma_x \\ \sigma_x & 0 \end{bmatrix} = \begin{bmatrix} 0 & 0 & 0 & 1 \\ 0 & 0 & 1 & 0 \\ 0 & 1 & 0 & 0 \\ 1 & 0 & 0 & 0 \end{bmatrix} \tag{12.22a}$$

$$\alpha_y = \begin{bmatrix} 0 & 0 & 0 & -i \\ 0 & 0 & i & 0 \\ 0 & -i & 0 & 0 \\ i & 0 & 0 & 0 \end{bmatrix} \quad \alpha_z = \begin{bmatrix} 0 & 0 & 1 & 0 \\ 0 & 0 & 0 & -1 \\ 1 & 0 & 0 & 0 \\ 0 & -1 & 0 & 0 \end{bmatrix} \tag{12.22b}$$

In the case where $m = 0$ or at very high energies, E where a particle of mass, m behaves like zero mass, only three anticommuting matrices instead of four are required. In this case the Pauli matrices are sufficient and the spinors require only 2 components which relate to the chiral representation [24]. The σ's satisfy the equation $\sigma_x\sigma_y = -\sigma_y\sigma_x = i\sigma_z$. In general, we can write, of 2 x 2 matrices,

$$\sigma_x\sigma_y = -\sigma_y\sigma_x = i\sigma_z \tag{12.23a}$$

$$\sigma_y\sigma_z = -\sigma_z\sigma_y = i\sigma_x \tag{12.23b}$$

$$\sigma_z\sigma_x = -\sigma_x\sigma_z = i\sigma_y \tag{12.23c}$$

$$\sigma_x^2 = \sigma_y^2 = \sigma_z^2 = 1 \tag{12.23d}$$

Where $\underline{s} \times \underline{s} = is$ and $\underline{\sigma} \times \underline{\sigma} = 2i\sigma$.

The Pauli spin matrices are unitary $\sigma_x = \sigma_x^{-1}$. See Eq. (12, 19) for the 2 x 2 Dirac matrices. The Klein-Gordon equation is a 4D form where the wave function depends on (x,y,z,t) and is written as $\Box^2\Psi + \frac{2m}{\hbar}\Psi = 0$ where $\Box^2$ is the D'Alembertian operator, $\Box^2 = \frac{\partial^2}{\partial x^2} + \frac{\partial^2}{\partial y^2} + \frac{\partial^2 z}{\partial t^2} - \frac{1}{c^2}\frac{\partial^2}{\partial^2 t}$ and m is the mass of the particle under consideration. Note that this equation is second order in space and time as is the classical wave equation whereas the Schrödinger equation is second order in space and first order in time in part the reason for the $i = \sqrt{-1}$ term in the equation. The first order in time term requires the *I* term in it.

We now write the Dirac equation in terms of the γ matrices. For a spin, s= ½ particle, the spin vector $u(p)$ is written as $\begin{pmatrix} 1 \\ 0 \end{pmatrix}$ and $\begin{pmatrix} 0 \\ 1 \end{pmatrix}$ for spin up and spin down respectively where p is momentum. For a particle with mass we have $c \neq 1$. For the independent form of Eq. (12.21),

$$\left(-i\hbar c\alpha_\mu \frac{\partial}{\partial x_\mu} + \beta mc^2 \right)\psi = 0 \tag{12.24}$$

for the time independent equation, and we can divide Eq. (12.24) by $i\hbar c$ and have,

$$\left(\gamma_\mu \frac{\partial}{\partial x_\mu} + \frac{mc}{\hbar} \right)\psi = 0 \tag{12.25}$$

where $k_0 = p/\hbar$ or $k_0 = mc/\hbar$ and $\gamma_\mu = i\hbar c\alpha_\mu$ where indices μ run 0 to 3. The dependent Dirac equation is given in Eq. (12.21).

Consider spinors as basic geometrical entities that apply at a deeper level of spacetime. Spinors are complex and have real fields and real manifolds have on underlying complex nature. An essential description of nature involves complex numbers and holomorphic functions. Spinors can be mapped to twisters and vice versa. Spinors are two component entities involving the isomorphism of the conformal group and SU(2,2) which can be related to the Yang-Mills theory. The solution to the Dirac equation is in terms of spin $u(p)$ as

$$\psi = u(p)e\frac{i}{\hbar}(p\cdot\underline{x} - Et) \tag{12.26}$$

the Dirac spin matrices $\gamma_\mu = i\hbar c\alpha_\mu$. The spinor calculus is related to the twistor algebra, which relates a 2-space to an associated complex 8-space (see reference [25]).

An example of the usefulness of spinors is in the Dirac equation. For example, we have the Dirac spin matrices, $\gamma_\mu = \begin{pmatrix} 0 & \sigma_\mu \\ \sigma_\mu & 0 \end{pmatrix} = -i\beta\alpha_k$ where terms such as $\gamma_\mu(1-\gamma_5)$ come into the electroweak vector-axial vector formalism. The three Dirac spinors are given as,

$$L_x = \frac{\hbar}{2}\sigma_x = \begin{bmatrix} 0 & 1 \\ 1 & 0 \end{bmatrix},\ L_y = \frac{\hbar}{2}\sigma_y = \begin{bmatrix} 0 & -i \\ i & 0 \end{bmatrix},\ L_z = \frac{\hbar}{2}\sigma_z = \begin{bmatrix} 1 & 0 \\ 0 & -1 \end{bmatrix}. \tag{12.27}$$

Then $\gamma_5 \equiv i\gamma_0\gamma_1\gamma_2\gamma_3 = i\gamma^0\gamma^1\gamma^2\gamma^3$ for $\gamma_0 = \beta$ is given as

$$\gamma_0 = \beta = \begin{bmatrix} 1 & 0 & 0 & 0 \\ 0 & 1 & 0 & 0 \\ 0 & 0 & -1 & 0 \\ 0 & 0 & 0 & -1 \end{bmatrix} \tag{12.28}$$

for trace $tr\beta = 0$, that is Eqs. (12.21) and (12.28) can be written as,

$$\gamma_0 = \beta = \begin{pmatrix} I_2 & 0 \\ 0 & -I_2 \end{pmatrix} \tag{12.29}$$

where we have the 2 x 2 spin matrix as $I_2 = \begin{vmatrix} 1 & 0 \\ 0 & 1 \end{vmatrix}$ for trace $I_2 = 2$. The Dirac spinors are the standard generators of the Lie algebra of SU_2. The commutation relations of the Dirac spin matrices is given as

$$\{\gamma^\mu, \gamma^\nu\} = \gamma^\mu \gamma^\nu + \gamma^\nu \gamma^\mu = i g^{\mu\nu} I \tag{12.30}$$

where I is the identity element and $\det\left|\gamma_{\mu\nu}\right| = \det\left|\left|g_{\mu\nu}\right|\right.$ where $g_{\mu\nu}$ is the metric tensor. The Dirac spin matrices come into use in the electroweak vector-axial vector model as $\gamma_\mu(1-\gamma_5)$ for γ_5 as,

$$\gamma_5 = i\gamma_0\gamma_1\gamma_2\gamma_3 = i\gamma^0\gamma^1\gamma^2\gamma^3 \tag{12.31}$$

where indices run 0 to 3. We can also write,

$$\gamma_{\mu\nu}(x^5, x^\mu) = \sum_{n=-\infty}^{\infty} \gamma_{\mu\nu}^{(n)}\left(x^\nu\right) e^{inx^5} \tag{12.32}$$

which expresses some of the properties of the 5D Kaluza-Klein space, having $\gamma_0, \gamma_1, \gamma_2, \gamma_3$ and γ_5. See Chap. 4.

As before stated 4 x 4 the γ matrices are Hermitian, $\gamma_\nu^* = \gamma_\mu$ and $\gamma_\mu\gamma_\nu = -\gamma_\nu\gamma_\mu$ where $\mu \neq \nu$ and $\gamma_\mu^2 = 1$. The form of the Dirac equation in Eq. (12.25) is the covariant form of the wave equation. The 4-vector form for spin ½ fermions for s = ½ and $m_e = m$, the mass of the electron. The γ matrices are 4 x 4 matrices with 16 elements which obey the following relations $\left(\gamma_\mu, \gamma_\nu\right)_x = \gamma_\mu\gamma_\nu + \gamma_\nu\gamma_\mu = 2\delta_{\mu\nu}$ where $\delta_{\mu\nu}$ is the Kronecker delta function. The Dirac spin matrices obey Fermi-Dirac statistics, where particles such as photons obey Bose-Einstein statistics.

The γ_5 matrix is associated with a 5D metric tensor. See Chap. 11. This 5D space passes exactly one geodesic curve which returns to the same point with a continuous direction. Note that this is a similar formalism to that of the Dirac string trick 720° path. A connection can also be made to the electromagnetic potential and the metric of the Kaluza-Klein geometry. We can express $\gamma_{\mu 5}$ in terms of a potential ϕ_μ so that

$$\gamma_{\mu 5} = \sqrt{2k\phi_{\mu}} \tag{12.33}$$

Where $k \equiv 8\pi / F$ and where $F = c^4 / G$ or the quantized cosmological force [8-10] (also see Eq. (12.14)). Then we have a 5-space 3-vector as,

$$\gamma_{\nu 5} = \begin{pmatrix} 0 \\ 0 \\ 0 \\ 0 \\ 1 \end{pmatrix}. \tag{12.34}$$

Through this approach, we can relate covariance and gauge invariance. See section 12.2 and Chaps. 5 and 6.

For the covariant equation of motion in terms of ψ^{ψ}

$$\frac{\partial \psi^*}{\partial x_{\mu}} \gamma_{\mu} - \frac{\partial \psi^*}{\partial t} \gamma_0 + \frac{mc}{\hbar} \psi^* = 0 \tag{12.35}$$

Then $\overline{\psi} = \psi^* \gamma_0$ and $\gamma^* = \overline{\psi}\gamma_0$ and using Eqs. (12.21), (12.22), and (12.32) we can write the matrix for ψ as the complex conjugate of ψ^* for two spin states of electrons. The corresponding wave function can be written as the bispinor or 4-spinor. The 4-component function transform under rotations in exactly the same manner as the Pauli spinors. The wavefunction, ψ is four rows and one column, 4 x 1 vector matrix.

$$\psi = \begin{pmatrix} \psi_1 \\ \psi_2 \\ \psi_3 \\ \psi_4 \end{pmatrix} = \begin{pmatrix} \psi_u \\ \psi_{\ell} \end{pmatrix}, \tag{12.36}$$

$$\psi^u = \begin{pmatrix} \psi_1 \\ \psi_2 \end{pmatrix}, \quad \psi^{\ell} = \begin{pmatrix} \psi_3 \\ \psi_4 \end{pmatrix}. \tag{12.37}$$

Where the indices u and ℓ correspond to upper and lower respectively and are each 2-component spinors.

The γ_5 matrices are utilized in the formulation of the electroweak theory. The weak interaction Hamiltonian is formed in analogy to QED in which the Hamiltonian H is given as,

$$H_{e+m} = ie(\bar{\psi}_i \gamma_\mu \psi_j) A_\mu \,. \tag{12.38}$$

Where $\bar{\psi}_i$ is the Hermitian conjugate of ψ_j which are the eigenfunctions of the Hamiltonian and A_μ is the electromagnetic potential. In analogy to Eq. (12.38), the weak interaction Hamiltonian

$$H_{\text{weak}} = \bar{u}\gamma_5(1-\gamma_5)u \tag{12.39}$$

where $\gamma_5\gamma_u$ is the axial vector part and the wave function is u.

12.5 The Dirac Equation in Complex 8-Space

We examine the formalism for the Dirac equation in the complex 8D space where the additional nonlinear terms arise from the imaginary components of the 8D space. The approach here is similar to that which we performed for the Schrödinger equation solved in 8D space; see Chap. 10. We proceed from the complexification of the Minkowski spacetime in which we formulated Maxwell's equations, Chaps. 5 and 6 as well as the Schrödinger equation. We identify the spinors as acting in a spin space in which spin is a conserved quantum number. Such a picture gives us understanding of the properties of spin but not its origin or source. This point is similar to that we made about charge. Physicists currently discuss the properties of charge as a conserved quantum number but the manner in which it arises is not addressed as we previously discussed. However, the origin of mass is formulated in terms of the elusive Higgs particle which may be an artifact of Gauge Theory being an approximation and might not exist.

The complex conjugate of spin space can be made since the Dirac 2 x 2 and Pauli 4 x 4 matrices are real and imaginary; hence the matrices in Eqs. (12.21) and (12.22) and their commutation relations will be effected by Eqs. (12.23a), (12.23b), (12.23c) and (12.23d). The angular momentum space will also be effected by a transformation in complex L space; see Eq. (12.19). Essentially formulating the Dirac equation in complex space and time utilizes the complex Minkowski formalism presented in Chap. 2. We proceed along the approach we have taken in Chap. 10 for the Schrödinger equation.

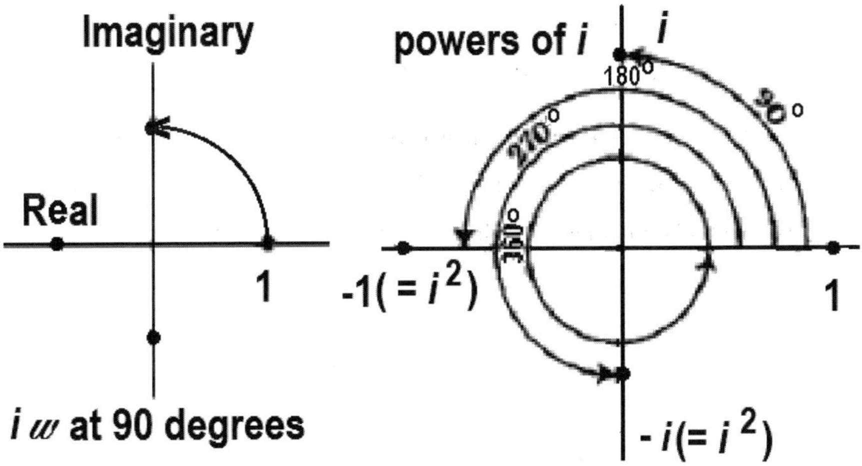

Fig. 12.2. Through a 90° transformation Re →Im and 180° Re→-Re, for a 270° rotation Re→- Im and for a 360° rotation + Re comes back to +Re. These comprise conditions in which the 360° case is relevant to the 0° case.

12.5.1 *Complexifying Spin Space*

Complexifying spin spaces effects the Dirac spinor and Pauli matrices. These are formulated in angular momentum space, see Eq. (12.19). For example, the SU_3 octet with the mass splitting of the p^+ and N^0 and octet is plotted in Y spin and I_2 space. For example from Eq. (12.19), the Pauli matrices $\sigma_x \to i\sigma_{ix}$ $\sigma_y \to i\sigma_{iy}$ and $\sigma_z \to i\sigma_{iz}$ which satisfy the commutation relations $i\sigma_{ix} i\sigma_{iy} = -i\sigma_{iy} i\sigma_{ix} = i\sigma_{iz}$ for $i = \sqrt{-1}$ so that $-\sigma_{ix}\sigma_{iy} = +\sigma_{iy}\sigma_{ix} = -\sigma_{iz}$ therefore $\sigma_{ix}\sigma_{iy} = -\sigma_{iy}\sigma_{ix} = \sigma_{iz}$. This commutation relation for the imaginary components of the σ's give a new commutation relation, that is, instead of $i\sigma_z$ we have σ_{iz}. The real components of the 2 x 2 matrices given in Eq. (12.19) become

$$i\sigma_{ix} = \begin{bmatrix} 0 & i \\ i & 0 \end{bmatrix}, \quad i\sigma_{iy} = \begin{bmatrix} 0 & 1 \\ -1 & 0 \end{bmatrix} \quad \text{and} \quad i\sigma_{iz} = \begin{bmatrix} i & 0 \\ 0 & -i \end{bmatrix} \tag{12.40}$$

so that now σ_{ix} and σ_{iz} become imaginary and σ_{iy} becomes real as opposed to the expression in Eq. (12.19) where σ_x and σ_z are real and σ_y is imaginary. We can expand this approach to Eqs. (12.23a), (12.23), (12.23c) and (12.23d).

We can term the 4 x 4 Dirac matrices α_μ and β for Eqs. (12.21) and (12.22), as real and so is α_x and α_z but α_y is imaginary. These matrices comprise the real components of the complex 8D space. For $i\sigma_D = i\beta$ then

$$i\beta = \begin{bmatrix} i & 0 & 0 & 0 \\ 0 & i & 0 & 0 \\ 0 & 0 & -i & 0 \\ 0 & 0 & 0 & -i \end{bmatrix}. \tag{12.41}$$

Where the trace, $tri\beta = 0$ is the real form of β.

For the imaginary part of the 4 x 4 α_μ matrices, from Eq. (12.22) we have,

$$i\alpha_{ix} = \begin{bmatrix} 0 & 0 & 0 & i \\ 0 & 0 & i & 0 \\ 0 & i & 0 & 0 \\ i & 0 & 0 & 0 \end{bmatrix}, \; i\alpha_{iy} = \begin{bmatrix} 0 & 0 & 0 & 1 \\ 0 & 0 & -1 & 0 \\ 0 & 1 & 0 & 0 \\ -1 & 0 & 0 & 0 \end{bmatrix}, \; i\alpha_{iz} = \begin{bmatrix} 0 & 0 & i & 0 \\ 0 & 0 & 0 & -i \\ i & 0 & 0 & 0 \\ 0 & -i & 0 & 0 \end{bmatrix} \tag{12.42}$$

Note that none of the matrix in Eq. (12.28) or Eq. (12.30) are the same as Eq. (12.19) or Eqs. (12.21) and (12.22). The notation for the imaginary part of $i\sigma_{ix}$ is the same as $i\sigma_{xim}$, etc. [20]. Consider Eq. (12.30), for γ_5, we chose the imaginary components of the γ matrices so that,

$$i\gamma_{5\mathrm{Im}} = -i\gamma_0\gamma\gamma_2\gamma_3 = -i\gamma^0\gamma^1\gamma^2\gamma^3 \tag{12.43}$$

In which $\gamma_{5\mathrm{Re}} = -\gamma_{5\mathrm{Im}}$. From Eq. (12.34) we have the imaginary component as

$$i\gamma_{v5} = \begin{pmatrix} 0 \\ 0 \\ 0 \\ 0 \\ i \end{pmatrix}. \tag{12.44}$$

The imaginary components of the γ matrices remain covariant under the transformation to the imaginary light cone.

Writing out the components of the γ matrices in the Dirac equation, we have

$$\left(\gamma_1\frac{\partial}{\partial x}+\gamma_2\frac{\partial}{\partial y}+\gamma_3\frac{\partial}{\partial z}-\gamma_0\frac{i}{\hbar}\frac{\partial}{\partial t}\right)\psi+\frac{mc}{\hbar}\psi=0 \tag{12.45}$$

for the time dependent form. Equation (12.45) is first order in space and time. If we consider the complexification of the bispinor space and spacetime, the imaginary forms of the γ_μ functions and the spatial and temporal derivatives remain the same under a transformation, however the mass term in Eq. (12.45) goes from $\frac{mc}{\hbar}$ to $\frac{imc}{\hbar}$. That is the signal becomes tachyonic. Complexification produces more changes in the Schrödinger equation because it is second order in space and first order in time but since the Klein-Gordon equation and Dirac equation are the same order in space and time so that only the mass terms are effected. This holds true for the linear approximation of these equations. Nonlinearized forms can lead to distinctly different results. See next subsection and Chap. 10.

We can write the imaginary form of Eq. (12.45) as,

$$\left(\gamma_1\frac{\partial}{\partial x}+\gamma_2\frac{\partial}{\partial y}+\gamma_3\frac{\partial}{\partial z}-\gamma_0\frac{i}{\hbar}\frac{\partial}{\partial t}\right)\psi+\frac{+imc}{\hbar}\psi=0 \tag{12.46}$$

for the tachyonic mass, *im* summing real and imaginary components yields a factor of 2 times the components of Eq. (12.46) except we have the sum of tardyon and tachyon mass terms as

$$\frac{(m+im)c}{\hbar} \tag{12.47}$$

and the interpretation of such a term requires further examination such as the imaginary component relating to the particle decay time of mass, *m*. Electrons are stable but other fermions, such as electron, muon and tau neutrinos, muon and tau can decay. This approach will affect our solutions to the Dirac equation; see Eqs. (12.36) and (12.37).

12.5.2 *Nonlinear Formalism of the Dirac Equation*

In this subsection, we examine some of the properties of the Dirac equation by considering the introduction of a small nonlinear term arising from a projective geometry from the full complex 8-space, $\mathbb{C}_4$ into the 4D Minkowski space, M_4 such that the imaginary components of $\mathbb{C}_4$ are expressed in terms of a nonlinear term $g^2(\bar{\psi}\psi)\psi$ for the wave function ψ [21]. The essential properties of the complex 8-space is nonlocality and by introducing the additional imaginary components of the $\mathbb{C}_4$ space, remote spacetime connections are allowed for microscopic connections (see Chaps. 4 and 10) and macroscopic phenomena such as in Chaps. 2,5,6 and 7. It is interesting to examine the Dirac equation in this light because it is a quantum expression which is relativistically invariant.

We can write the equation of motion for a nonlinear system

$$\left(i\frac{\partial}{\partial x}-m\right)\psi+g^2\left(\bar{\psi}\psi\right)\phi=0\ . \tag{12.48}$$

Where $\bar{\psi}$ and ψ^* are the Hermitian and complex conjugate of ψ respectively; ψ^+ is also used for Hermitian conjugate the nonlinear term is expressed as the coupling term g^2. For the associated action variable, S, expressed in terms of a field $\phi(x,t)$ and its conjugate $\phi^+(x',t)$, we write

$$S=\int dt\ \ dx\left(i\phi^+\right)\phi-H \tag{12.49}$$

and where $\left[\phi,\phi^+\right]=\gamma\left(x'-x\right)$ and ϕ and ϕ^+ are orthogonal to each other. The Hamiltonian, H for this system is given in terms of our nonlinear term g^2

$$H=\int dx\,H=\int dx\ \ \frac{\partial}{\partial x}\phi^+\frac{\partial}{\partial x}\phi-\alpha\phi^+\phi-g^2\left(\phi^+\phi\right)\ . \tag{12.50}$$

The solutions for this equation of motion, Eq. (12.48) are

$$\phi(x,t)=\mathrm{A}(x,t)e^{-i\alpha t} \tag{12.51}$$

$$\phi^+(x,t)=\mathrm{A}^+(x,t)e^{+i\alpha t} \tag{12.52}$$

where $\mathrm{A}(x,t)$ is the wave amplitude.

We form an expression for the Dirac equation for $g^2\left(\psi^+\psi\right)$ as a small additional term as,

$$\left(i\partial_x - m\right)\psi + g^2\left(\psi^+\psi\right) = 0 \tag{12.53}$$

where we use the notation, $\partial_x \equiv \partial/\partial x$. We can now write the charge density Hamiltonian as

$$H = \bar{\psi}\left(-i\gamma\frac{\partial}{\partial x} + m\right)\psi - \frac{g^2}{2}\left(\bar{\psi}\psi\right)^2 \tag{12.54}$$

The Lagrangian for plane wave solution is given as

$$L = \bar{\psi}\left(i\partial_x - m\right)\psi + \frac{g^2}{2}\left(\bar{\psi}\psi\right)^2 \tag{12.55}$$

where the γ Pauli spin matrices and ρ, density matrices $\rho = \psi^+\psi$. Then the lowest energy plane wave solutions are expressed in terms of spinors

$$\psi = \exp - i\omega t\begin{pmatrix} u \\ - \\ v \end{pmatrix}; \tag{12.56}$$

where the spinors are $\begin{pmatrix} u \\ - \\ v \end{pmatrix}$.

For the case where the coupling constant g^2 small $g \geq 0$, the attractive force for nonlinear term and ψ is the quantized Fermi field. The small nonlinear term $g^2\left(\psi + \psi\right)$ can be identified with the imaginary part of the mass, where in the linear approximation, $m_T = m = m_{Re} + im_{Im}$ where $m_T = m$ is the total mass. In Eq. (12.48) we associate m with the real part of the mass, m_{Re} and the additional imaginary component of the mass with m_{Im}. The imaginary component of mass may be associated with particle decay times for fermions in general.

We consider the solutions to two mass free coupled equations, where the coupling constant is expressed in terms of the nonlinear term g^2 where g^2 has two eigenvalues, g and ϕ. For our coupled equation formalism, we have wave amplitude eigenfunctions u_1 and u_2. We have considered the coupled channel formalism in nuclear physics applications with good success [22-24].

$$\frac{\partial u_1}{\partial x} + igu_1 = ig\phi u_2 \tag{12.57a}$$

$$\frac{\partial u_2}{\partial x} + igu_2 = ig\phi^+ u_1\,. \tag{12.57b}$$

The boundary conditions in the asymptotic limit on ϕ and ϕ^{+} is given as ϕ and $\phi^{+} \to 0, \lim_{x\to\infty}$.

The solutions take the form of

$$\phi(x,g) = \exp - igx + ig\int ds \;\; \exp - ig(x-s)\phi(s)\phi(s,g) \qquad (12.58)$$

for $g^2 \to 0$, then we have,

$$\phi \to a(g)e^{-igx} \qquad (12.59a)$$

$$\phi^{+} \to b(g)e^{igx} \qquad (12.59b)$$

For the case where g small perturbation expansion can be made for g^2 related to $Jm = mJm = m^*$. There is much more to explore in the richness of the Dirac theory. The Fermi-Dirac model is significant in the considerations of nonlocal coherences in plasmas and other material media and the possible relation of the vacuum concept to advanced potentials and hidden variable theories related to nonlocality such as presented in Chap. 4.

12.5.3 *Generalized Wave Equations, Classical, Quantum, Nonrelativistic And Relativistic in Linear and Nonlinear Forms*

We present a detailed comparison of the form of a number of wave equations in linear and nonlinear forms and we demonstrate their interrelationship. We summarize and discuss the structure of the Schrödinger, Klein-Gordon and Dirac equation. The uniqueness of the properties of spin and chirality of the Dirac string trick is presented, which is unique to the Dirac formalism [25-29]. The standard wave equation is second order in space and time

$$\frac{d^2U}{dx^2} = \frac{1}{v^2}\frac{d^2U}{dt^2}, \qquad (12.60)$$

where the amplitude, U is a function of space and time, $U(x,t)$, v is the wave velocity and the amplitude, U is expressed in terms of oscillatory solutions.

The Klein-Gordon equation is also expressed as second order in space and time as

$$\square^2\psi + \frac{2m}{\hbar}\psi = 0 \qquad (12.61)$$

where the D'Alembertian operator is given as

$$\square^2 = \frac{\partial^2}{\partial x^2} + \frac{\partial^2}{\partial y^2} + \frac{\partial^2}{\partial z^2} - \frac{1}{c^2}\frac{\partial^2}{\partial t^2} \tag{12.62}$$

We can write the wave mechanical treatment by revising the relativistic Klein-Gordon equation for the zero rest mass of the photon, $m = 0$ so that $\left(\nabla^2 - k^2 \frac{\partial^2}{\partial t^2}\right)\phi = 0$. For $m \neq 0$, we have $m_{\mathrm{Re}} = m$ and $m_{\mathrm{Im}} = im$ terms, which may add further to the understanding of the strong force [14]. Under the action of a potential goes as 1/r for a particle of mass, *m* such as the pion, π particle mass $\left(\nabla^2 - k^2 \frac{\partial^2}{\partial t^2} + m^2\right)\phi = 0$ then $\phi \to \left(\frac{e - cr}{r}\right)$ which yields the Yukawa potential for nuclear forces. The key is the richness of the quantum theory approach and perhaps its universality as exemplified by the Heisenberg uncertainty or indeterminacy relations and the conditions of the EPR paradox. See Chap. 4. The Sommerfeld quantization condition $\oint pdq = n\hbar$ is to the Heisenberg relations and to phase space analysis in terms of (*p,q*). The duality of p and x and *E* and *t* both form phase spaces. Note that we denote q generalized spatial parameter such as *x* and *p* as momentum. This phase space (*p,q*) approach leads to the Heisenberg indeterminacy or uncertainty principle. We may be able to relate the "phase spaces" such as (*x,t*), (*p,E*), and (*x,p*), (*E,t*), to multidimensional Fourier transforms and some physical processes [30].

The Schrödinger hypergeometric equation is formulated in terms of the second order in space and first order in time as for the potential free case,

$$\frac{\hbar\nabla^2\psi}{2m} = \frac{1}{i}\frac{\partial\psi}{\partial t}\ . \tag{12.63}$$

In the case where a potential of a force is present, we have

$$\frac{\hbar\nabla^2\psi}{2m} + V\psi = \frac{1}{2}\frac{\partial\psi}{\partial t} \tag{12.64}$$

where we have the potential, *V* and $\frac{\partial\psi}{\partial t}$ is the time dependent term, where ψ is a function of the independent variable *x,t* as $\psi(x,t)$. For the term

$\frac{1}{i}\frac{\partial\psi}{\partial t}=0$, then we have the time independent Schrödinger equation. In general the time dependent solution is of the form

$$\psi \propto e\frac{i(kx-\omega t)}{\hbar} \tag{12.65a}$$

and

$$\psi^* \propto e\frac{-i(kx-\omega t)}{\hbar}. \tag{12.65b}$$

The quantum theory is formulated in terms of probabilities, $\psi^*\psi$ but the equations of quantum mechanics are analytic.

The Dirac equation is formulated in terms of a first order in space and time. We write the time independent Dirac equation as

$$\left(\gamma_\mu \frac{\partial}{\partial x_\mu}+\frac{mc}{\hbar}\right)\psi=0\ . \tag{12.66}$$

The γ_μ matrices are expressed in terms of the Dirac matrices, σ which are 2 x 2 matrices and the indices run 0 to 3 and the γ_μ matrices are 4 x 4 matrices. The solution to the Dirac equation takes the form $\psi=u(p)\exp\left[\frac{i}{n}(px-Et)\right]$. The quantity $u(p)$ is a spinor with components $\begin{pmatrix}1\\0\end{pmatrix}$ and $\begin{pmatrix}0\\1\end{pmatrix}$ for spin up and spin down respectively. See Chap. 11. Since we can express the $P\mu$'s in terms of the Pauli spin matrices, α_μ which we can express in terms of the Dirac matrices, σ, we then express the Dirac equation as

$$\left(-i\hbar c\alpha_\mu \frac{\partial}{\partial x_\mu}+\beta mc^2\right)\psi=0\,. \tag{12.67}$$

The Pauli spin matrices, α_μ are expressed in terms of the Dirac 2 x 2 matrices, σ as

$$\sigma_x = \begin{vmatrix} 0 & 1 \\ 1 & 0 \end{vmatrix}, \quad \sigma_y = \begin{vmatrix} 0 & -i \\ i & 0 \end{vmatrix} \text{ and } \sigma_z = \begin{vmatrix} 1 & 0 \\ 0 & -1 \end{vmatrix} \tag{12.68}$$

Table 12.2 Linear and Nonlinear Wave Equations

- Non-dispersive – Non-diffusive wave equations $\dfrac{\partial\psi}{\partial t}+\dfrac{\partial\psi}{\partial x}=0$.
- Dispersive wave equations, where the third order term has dispersive losses $\dfrac{\partial\psi}{\partial t}+\dfrac{\partial\psi}{\partial x}=-\beta\dfrac{\partial^3\psi}{\partial x^3}$.
- Diffusive wave equation where the second order term has diffusional losses $\dfrac{\partial\psi}{\partial t}+\dfrac{\partial\psi}{\partial x}=\dfrac{\partial^2\psi}{\partial x^2}$ where β,δ are constants.
- Korteweg-deVries equation is nonlinear and is dispersive but not diffusive where the nonlinear term $\psi\partial/\partial x$ overcomes dispersive losses $\dfrac{\partial\psi}{\partial t}+\psi\dfrac{\partial\psi}{\partial x}=-\beta\dfrac{\partial^3\psi}{\partial x^3}$ and has soliton solutions.
- Burger's equation is nonlinear and diffusive but is not dispersive, where the nonlinear term $\psi\dfrac{\partial\psi}{\partial t}$ overcomes dispersive losses $\dfrac{\partial\psi}{\partial t}+\psi\dfrac{\partial\psi}{\partial x}=\delta\dfrac{\partial^2\psi}{\partial x^2}$.
- Nondispersive and nondiffusive, nonlinear wave equation $\dfrac{\partial\psi}{\partial t}+\psi\dfrac{\partial\psi}{\partial x}=0$.

If we express the Klein-Gordon equation in complex 8-space, the complexification of the spatial and temporal components remain unchanged. Thus, the Klein-Gordon equation does not form extra imaginary components for the spatial and temporal second order derivatives [24]. The Dirac equation is first order in space and time. Essentially one can express the Klein-Gordon equation as a dual Dirac equation, except of course, the Dirac equation is expressed in terms of spinors, which the Klein-Gordon equation is not. Because of the electron spin symmetry conditions or the Dirac string trick in which the rotation of the system must pass through a 270° rotation [25-27,30].

The so termed Dirac string trick involves tracing the spin of an electron in space. The requirement for the electron spin and chirality to be aligned or anti-aligned along the particles direction of motion requires a 720° twist or rotation. If we rotate a 90° spin change we move from the real to imaginary axis so that a variable, ζ has a real and imaginary part, then $\zeta_{\text{Re}} \rightarrow \zeta_{\text{Im}}$. Through a rotation of 180° then, ζ_{Re} comes back to real again and without chirality considerations, only a phase sign charge has occurred. In the case of the Dirac spinors, symmetry requirements lead to the 270° rotation so that ζ_{Re} is now mapped into ζ_{Im} as $\zeta_{\text{Re}} \rightarrow \zeta_{\text{Im}}$ and hence the Dirac equation does not remain uncharged under the transformation from real spacetime, x_{Re}, t_{Re} to imaginary spacetime, x_{Im}, t_{Im}. See Fig. 12.2.

Table 12.3 TYPES OF WAVE EQUATIONS: CLASSICAL, QUANTUM AND RELATIVISTIC

- Time dependent classical wave equation in 1D $\dfrac{\partial^2 u}{\partial x^2} = \dfrac{1}{v^2}\dfrac{\partial^2 u}{\partial t^2}$ for wave amplitude solution $u(x,t)$ and v is the classical velocity, $v << c$.
- Time dependent Klein-Gordon equation [32-35] in 3D with $m \neq 0$ $\square^2\psi + \dfrac{2m}{\hbar}\psi = 0$ for $\square^2 = \dfrac{\partial^2}{\partial x^2} + \dfrac{\partial^2}{\partial y^2} + \dfrac{\partial^2}{\partial t^2} - \dfrac{1}{c^2}\dfrac{\partial^2}{\partial t^2}$.
- Time independent Dirac equation with $m \neq 0$. $\left(\gamma_u \dfrac{\partial}{\partial x_u} - \dfrac{mc}{\hbar}\right)\psi = 0$. The time dependent Dirac equation with $m \neq 0$: $\left(\partial_\mu \dfrac{\partial}{\partial x_\mu} - \dfrac{mc}{\hbar}\right)\psi = \dfrac{i}{\hbar}\dfrac{\partial \psi}{\partial t}$
- Time dependent Schrödinger equation $-\dfrac{\hbar^2}{2m}\nabla^2\psi + V\psi = -\dfrac{\hbar}{i}\dfrac{\partial \psi}{\partial t}$ or $\nabla^2 = \dfrac{\partial^2}{\partial x^2} + \dfrac{\partial^2}{\partial y^2} + \dfrac{\partial^2}{\partial z^2}$ time independent Schrödinger equation for $H\psi = E\psi$ where $H = T + V$ and V is the potential energy.

In the following tables we present a summary of structure of the major wave equations of physics. We enumerate a set of wave equations having classical properties. These are better linear and nonlinear equations and are classical in nature. These equations have various properties of dispersive and diffusive energy and information losses. Nonlinear terms can overcome these loss mechanisms and form coherent, non-dispersive and non-diffusive states. See Table 12.2.

In Table 12.3, we enumerate types of time dependent, time independent classical, quantum and quantum relativistic equations. All these equations are linear. If we consider a small deviation from linearity, we formulate nonlinear equations that take forms that overcome dispersive and diffusive losses. Essentially in the Everett-Graham-Wheeler Multiverse picture or in the infinite possible string theory vacuum solutions, the number of possibilities may be reduced, see Chaps. 2 and 10 [6,23]. Selection of higher probability terms is made by inclusion of nonlinear terms in the wave equations, in some cases yielding solitary wave or soliton solutions (Chap. 10).

In Table 12.4 we present the nonlinear forms of the Schrödinger and Dirac equation for both time dependent and time independent forms.

Table 12.4 Nonlinear Quantum Wave Equations

- Nonrelativistic nonlinear time dependent Schrödinger equation $\frac{\hbar\nabla^2\psi}{2m}+g^2\left(\psi^+\psi\right)\psi=\frac{1}{i}\frac{\partial\psi}{\partial t}$ where $g^2\left(\psi^+\psi\right)$ is the nonlinear term and ψ^+ is the Hermitian conjugate of ψ.
- Relativistic time independent Dirac equation $\left(i\frac{\partial}{\partial x}-m\right)\psi+g^2\left(\psi^+\psi\right)\psi=0$ for the nonlinear term $g^2\left(\psi^+\psi\right)$.
- Relativistic time dependent Dirac equation $\left(i\gamma_\mu\frac{\partial}{\partial x^u}-m\right)\psi+g^2\left(\psi^+\psi\right)\psi=\frac{i}{\hbar}\frac{\partial}{\partial t}\psi$

References and Notes

[1] Dirac, P.A.M. (1928) The quantum theory of the electron, Proc. Roy. Soc. London, Series A, 117, 778, 610.
[2] Dirac, P.A.M. (1948) Phys. Rev. 74, 817.
[3] Dirac, P.A.M. (1958) The Principles of Quantum Mechanics, 4th Edition, Oxford: Clarendon.
[4] Dirac, P.A.M. (1965) Phys. Rev. B139, 684.

[5] Feynman, R.P. (1963) *Acta. Phys. Polon.* 24, 697, Feynman developed his model when J.A. Wheeler, his Ph.D. thesis adviser suggested that all the electrons in the Universe could be considered as one electron moving rapidly forward and backward in time.
[6] Feynman, R.P. (1949) Phys. Rev. 76, 769.
[7] Rauscher, E.A. (1968) Collective electron states in quantum plasma physics, J. Plasma Phys. 2, 517.
[8] Haramein, N. & Rauscher, E.A. (2005) Collective coherent oscillating plasma models in surrounding media of black holes and vacuum structure, in R.L. Amoroso, B. Lehnert & J.P. Vigier (eds.) pp. 279-331, Orinda: The Noetic Press.
[9] Corben, H.C. & Stehle, P. (1960) Classical Mechanics, 2nd Edition, San Francisco: John Wiley and Sons.
[10] French, A.P. (1958) Principles of Modern Physics, San Francisco: John Wiley and Sons.
[11] Rauscher, E.A. (1971) A Unifying Theory of Fundamental Processes, Berkeley: UCRL-20808, LBNL-UCB Press.
[12] Westervelt, P.J. (1970) The Theory of General Relativity, Report NYO-2262-TA-218, Brown University.
[13] Rose, M.E. (1961) Relativistic Electron Theory, San Francisco: John Wiley.
[14] Messiah, A. (1999) Quantum Mechanics, Vols. 1-3, Mineola: Dover.
[15] Bohm, D. (1989) Quantum Theory, New York: Dover.
[16] Brink, D.M. & Satchler, G.R. (1968) Angular Momentum, Oxford: Oxford Univ. Press.
[17] Peebles, P.J.E. (1992) Quantum Mechanics, Princeton: Princeton Univ. Press.
[18] Schift, L.I. (1955) Quantum Mechanics, New York: McGaw Hill.
[19] Hatfield, B. (1992) Quantum Theory of Point Particles and Strings, Reading: Addison Wesley.
[20] Rauscher, E.A. (1984) The formulation of the Dirac equation in complex spacetime, Berkeley: TRL Report, PSRL-61289, June.
[21] Rauscher, E.A. (1982) Coherent, nondispersive nonlinear states for the Dirac equation, Berkeley: TRL Report, PSRL-52614.
[22] Rauscher, E.A., Rasmussen, J.O. & Harada, K. (1967) Coupled channel alpha decay beta theory applied to Po^{212m}, Nucl. Phys, A94, 331.
[23] Harada, K. & Rauscher, E.A. (1968) Unified theory of alpha decay, Phys. Rev. 169, 818.
[24] Rauscher, E.A., Seinski, A.J. & Rasmussen, J.D. (1977) Coupled channel alpha decay theory of odd-mass nuclei, Eo^{253} and Fm^{255}, Nucl. Phys, A291, 386.
[25] Dirac, P.A.M. (1935) Ann. Math. 36, 657.
[26] Dirac, P.A.M. (1952) Is There an Ether?, Nature 169, 172.
[27] Bergmann, J. (1957) Two component spinors in general relativity, Phys. Rev. 107, 624.
[28] Amoroso, R.L. (2008) Universal quantum computing: Anticipatory parameters predicting bulk implementation, in D.M. Dubois (ed.) Intl. J Computing Anticipatory Systems, 22: 254; 254-282, Proceedings CASYS07, Liege, Belgium.
[29] Amoroso, R.L. & Rauscher, E.A. (2008) Emergence of generalized F-theory 2-branes from SUSY spacetime parameters of the discrete incursive oscillator, in D.M.

Dubois (ed.) Intl. J Computing Anticipatory Systems, 22: 283; 283-291. Proceedings CASYS07, Liege, Belgium.
[30] Rauscher, E.A. (1972) Generalized Heisenberg relations, Il Nuovo Cimento, 4, 757.
[31] Amoroso, R.L. & Rauscher, E.A. (2009) The Holographic Anthropic Multiverse: Formalizing the Geometry of Ultimate Reality, Singapore: World Scientific.
[32] Klein, O. (1926) *Z. Physik*, 37, 895.
[33] Gordon, W. (1926) *Z. Physik*, 38, 117.
[34] Fock, V. (1926) *Z. Physik*, 38, 2421.
[35] Fock, V. (1926) *Z. Physik*, 39, 226.

Chapter 13

Speculation on a Unified Field Theory (UFT), Grand Unification Theories (GUT) and Supersymmetry and Superstring Theories

Sit down before fact like a little child and be prepared to give up every preconceived notion. Follow humbly wherever and to whatever abyss nature leads, or you shall learn nothing. – Thomas H. Huxley

Eugene Wigner's curiosity about mathematics and its meaning, like many physicists, have put forward the idea that some of the most important concepts in physics, including that of quantum theory, owe their success to mathematical systems that have been devised without any idea as to what they would someday be applied to. "It is difficult to avoid the impression that a miracle confronts us here." Wigner wrote this comment in his paper entitled, The Unreasonable Effectiveness of Mathematics in Natural Science, 1980 – E.A. Rauscher

The search for a comprehensive theory of the Universe, i.e. a theory of everything (TOE) has been the desire and attempted fulfillment throughout history. In this chapter, we cannot give a comprehensive description of this unfolding story but will present some of the lights from our perspective and particularly that of (EAR) from her Lawrence Berkeley National laboratory (LBNL) days when she was a graduate student and then a staff member in the Nuclear Science and Department of Theoretical Physics. She was one of the very few staff members, not only interested in particle physics, general relativity, astrophysics and cosmology, but also the nature of consciousness. Both of us have a passion to ken the Universe, which clearly includes conscious sentient beings. Here we address the issues in unified field theories that relate to our work in this volume on multidimensional geometries and their relationship to describing nonlocal, nonlinear, anticipatory phenomena. One of our motivations is to include a domain for the action and participation of the conscious observer and consciousness in general. The process of scientific discovery involves observation, data systemization and theoretical hypothesis and formulation about the observed data. The process of experimental examination and theoretical hypothesis and further tests of these concepts, leads to the development of scientific rules or laws.

13.1 The Observer in the Quest for Truth

In any unified theory a reality, it appears to us its observer must be included (as, in fact, also chemistry biology and life). Does the unobserved universe really exist or as in Descartes' and Wheeler's views are observation and perception required for existence to exist. How are the observer and observed connected and what is observed and how is it observed? And what of the role of the observer / participator in the collapse of the wave function in the case of linear quantum theory, in the Schrödinger's cat paradox? In the Schrödinger cat paradox arose out of heated discussions in the 1930's as to what is quantum measurement and how the entangled wave function collapses to a single state solution. In this *gedanken* experiment, a cat is contained in an enclosed cage. A radioactive source decay, which is a form of random number generator, (RNG) will determine the time at which a pellet containing a lethal gas such as cyanide is broken by a hammer set into action by the RNG that breaks to containment vile and kills the cat. The unobserved system has the wave function $\Psi = a\ \Psi_{alive} + b\ \Psi_{dead}$ that is two states for the cat's condition of cat alive and cat dead exist simultaneously which is the paradox. See figure 13.1. It is through an observation or measurement that collapses the wave function to either state Ψ_{alive} or Ψ_{dead}. The longer time goes on the more likely that Ψ_{dead} will be the observed state. But who really measures the state? The cat "knows" if it is alive and facetiously the cat "knows" if it is dead, if there is life after death, if there is life after death for a cat. What of the rat or gnat paradox? Is there a mutual subject – experimenter wave function collapse?

Who or what is conscious? Standard quantum mechanics deals with the additions summations of probabilities. If there are small nonlinear terms in the quantum theory, such as solving the Schrödinger equation in complex 8 space, see Chap. 10, such that a shift in probabilities occurs which increases the likelihood of say b Ψ_{dead} if b is larger than a in the earlier equation. As Wheeler suggests, it is a participatory universe; the observer no longer remains passive. It is clear what one does to the Universe, such as in the photoelectric effect, the Compton Effect, etc. implies that one effect a system to observe it. Wheeler suggested that the Universe may not be able to exist without an observer as well as the observed [1].

So how does the relative scale of a (human) observer effect the collapse of the wave function? The path towards a comprehension of physical reality involves the scale from the mini black hole to the size of the Universe and all in between. In Fig. 13.2, we present a scale from the size of the Universe, as it is conceived of in current physical theories. If the Universe is 10^{28} cm in size with a mass of 10^{56} gms, it fits the Schwarzschild condition of $r_s = 2GM/c^2$ where G is the gravitational constant, for a mass, M undergoing the gravitational collapse. The Universe, in this sense fits the Schwarzschild condition even though the Schwarzschild solution is an exterior solution. If the quantum form of space has a lower limit of a mini Planck black hole of $\ell = \left(\hbar G/c^3\right)^{\frac{1}{2}}$ also fits the Schwarzschild condition. The length ℓ is 10^{-33} cm.

$$\Psi = a\psi_{\text{alive}} + b\psi_{\text{dead}}$$

Figure 13.1. is a depiction of the Schrödinger cat paradox. A cat in a sealed box with air and a hammer that can break a cyanide pellet. The trigger for the hammer is set off by the nuclear decay of a radioactive element, which acts as a random number generator, RNG.

An interesting consideration in terms of interconnectedness in scale is Mach's principle. See Chap. 4. Mach's principle is concerned with the relationship of local phenomena to cosmic, large-scale phenomena. If a bucket of fluid is rotated, the meniscus (surface of the fluid) changes shape, from flat to parabolic. The faster the rotation, the more parabolic the surface becomes. Ernst Mach stated that the rotation of the bucket effects the surface of the liquid because the rotation is defined and occurs relative to the fixed star system (or the whole Universe). It appears that Mach's principle and other nonlocal phenomena display macroscopic as well as microscopic remote interconnectedness.

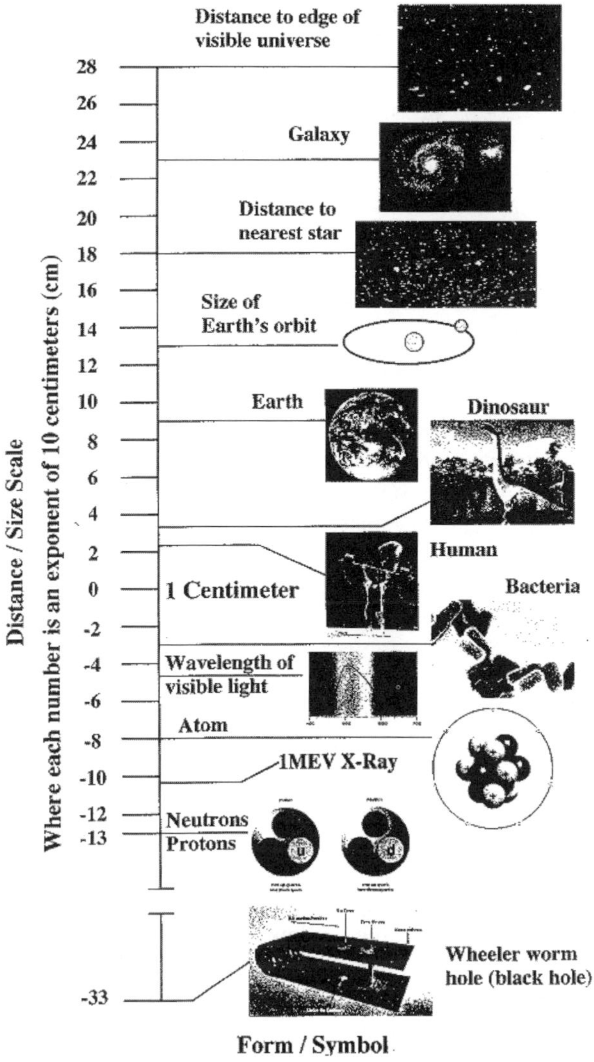

Figure 13.2. Represents scale of various small to large objects in the Universe. On the vertical axis, we denote the exponent x factor of 10^x (in cm) and corresponding to the vertical axis size scale is given some examples of objects of various sizes from the mini Planck black hole to the observable size of the Universe. Note that the observer human has a good vantage point of being roughly between $\ell_{mp} \sim 10^{-33}$ cm and $U_R \sim 10^{28}$ cm. (The center of which would be $10^{2.5}$ smaller than a human.) The size scale of a form determines what symbology we give to create a thought content relative to our size.

Knowledge is an accumulation of facts, theories, concepts, ideas and beliefs. Throughout history there are those who pursued the path of reason (originating from the Greek word for ratio or relationship) and logic (from the Greek word for Logos or "the word"), which are the foundation of natural philosophy, currently known as physics. The origin of the word philosophy is lover of knowledge. Many scientists have contributed to our body of knowledge. We do not know of the useful knowledge that has been lost such as the burning of the Library of Alexandria. From the Greek, Egyptians, Ionians and other early civilizations have built the knowledge base of our social, political, military and academic institutions. The basis of science and the scientific method have evolved over the centuries [4,5].

It is assumed that physics is the most fundamental of all sciences and its perhaps the basis for all human knowledge, using the precise and logical language of mathematics. Our current understanding of physics grow out of our attempt to understand the natural world and has been the result of accumulated knowledge by a succession of inductive and deductive inferences derived from observation and theoretical hypothesis and theoretical explanation and prediction and experimental confirmation. The concept of a unified theory of physics or a theory of everything (TOE) assumes there is a point at which the origin of everything is explained and also it can be explained in terms of a single obvious source. Since knowledge can occur when one does not know "things", that is, one can only learn what we do not already know or think we know. So is it possible for the dream of a "final theory of everything" possible or, as in the past, the process of learning is an ongoing process.

One can consider, as in the past, that objectivity involves what one knows, because that is all on knows what is in one's mind and that external reality is the subjective reality, i.e. the subject of one's study. Currently, we define objective reality as what is "out there" as fixed and immutable and that what one knows is subjective, incomplete and inaccurate knowledge of the objective reality. The objective reality, physics, biology, etc. but what of the mind/brain issue? Is consciousness just an epiphenomenon of brain neuronal activity [2]? We do not think it is just collective neuronal, glial cell activity and there is evidence it isn't. [3] Some features of conscious affection appear to act nonlocally. This in one of many attributes of consciousness as well as doing math, physics, science and creating civilization that leads one to an inevitable and reasonable conclusion that there is more to consciousness than the firing of a programmed set of neurons.

In Fig. 13.3 we represent the objective view of reality as a "contour integral" of which the inside contains objective reality. The reality of the mind is excluded as some subjective entity not fully describable by that which lies within the contour integral. In the complex hyper dimensional space, the contour integral includes the mind.

- What we perceive as real depends on our State of Consciousness
- Barriers to new knowledge are artificial constructs of mind
- What is real must necessarily include that aspect called mind reality to be complete or unified

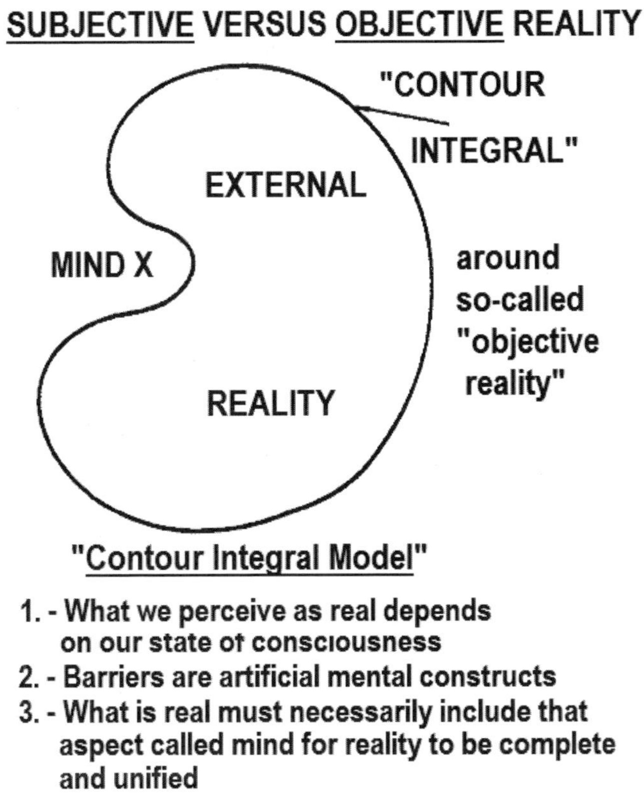

Figure 13.3. Subjective versus objective reality in the contour integral model.

13.2 GUT, TOE Supersymmetry Models, Complex 8-Spacetime and Kaluza-Klein Theory

This chapter in not a comprehensive discourse on group theory and a detailed formulation of the foundations of physics, our purpose is to present enough background to the current approaches to fundamental physics and to demonstrate the manner in which the complex eight space and the Kaluza-

Klein geometry, as well as spinors and twistors, occupy a basic role in the foundations of physics.

The Kaluza-Klein approach lead to the concepts of additional dimensions, XD which required consideration in developing a UFT or TOE theory. The foundation of the standard theory is group theory and, in the context of this approach, and that of the supersymmetry models, is group theory, which has greatly expanded our understanding of the vast accelerator particle physics "zoo" data [6]. In the supersymmetry models and superstring theory, the attempt is to include gravity to the GUT theories. The so termed standard model involves the three stronger forces, the strong, electromagnetic and weak forces of SU_3, U_1 and SU_2 respectively. Particles are split by their quantum number properties, that is for example, symmetry breaking effects are those that violate the invariance of a given symmetry scheme. For example, is that of change independence of the strong force which is broken by the electromagnetic and weal (electroweak) interactions of SU_2 x U_1. The supersymmetry models include gravity and comprise the TOE approach.

Current unification theories involve multidimensional geometries. Physicists currently are attempting to develop a unified theorem of the four major force fields, strong nuclear force, electromagnetic force, weak nuclear decay force, and the gravitational force. The mathematical model involving partial group symmetries leads to an 11D space in the GUT scheme, combined with the gravitational force, leading to the supersymmetry models. In some models, scientists consider a 24-element group is given in terms of spacetime and strong and weak interactions and the electromagnetic fields.

The GUT (grand unification theory) involves strong, electromagnetic, and weak force, and involves ten or more dimensions and ten or more Vector Bosons exchanged as a subset of the superstring model as $SU_2 \otimes U(1)$ and SU_3 where U(1) is the group of electromagnetic interactions with the exchange of a photon and electroweak interaction with an exchange of the three Vector Bosons $W_{\pm}$, Z^0 and the eight gluons of the strong nuclear force (holding nucleons together), obeying the group SU_3 so that we have the group of elements of 1 (of U(1), 2 (of SU_2), and 8 (SU_3) so we have 1+2+8 comprising an 11D GUT theory. To summarize, electromagnetic and weak interactions are combined as the electroweak interaction force, which combined with the strong force, or quantum chromodynamics (QCD) and this model, is termed the GUT model, which when unified with the gravitational field is termed the supersymmetry unified model.

In Table 13.1 lists various current formulations of the foundations of physics of particular interest is the U_4 are the 4-spacetime dimensions of $U_2 \times \tilde{U}_2$ for the complexification of M_4 space, the U_2 group is that of the 4D real spacetime and $\tilde{U}_2$ constitutes the imaginary 4D spacetime which comprise the 8D space, $\mathbb{C}_4$ which is a subset of the GUT scheme. For the 5D Kaluza-Klein geometry the electromagnetic space for U_1 is related to the proper orthogonal group of rotations SO_3^+ as a rolled up dimension of the order of the Planck length $\ell = \left(\hbar G / c^3\right)^{1/2} \sim 10^{-33} cm$. Table 13.1 lists some of the major tenets of group theoretical features for the unification of the four fundamental force fields. The GUT and TOE theories are examined. The complex $\mathbb{C}_4$ space, Kaluza-Klein and twistor spaces are examined as subsets of the GUT and TOE theories.

Table 13.1
ALTERNATE GROUP AND GUT THEORIES

- $S_4 = U(1) \otimes U(1) \otimes SU_2 \otimes SU_3 \otimes SU_3 = 11$
- $SU_5 = SU_2 \otimes SU_3 \otimes U(1)$
- 25-1=254 Permutations in Eddington-Klein Group
  $\bar{S}_4 = U_2 \otimes \tilde{U}_2 \otimes U(4)$ AND 24=4+4+16
- $\bar{S}_4 = SU(2,2/1)$ Penrose Twistor and Supergravity Matter Fields
- The U(4) are 4-Space and Time $U(2) \otimes \tilde{U}(2)$ for the Complexification of 4-Space

  $U(2)$: 4 Real Space and Times

  $\tilde{U}(2)$: 4 Imaginary Space and Times

  ] Complex Space as a Penrose Twistor
- Kaluza-Klein 5-space, U(1) Relates to SO_3 + Electromagnetic and Gravitational Fields and the Spinor Calculus
- The 8-Space Twistor Algebra is Mapable to the Spinor Calculus of the 5D Space

The approach is the supersymmetry models for the 11D space unifying the four force fields includes the strong force with SU_3 symmetry of the gluons or 8D and strongly interacting particles, the electromagnetic force, with U_1 symmetry of 1D and the weak force nuclear decay symmetry of SU_2 symmetry with 2D which comprises the 11D GUT picture. See Table 13.1. In Table 13.1 we present some of the recent approaches to the unified field theories such as the grand unification theory, GUT, and supersymmetry models. These approaches are related to those of Sirag [7], and Penrose [8] as well as the complex Minkowski space of Hanson and Newman [9] and Rauscher [10, 11]. See Chaps. 2 and 4. Rauscher has demonstrated the relationship of the Kaluza-Klein geometry and the complex eight space, which is related to the twistor algebra of the complex eight space through the spinor calculus of the Kaluza-Klein geometry or other spinor models. See Chap. 11. It is interesting to note that the triple torus, which comes out of the Calabi-Yau string theory, can also be constructed from the Penrose model [12-16]. See Table 13.2 for a schematic of the strategy for relating superstring theory to models of reality.

Table 13.2 the Strategy for Superstring Theories

- Superstrings in 11D (4D noncompactified and 7D compactified) relates the strong, electroweak and gravitational forces.
- The 10D supergravity involves the compactification on a Calabi-Yau manifold which is a 3-torus, $U_2 \times U_2 \times U_2$.
- The 4D "no-scale" supergravity and the Gauge group $\subseteq E_6$ predict Gaugino condensation in hidden vector space.
- Local supersymmetry, that is, broken gravitino rest mass, $\neq 0$ and rest mass $\neq 0$ produced by quantum vacuum state polarization.
- Effective low-energy theory in which Supersymmetric Gauge Theory yields non-zero particle masses, i.e. having gravitational effects which would require an expansion of Gauge conditions.
- Additional neutral currents and additional matter particles may require additional dimensions (XD).
- We await empirical results from the CERN supercollider, but more recent theories suggest 11 noncompactified dimensions, no superpartners (sparticles), no Higgs Boson, and no quantum gravity in the form currently sought. These parameters instead become topological conditions of the complex space.

Table 13.3 presents some of the superstring theories, in particular, the Calabi-Yau manifold and its relation to torus topology and the quantum vacuum. The concept of small but finite rest mass of the photon, m_γ is suggested by Vigier, Rauscher and others. Rauscher has demonstrated that the solutions of Maxwell's equations in a complexified Minkowski space, $\hat{M}_4 \subset \mathbb{C}_4$ also requires generalized gauge conditions and a finite rest mass of the photon, m_γ. The vacuum state polarization produces effects on quantum systems and perhaps larger non-quantum systems. Mathematical formulations which include material particles may require additional dimensions.

	FORCE	COUPLING CONSTSNT	RANGE (fm)	INTERACTION TIME
8 SU(3)	Strong (Nuclear)	1 to 10	1 fm	10^{-23}
1 U(1)	Electromagnetic	10^{-2} to 10^{-9}	∞	10^{-21}
2 SU(2)	Weak (Decay)	10^{-11} to 10^{-14}	Short Range	10^{-6} to 10^{-9}
G (2)	Gravitational	10^{-39}	∞	∞

Table 13.3. Relative force strengths. The relative magnitude and range of the four major forces are given.

In Table 13.3 we list the four force fields of physics, the strong nuclear, the electromagnetic, the weak nuclear and gravitational force and their relative strengths or coupling constants. Also the interaction range and decay times are given. The relevant group theory is also noted in the first column leading to the supersymmetry 11D space theory [18].

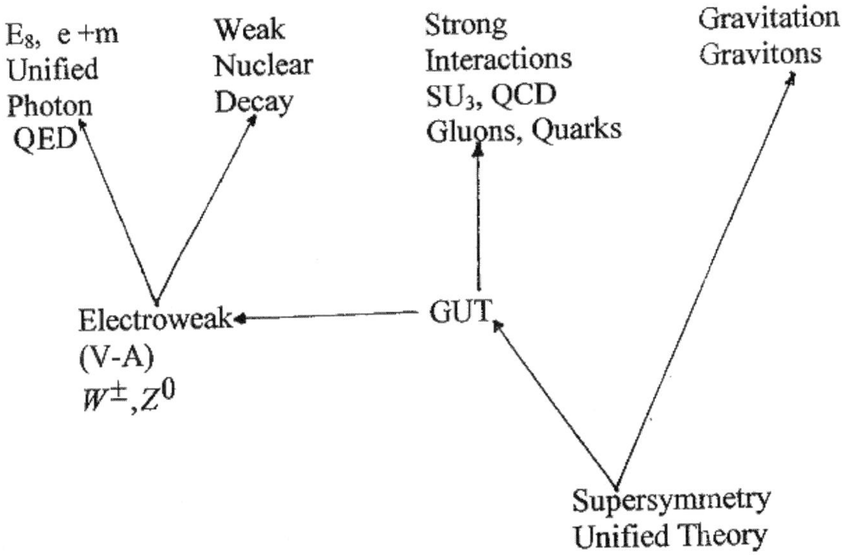

Figure 13.4. Unification schemes: The strong color gauge is $SU(10)_c$ can be decomposed into $U(1)_c \times SU(9)_c$. Symmetries is associated with conserved color components of the electric charge of fermions.

Invariance under the local gauge group SU_2 can be extended to larger groups SU_n. The Yang-Mills theory is a theory which is invariant under the local gauge group SU_n. Quantum chromodymics (QCD) is an extension of QED to include the strong force and is a Yang-Mills theory with the gauge group SU_3 [17]. That is QCD is defined as the Yang-Mills theory with SU_3 as local gauge invariance. In this scheme, it is well known that the quarks are fermions and each flavor or type of quark has three internal degrees of freedom i.e. color. There are $n^2 - 1 = 8$ vector gauge fields or gluons. For the SU_n gauge group, each fermion has n internal degrees of freedom and is coupled to $n^2 - 1$ vector gauge fields, g. As we know, hadrons, which include strongly interacting particles which includes all baryons, π and K mesons but not the muons or electrons, μ or τ and neutrinos, ν_e, ν_m and ν_τ which are leptons or fermions.

In Fig. 13.4 is another manner of examining the relationship the four force fields (FFF) and current physical theories. Their interrelationship is denoted relevant to the current unified physics theories, under the supersymmetry scheme. The QED, electroweak, and strong force are unified under the GUT scheme. The QCD formalism is a strong force theory which allows the GUT theory, to be combined with gravitation, leading to the supersymmetry unified theory.

We enumerate the details of the standard 11D theory that incorporates the four force fields. This approach is basic to current the emergent TOE theories. In these theories the concept of dimension involves the usual extended dimensions of 4D spacetime and rolled up 7D spinor like dimensions such as in the Kaluza-Klein geometry. The essential aspects of the theory are:

- Kaluza-Klein geometry relates gravity and electromagnetics the additional 5th dimension as an internal dimension acts to create a charge, for a clockwise loop as a plus charge and counterclockwise a minus charge. One dimension is for the electromagnetic field and the other four is for the gravitational Minkowski field.
- The weak force involves the exchange of intermediate Vector Bosons $W^{\pm}$ and Z^0 in the Vector minus axial vector theory having 2D.
- The strong nuclear force, 8 charged particles for quarks, 8D for 8 gluons.
- Hence, the total dimensions is 1D for the U_1 (electromagnetism), 2D for the SU_2 (weak interactions), 8D (gluons and quarks for the octet and triplet of SU_3) is 11D, that is 7D plus spacetime (4D) for the four forces. In the GUT, the scheme implies a 10 or more Vector Bosons exchanged which is a subset of substring.

The following tenets of a TOE theory are listed in terms of some of the theoretical approaches that are utilized in developing a unified field theory.

- The properties and characteristics of the four force fields.
- Strong nuclear processes that involve strongly interacting hadrons (Baryons) with substructure, quarks and near massless gluons which transmit hadronic energy between the quarks.
- Symmetry triplet of quark, color SU_3. From the SU_3 type model, five flavors, three families as the four following points.
- Up (u) and down (d)
- charm (c), strangeness (s)
- Top (t), and bottom (b)
- Left (L) and right (R) or dextral.
- Quark confinement in quantum chromodynamics (QCD), white color singlet are found in independent particle entities.
- Three primary color values of quarks which represent hadronic charge, threefold, triplet, R (red), G (green), and B (blue) and anti color counterparts, $\overline{R}$, $\overline{G}$ and $\overline{B}$. The three quarks make up a white particle or baryons as R, G, B and mesons are made up of quark – antiquark pair.
- Photons act as the transmitters electromagnetic energy; the symmetry group is U_1 where all phase factors, that is, complex numbers of unit magnitude, and we have electric charge.
- spin 2 graviton, mediator of the gravitational force.

The foundation of the standard model is a fiber bundle which is a union of the usual 4D spacetime with an extended version of isotopic spin space. This picture is exemplified by the usual gauge conditions. The gauge transformation is characterized by gauge parameters. In the standard model, the four vector potential is quantized as the spin 1 bosomic field, the photon. The fiber bundle is built from the usual 4D spacetime in which each fiber represents one of the groups U_1, SU_2. and SU_3 which include the strong, electromagnetic and weak forces. See Chap. 8.

Superstring theory accommodates a supersymmetry which relates bosons and fermions. Superstrings contain the supersymmetry model. The hope is in a “final theory” or theory of everything (TOE). When Einstein attempted to unify gravity and quantum mechanics, his concept and others was to find a unique set of equations with a unique solution. However, string theory allows 10^{100} or so possible vacuum solutions. So why does it appear we “live in” a unique universe? Maybe we do not according to some superstring models incorporating the Multiverse concept [16].

Figure 2.2 in Chap. 2 also represents the current Multiverse that comes out of the superstring theory that accommodates supersymmetry (relating bosons and fermions) to string theory. Since string theory and superstrings do not provide a unique vacuum solution, this view is akin to the Copenhagen quantum view in which solutions remain unmeasured or entangled or as a Everett, Graham, Wheeler multiple universe or Multiverse model. This approach is one in which the does not require one of the vast multitude of string theories to be the only unique "our universe" correct string theory. String theory is current leading proposal for unifying quantum theory and relativity. In order to have a diffeomorphic manifold for relativity, particles are not considered as point particle, creating singularities, but a tine vibrating strings. The size of these rolled up dimensions, which are like the Kaluza-Klein electromagnetic, dimension are of the order of the Planck length $\ell \sim 10^{-33} cm$.

13.3 Lorentz Transforms and the Universality of the Laws of Physics, Analyticity and Unitarity

In Chap. 3 we presented some of the major principle of modern physics. These are Lorentz invariances, analyticity and unitarity. Since we are address GUT and TOE theories, the principles are most applicable.

13.3.1 *Lorentz Invariant Conditions*

The principle of covariance states that the general laws of physics can be expressed in a form which is independent of the choice of spacetime coordinates and the essential physical contents of these laws are unchanged by a proper Lorentz transformation. The Lorentz group is defined as all real linear transformations,

$$X^{u'} = L^{\mu}_{\nu} X^{\nu} \text{ and } \det\left|L^{\mu}_{\nu}\right| = \pm 1 . \tag{13.1}$$

The laws of physics are independent of the choice of spacetime coordinates or the basis set of vectors of the space. For a Lorentz transformation we can write for $X^{'}_{\nu} = L_{\nu\mu} X^{\mu} + a_{\nu}$ for the inhomogeneous transformation or translation and rotation or for the homogeneous transformation $X^{'}_{\nu} = L_{\nu\mu} X^{\mu}$ which can also be written $\underline{u}^{'}_{\nu}\left(X^{'}\right) = L^{\mu}_{\nu} u_{\mu}\left(X\right)$ for covariant and contravariant tensors and mixed tensors another manner to state the above conditions is conformal invariances is the mathematical property which permits the using to be written in new equivalent ways.

For the metric of the space, $g_{\mu\nu}$ we have $ds^2 = g_{\mu\nu}X^{\mu}X^{\nu}$ and for the Lorentz invariance of the metric

$$g_{\mu\nu}L^{\mu}_{\alpha}L^{\nu}_{\beta} = g_{\alpha\beta}\,. \tag{13.2}$$

For gauge invariance we define a gradient

$$A_{\mu} = \frac{\partial \Lambda}{\partial x_{\mu}} \tag{13.3}$$

so that the potential does not change the fields by anything physically observable. See Chap. 8 for the generalization of gauge invariance conditions.

Let us define a vector potential, A_{μ} form the electromagnetic fields,

$$F_{\mu\nu} = \frac{\partial A_{\nu}(x)}{\partial x_{\mu}} - \frac{\partial A_{\mu}(x)}{\partial x_{\nu}} \tag{13.4}$$

the gauge transform is given as

$$A'_{\mu}(x) \to A_{\mu}(x) + \frac{\partial A_{\mu}(x)}{\partial x_{\mu}}\,. \tag{13.5}$$

The Lorentz condition yields $\frac{\partial A_{\nu}(x)}{\partial x_{\nu}} = 0$ and then the equation of motion becomes

$$\Box A_{\mu}(x) = 0 \quad \text{and} \quad \Box\frac{\partial A_{\mu}(x)}{\partial x_{\nu}} = 0 \tag{13.6}$$

for free potentials, A_{μ}. Then we have the commutation relations

$$\left[A_{\mu}(x), A_{\nu}(x)\right]_{x=x'_0} = 0 \tag{13.7}$$

where we define $\partial_{\mu} \equiv \partial/\partial x_{\mu}$ as that $\partial_{\mu}\partial^{\mu} \equiv \nabla^2$.

13.3.2 *The Analytic S-Matrix in Particle Physics*

In analytic function is analytic throughout a region in a plane if its derivative exists at every point in the region. Analyticity is one of the principles of physics and particular relates to the solutions of equations of motion in the complex plane. Two major approaches to particle physics were at sway in the 1960's, 1970's and 1980's at LBNL where the S-matrix approach [19] and other was the field theoretic approach [20]. The analytic S-matrix and the field theoretic approach both had their advantages and disadvantages such as divergencies and singularities which the string theories attempting to resolve, but both have their usefulness and also their incompleteness. From the S-matrix days arose the Veneziano model, which led to the string theories. The field theoretic approach, using creation and destruction operators to describe particle interactions gave rise to QED and later QCD. The group theory formalism yielded great progress in understanding particle properties and their interactions.

The S-matrix is a unitary matrix which relates Ψ_i to Ψ_f so that $\Psi_f = SU_1\Psi_i$ where Ψ_i is the initial state and Ψ_f is the final state. The S-matrix analyticity properties, as well as Poincairé invariance and unitarity are implied by the observed characteristics of the real world, then there may exist one possible S-matrix compatible with Euclidean spacetime unlike string theory with its multiple possibilities. In view of the S-matrix theory of the 1970's, the partial bootstrap hypothesis [21] is that the observed hadron phenomena correspond to the unique Lorentz invariant, unitary, analytic S-matrix containing only singularities that correspond to particle states or only Regge Poles. The Regge trajectory and recurrence are paths of spin as a function of mass, $s(m)$ that connects particles having the same baryon number, B and I and Y spin numbers. The quantum number strangeness, S, hypercharge Y are quantum numbers assigned to hadrons to assure the law of conservation of strangeness to be formulated. Both S and Y are conserved by strong interactions as $Y = B + S$ and $Y = \frac{1}{2}Q + B$ where Q is the charge for the I spin triplet n, $p+$ and Λ and for charge conjugation to antiparticles $I_z = Q - \frac{1}{2}B$.

Experimental research indicates that S-matrix elements are analytic functions of the hadron momenta on which they depend. In the absence of zero-mass hadrons there is no bar to analytic continuation of complex values of momentum, energy, and mass, apart from isolated singularities required by unitarity. Among these singularities, simple poles correspond to particles, the pole position determining the particle mass, the residue determining a partial width which corresponds to the imaginary part of complex mass which means that the hadron is unstable, the imaginary part of the mass corresponding to the lifetime. There also occur various branch point associated Landau's work and related to the possibility that complicated reactions proceed through a succession of simpler reactions. Causality is ensured by the proper location of the Landau branch points [22]. The S-matrix constraint of "first-degree analyticity" requires postulation of no momentum singularities other than particle poles and Landau branch points. This third constraint has substantial experimental support, although its basis is not as compelling as that for the constraints of Lorentz invariance and unitarity.

The concept is that the fundamental constituents of matter are entities with distinct properties (elementary and composite particles) connected as a series of events in spacetime. The properties of particles can be described in terms of their quantum numbers. The location of the poles and cuts in complex momentum space of p_{Im} vertical and p_{Re} longitudinal coordinates yielded the particle mass and the residue of the pole gave the irradiative decay time. In formulating equations of motion, the manifold is diffeomorphic was plagued by infinities from the expanded series such as the e^{alg} = group elements and singularities in the domain. Then Veneziano model was the first attempt to develop a singularity free theory in the 1970's which led to the current Witten et al. [16,23-25] string theories. The mathematics of the Veneziano model is similar to string theory. For hadrons the string size is similar to 10^{-13}cm but for gravity, the scale is of the order of 10^{-33}cm.

The nice idea of a unique solution such as in the S-matrix approach was soon lost in a morass of solutions and possibilities so that the concept of finding a single solution to a set of equations, which describes our unique only universe was lost and the Multiverse concept was born. This view also captures the essence of one way out of the Schrödinger cat paradox, that is it leaves the cat alive, Ψ_{alive} and cat dead, Ψ_{dead} but in alternate universes so infinitum! See Fig. 13.1.

13.3.3 *Conservation Principle in Quantum Mechanics and Relativity Theory*

We briefly present some basic concepts related to conservation principles in the quantum and relativity theory. The U matrix is unitary and then $UU^{+}=1$ or $U^{-1}=U^{+}$ so that when

- Ψ is transformed as $\Psi' = U\Psi$ leaves the operators the same so as $A' \leq \Psi \left| U^1 A U^{-1} \right| \Psi >$ or

- the transform $A' = UAU^{-1}$ where A is a diagonal matrix which leaves the wave functions unchanged but both cannot be true.

So U = U^+ = U^{-1} and U is the identity element, $\underset{\sim}{I}$ where U^+ is the Hermitian conjugate of U and U^{-1} is the inverse of U. A unitary matrix assures that the total probabilities $<\Psi|\Psi>$ are conserved [28]. In the relativity theory, we consider the conservation principles for mass energy. In terms of the stress energy tensor $T_{\mu\nu}$ we have $\frac{\partial T_{\mu\nu}}{\partial x^{2\nu}} = 0$ that is, the divergence of the field vanishes. This can also be written in more compact notation as $T^{\mu\nu}_{2\nu} = 0$. Note that the notation also used is $\partial_\mu \equiv \partial / \partial_\mu$ and $\partial_\mu \partial^\mu = \nabla^2$ are previously stated.

The principle of invariance leads directly to the conservation laws of physics such as the conservation of energy or mass – energy and the continuity equation. The continuity equation can relate to change, current and mass conditions such as $\frac{\partial \rho}{\partial t} + \nabla \bullet (ev) = 0$ where ρ can stand for mass density or current density Also for 4 spacetime $\iiiint \sqrt{-g}\, d\tau$ is an invariant for $d\tau = \prod_n dx_n = dx_1 dx_2 dx_3 dx_4$ where index n runs 1 to 4. In relativity theory c, the velocity of light is an invariant or a constant. Under affine connections, transformations are linear and rotational in a uniform manner. Straight lines are carried into straight and parallel lines into parallel lines but distances between points and angle can be altered. Orthogonal transformations in five dimensions with metric (1,1,1,1,-1) and transformation matrix determinate, det = +1 is the SO(1, 4) group. The Lorentz group SO(1,3) is similarly defined relative to the 4D Minkowski space with the metric (1,1,1,-1).

13.4 A Brief Background on Group Theory

Unification theories have their expression in group theoretical terms. Group theory comprises a powerful tool for codifying the basic symmetry observed in physics from elementary particle interactions to crystals, to flowers, to tornados and hurricanes to galaxies. The basis of physical law is the formulation, in terms of variables and constants of the relationship, constancy (conservation),and symmetry of the physical universe. Both finite groups with finite algebraic generators of the group and Lie groups with infinitesimal generators of the group, which comprise the algebra are extremely useful mathematical tools.

Group theory is based on point group sets, continuous groups which the foundation of the GUT scheme. There are three types of simple reflection spaces; these are A, D, and E and there are two infinite series reflection spaces A and D. We also have the finite discrete point groups or crystallographic groups. It may be possible to form an analogy between the formation of such a crystal as an example of symmetry breaking which is analogous to the spontaneous symmetry breaking in the GUT theory. Let us form some analogies,

- Symmetry for the crystal of rotational invariance analogous to GUT is electron, photon and quark indistinguishablility under the three force interactions,
- Spontaneous symmetry breaking in the crystal which has axes chosen from three distinct directions and in the GUT theory, the Higg's field chooses out three distinct particles i.e. the electron, photon and quark also the three distinct interaction force fields as displayed in the SU_3 scheme
- At low energy, less than 100 GeV of ordinary particle interactions, for the crystal, three fundamental axis of space, for the GUT theory, yield three distinct particles, the electron, neutrino and quark, and
- High temperature "Big Bang" physics for the crystal-rotational invariance is restored, for the GUT theory, phase transitions at $T \approx 10^{27}$ degrees Kelvin ultimate symmetry is restored i.e. there is no separation of the four force fields. This is presumably the case at the time of the initial "Big Bang", the mass density is $\rho = 10^{93}$ gm/cm^3 at t~10^{-44} sec [29-31].

The basis for modern group theoretical description of based on the work of Sophus Lie in the 1890's [32]. His work was on infinite groups which have infinitesimal groups generators and infinite elements. Most group theory is involved with groups having finite elements and are based on the crystal group description [33]. A continuous group is defined as a system of objects called group elements which can be characterized by parameters varying continuously in a certain region. To every group element corresponds a set of values of the parameter within a specific region. These regions are called a group space. There is a one-to-one correspondence between group elements and points in group space. Group elements, whose parameters differ only slightly from one another, are said to be "adjacent". The products and reciprocals of adjacent elements must also be adjacent. The laws of associative law, identity element and inverse remain valid, being supplemented by the requirement of continuity. If the parameter changes continuously, we say that the group elements change continuously. Groups whose elements can be denoted by n parameters are known as n-parametric group. The region of variability of the parameters can be simply or multiply connected. The portion of the group to which the identity element is known as the infinitesimal group.

For particle physics Lie groups are particularly applicable [34,35]. We can use the short hand notation that $e^{\text{alg.}}$ = group, that is for example

$$e^{tA} = \sum_{n=0}^{\infty} \frac{t_n}{n!} A^n \tag{13.8}$$

where the group elements form a Taylor series and $\underset{\sim}{A}$ is a square matrix representation of a group or A_{nm} where indices n = m. If we have $e^{tA}e^{tB}$, e^{tA} can be used as a unitary transformation (such as to conserve the elements of the system) as $e^{tA}Be^{-tA}$ then e^{tA} = *U*, a unitary matrix of $U_{mn}^{-1} = U_{mn}^{+}$. In an *nxn* square matrix, *n* is the number of degrees of freedom, then *A* and *B* are matrices that obey commutation relations. For example $\Omega_n(A,B)$ form commutation relations such that the transformation preserve operators. Then $\Omega_0 = (A,B) \equiv B, \quad \Omega_1(A,B) = [A,B] \quad and \quad \Omega_2(A,B) = \left[A,[A,B]\right]$ or in general $\Omega_{n+1}(A,B) = \left[A, \Omega_n(A,B)\right]$ which are the formal power series as the group representations. We have

$$e^{tA}Be^{-tA} = \sum_{n-0}^{A} \frac{t^n}{n!} \Omega_n(A,B) \tag{13.9}$$

and as we said e^{tA} can comprise a unitary transformation as e^{tA}Be^{-tA} = B'. If B is a similarity transformation then B̀ = UBU$^+$ = UBU^{-1}. We can say that there are elements of *A*, $x_i \supset A$ and x_j are the elements of B, as $x_j \supset B$. In general terms e^x = 1 + x + x^2 + x^3 …x^n.

Unitary operators preserve normals of vectors; for $\langle U\Phi | U\Phi \rangle = \langle \Phi | \Phi \rangle$ where $<\Phi|$ is a bra vector and 1 Φ > is a ket vector. For A and B commuting we have [A,B] = 0 and we can write U = A + *i* B so that A = ½ (U + U$^+$) and B = = ½(U – U$^+$) for $U = \dfrac{1+iK}{1-iK}$ for U = *i*K where A is

Hermitian and generates a complex space, see Chaps 2 and 5. The SU_2 group is the symmetry of charge independence associated with isotopic spin rotations in charge space to account for the two charge states of the nucleon, n or p^0 and p^+ which is compared with the two spin states of the electron e^+ and e^-. The SU_3 symmetry group, the special unitary group in 3D correspond to the unitary unimodular transformation of charge and hypercharge. The quark triplet and antiparticle triplet, that transforms into each other under the SU_3 transformation which have fractional electronic charge and baryon number. Quarks have the same isospin and strangeness as the proton, neutron and lambda particle, Λ.

The conserved quantities of a system will prove to be invariant under the symmetry group considered such as the group of rotations U = $e^{i\Theta J}$ where J are the generators of the group. For example for $[J_+, J_-] = i\sigma_z$ which for the algebra of the group. Both J as $(J_\pm, J_z)$ and I or $(I_\pm, I_z)$ are generators of the SU_2 group, where O_3 + is the cover group of SU_3 with generators for the type $[I_+, I_-] = iI_z$. The triplets, octets and decaplets of particle groupings have mass splitting or mass difference because of the preserved of the all pervasive Higg's field. Note that in the 1970's other larger groups were considered but found to be insufficient for a unified field theory including SU_6 and $\tilde{U}_{12}$ [35]. The A_4 reflection space was utilized by Georgi and Glashow to unify the three main non-gravitational forces [13] The dimensions were allocated as electromagnetic 1D, weak force 1D and strong force as 2D in which were developed the GUT theory using SU_5. The SU_3 group theory predicts the mass splitting and quantum numbers of the octets and deceplets of the strong color force but is non relativistically invariant. The J and I spins SU_2 are conserved but are not completely conserved in SU_3. For example for a Hamiltonian operator, H then UHU^{-1} for U, a unitary transformation $[U, H] = 0$. For a unitary group for SU_3 then these transformations $[U, H] \neq 0$. If $H = H^+$, then H is a Hermitian operator and if $U^{-1}_{mn} = U^{+}_{nm}$ then U is unitary. See Chap. 3 on the principles of modern physics and in particular conservation laws and unitarity.

As we stated, fundamental geometrical forms and their group representation and interpretation are based on the relationship between finite and infinitesimal groups. Some descriptions of groups can stand for either finite point groups such as A_n and the exceptional groups, E_n but also these group labels can designate the infinitesimal Lie groups. For the finite groups we have the self dual tetrahedron A_4 group. The octahedron S_4 system is dual to the cube and the icosahedron has its own group A_5. Symmetry groups act on vertices as permutations into themselves thus describing the geometric figures' form. These solid geometric point group systems can be related to the formalism of modern physics. For example the stellar octahedral as a "mirror image" of the cube represents the eigenvalues of the strong color (quark) force. The eight vertices denote the three quarks and three antiquarks for the $\nu + e^{-} -\!\!\rightarrow \underline{\nu} + e^{+}$ reaction, yielding the correct electrical quark charge of 1/3 and/or 2/3's. The cube octahedron with 12 vertices contains the eigenvalues of the U(1) and SU_3 or the SU_4 group, which is a 15-element group i.e. the 16-element, 16-1=15 with the identity element. Particle mass splitting is associated with a reflection like space based on the lattice groups where the points are eigenvalues. Lattice spaces are related to the reflective groups, A_3, B_1, C_2, C_3 and G_2 to Lie groups, special unitary, SU_2, SU_3, and SU_5 and orthogonal groups O_2, O_3 and their geometric forms and the unitary U_1 group. For example, the G_2 group is associated with a dodecahedral structure and the orthogonal O_n group includes the two superimposed reversal triangles.

Symmetry groups and their operation are fundamental to modern physics. We examined the possible mappings, of compact Lie groups on a manifold to C_n. It is possible to choose a smooth C^∞ which is infinitely differentiable to compare or map to a crystal symmetry group lattice space C_n of E (the Euclidean group). The subgroup $K \subset E$, transforms C into itself, K is a discrete subgroup of E such that we have a homogenous space *EMK* or *E:K* where E is mapped onto K. That is we find a representation C^∞ that has a subgroup of C_n so that we can find a mapping that allows the smooth operation of a compact Lie algebra with all its entailed properties and infinitesimal generators to the crystallographic finite discrete algebra of C_n. It appears that this single mapping or morphism is possible because of the symmetry breaking in the Lie algebras of SU_2 since these groups then approximate finite discrete point group sets.

Current physics is based on symmetry principles and conservation laws which describe the objects and process. These symmetry principles and conservation laws are expressed in the algebras that generate the groups in both the broken symmetries of particle physics and the operations principles in crystallographic point group sets. These operations describe the structure of and processes in the space being considered. These operations involve translation, rotation, reflection, inversion and reciprocity or reciprocal operations. Note that this is the type of group theory "Bookkeeping" that also applies to crystallographic sets.

The symmetry properties of the Lorentz group of the Lorentz transformations is fundamental to the concept that translation (Lorentz group) and rotations (inhomogeneous Lorentz group) do not modify the laws of physics. This is related to the general principle of experimental physics and the observation of the results of experimentation and observation, that when and where an experiment is conducted leads to reproducibility. This is more difficult for complicated experiments which require knowledge of the "state of the art".

The group SL(n,c) is a subgroup of GL(n,c) formed by the matrices of a determinant one. This is where the Lorentz group of Lorentz space-time operations comes in. For n = 2 on the Minkowski light cone the operation of SL(2,C) on E_{n^3} = E (Euclidean). The center Z_2 of SL(2,C) acts trivially and the quotient group SL(2,C)/Z_2 acts effectively as the isomorphic group connected to the Lorentz group with the conjugate group SU_2. The Lorentz group algebra is the set transformations that allow the Laws of Physics to remain invariant under the set of space-time transformations under physical processes. The group generated by the translations and the connected Lorentz group transformations is the connected Poincaré group. See Chap. 3. Then space-time is one orbit of this group; the stabilizer of any point 0 is the Lorentz group leaving 0 the origin fixed.

Mappings on a sphere as an ultimate symmetry figure abstracts is of important. We can relate E, SO_n, SU_n and U_1 and U_2 type groups to GL(n,C) type groups. Consider the transformations or actions of our groups on a sphere. On a sphere we have a center, equator including the center and a vertical axis or pole through the center perpendicular to the plane of the equator. For the relationship of SO_2 (spherical rotation group) and S_2 (spherical group) of 2 dimensions their orbits one parallel circles is 2 dimensional. The two poles are fixed point and the so termed "little group" SO_2. The action SO_2 on S_2 has a symmetry through the center, having no fixed points, each orbit with 2 parallels with the same N (up), S (down) latitude, one orbit and 2 poles, and the equator.

Although superstring theories have their critics, due to the fact that those theories contain a large number of "free" parameters, there has been great interest in these theories by the physics community. Superstring theory has been related to the standard model. Some string theories contain gravity and others do not. One of the major features of superstring theory is to treat particles as tiny loops rather than as point particles so as to avoid the problem of singularities. In the string theory, particles are treated as vibrations of a membrane (M-Brane surface), which is swept out by the vibrating string occurring in 8D space. These 8D comprise 8D of the 10D standard model in which two of the dimensions are the string surface itself. This vibrational space carries the symmetry of the Lie group E_8. Superstring theory represents both bosonic and fermionic particle states. The usual string theories occupy a 26-dimensional spacetime, representing bosonic particle states. A quantum state of identical bosonic particles is symmetric under the exchange of any two particles. A quantum state of identical fermionic particles is antisymmetric under the exchange of any two particles which includes the photon and gravitation. Then we have 8 x 8 = 64 dimensional states in some superstring theories. The closed string theory is called a type II string theory, which has the doubly fermionic states included, for a total of 2 x 8 x 8 = 128 fermionic states.

In addition to the Type-II string theory, there are two heterotic superstring theories which involve closed strings. Out of the 26-L bosonic coordinates of the bosonic factor, only ten are matched to R-bosonic coordinates of the superstring factor, hence this theory effectively exists in 10D spacetime. Heterotic strings comes in two versions, that is E_8 x E_8 and the SO(32) type. The vacuum is included and E_8 is the highest dimensional exceptional group. The E_8 x E_8 superstring theory is derived from the compilation of M-Theory. One of the most promising superstring theories that unifies the four forces is the E_8 x E_8 reflection space. This is possible only because reflection embedding provides for an embedding of A_4 in E_8. the relationship between the SO(32) heterotic string theory also utilizes the E_8 x E_8 formalism [14].

In general, the Lie algebra A_n associated with a reflection space C^n has a compact Lie group SU_{n+1}. Sirag attempts to develop an interesting unified field theory in terms of U_1 x SU_2 x SU_3 x SU_4 where he identifies the SU_4 group with the tensor gravitational field [7]. The SO(32), or SO_{32}, is the group generated by 32-by-32 matrices that are orthogonal. For the strong force, gluons are described by a four dimensional SU_3 Yang-Mills theory [20]. The full set of standard model gauge bosons is described by the Yang-Mills theory with the gauge group SU_3 x SU_2 x U_1. Alternatively, for the U_5 = SU_2 x SU_3 Yang-Mills theory, the gauge group that emerges as U_3 x U_2 = SU_3 x SU_2 x U_1 x U_1 where U_1 x U_1 is the topology of the torus.

Complex spacetime is implied in the group algebra C(OD). Embedded in this algebra is the unitary group U_2 x U_2 which Sirag gives the name complexified conformally compactified spacetime [7]. This is the exact spacetime required for twistor theory [8] and there is a profound connection between twistor theory and superstring theory approach to quantum gravity [14,37]. Any quantum gravity theory is considered to exist in at least 10D. As we stated the superstring theory replaces the point particles with vibrational modes of strings. The most popular version of string theory is the E_8 x E_8 Lie group which is a 496 D space hyperspace. The dimensionality of E_8 is a 248 D hyperspace and the full symmetry group of the standard superstring theory is a product of two E_8's as E_8 x E_8 with a dimensionality double that of E_8 or 496 D. For the product of two groups, the dimensionality of the space adds.

Some of these theories will be put to some replicative test and new tests with the advent of the CERN Hadron Supercollider may soon come into extensive use. The LHC collides two 6×10^9 proton beams every second and is considered at 7 TeV to reproduce the conditions at $t \sim 10^{-12}$ sec. after the Big Bang which is the implicit model. This is the era when mass condensed out of thought to be the radiation fields. The LHC may also yield clues about the so-called extra curled up dimensions as well as our approach to hyperdimensional geometries. In the spring of 2010, the CERN large hadron supercollider (LHC) brought together two proton beams with sustained collections. The two proton beams, each having 3.5 TeV electron volts combined 7 TeV is by far the largest and highest energy collider in the world and is expected to produce the energy for particle creation. It is 27 km in circumference using superconducting magnets to steer the beam. The search is on for the Higgs boson where the matter creation era in the Universe's evolution occurred. Perhaps other dimensions may be experimentally explored through the determination of possible properties of the so termed dark matter. In addition it is hypothesized by some that mini or Planck unit black holes may be created and decay rapidly.

Phenomenological models treat the mini or Planck unit black holes with their self energy and Hawking like radiation effect so that the Tevatron LHC energy would be high enough to create short lived and detectable black holes with the Tevatron LHC. Their predicted size is about 10^{-6} or more times smaller than a proton. Various researcher debate the lifetime of the LHC produced mini black holes from extremely short to about 10^{-3} sec. The lifetime depends on the nature of the self energy and the rate of the evaporation by the emission of Hawking-like radiation which involves a γ going into an e^+e^- pair at the event horizon [38]. The bare Plank mass of $\sim 10^{-5}$ gm is exceedingly large compared to the free particle proton mass of $\sim 10^{-24}$ gm. Also the XD arising from the background metric is of interest as in these experiments. Many researcher conjecture that these experiments will confirm the existence of mini black holes and the predictions of Hawking radiation and thus hold clues about the very nature of the fabric of space itself.

Well over four-thousand papers have formulated and detail their concept of the LHC high energy collider factory of mini black holes. The Hawking radiation is expected to be observed as high energy photons (γ and x-rays) and leptons from the subcomponents quarks or partrons of the accelerated hadrons (protons) in the center of mass of the LHC colliding beams. Some researchers treat the produced mini black holes as the sudden decay of the Schwarzschild black hole state, and other researchers include charge and angular momentum, i.e., using the Reissner-Nordstrom or the Kerr black hole. The rotating charged Kerr-Newman black hole requires a much more complex calculation for the cross section of black holes production in the LHD collisions. Some of the original motivation for constructing super high energy accelerators was to find the massive Higgs boson. Now much of the current effort at the LHC is the search for mini black holes. Both the Higgs particle and mini black holes are fundamental to the construct of a unified model of the four force fields. The Higgs field elementary particles and quarks may be describable in terms of the mini black holes.

The standard unified model of current physics requires the existence of the elusive Higgs boson. Higgs suggested that space was filled with heavy molasses like substance, currently termed the Higgs field. [29, 30] This field may be associated with a massive boson particle around 80-120 GeV. It is hypothesized that such a particle gives all particle their mass as they interact through the mediator particle, the Higgs boson. This yet undetected particle was used, as is the mini black holes, to explain the missing dark matter in the universe and may be revealed in the LHC accelerator experiments. The existence of the mini black hole production from the vacuum energy may be observed through a process analogous to Hawking radiation from astrophysical black holes such as the Dirac vacuum. While the carriers of the electroweak force, i.e. the $W^{\pm}$ and Z^0 bosons have mass unlike the photon mediator of the quantum electrodynamics, photons are abundant whereas the electroweak bosons are not. Diligent work at the CERN-LEP accelerator has not revealed the elusive Higgs boson. This work has been part of a 30-year quest that, in part was the motivation for the construction of the LHC. But now a major emphasis of the work on the LHC has turned, in force, to the search for the mini black holes as the mediator particle that fills all space in analogy to vertical vacuum pair production (or mini black holes) from an energetic vacuum having a theoretical density of 10^{93} gm/cm^3 [37]. This picture yields a unification of the compatibility of QED and QCD from the early universe and as yet may be revealed in the LHC high energy experiments in the current universe.

The revolution of the prediction and observation of the mini black holes in accelerator physics will yield a much more exciting and rich find, in that now gravity can be taken into account in terms of quantum gravity, but also that gravity is unified with the strong and electroweak force as in the supersymmetry models. A new basis of a unified theory model through the mini black hole model relate, to the massive black hole of stellar and interstellar space, exploring the so called missing mass problem in cosmology [39]. Through this theory, incorporating the existence of a vacuum field of mini black holes, a unified view can incorporate gravity which utilizes the plenum of a full vacuum concept. We suggest a dual 3-brane model in which one contains the standard model fields and the other model incorporates the spin-2 graviton excitation including the vacuum energy which acts as part of the gravitational field force fields. In the quantum gravity picture, the universe may be full of HD string theory objects. These dimensions are termed branes from membranes or 2D+ spatial strings.

It is important to consider Gödel's incompleteness theorem when considering the possibility of the development of a TOE theory. Loosely stated no mathematical system can be completely self described since all the rules necessary for describing the system cannot be stated within the system. because there are more truths of a mathematical system than axioms for an algebraic system. Gödel's theorem has been demonstrated to apply to algebraic systems and geometric systems which are open and incomplete systems within themselves. Since mathematics is our tool for describing a theory of everything (TOE) as a complete theory and perhaps a complete truth, what happens if our tool, mathematics is necessarily incomplete? It is clear that structure of a TOE theory in the form of a complete theory must have within it the manner in which to address Gödel's incompleteness theorem [40].

The authors would caution to not be too hasty and quick to judge that we are soon to reach a final theory. There are many conceptual and mathematical issues to be resolved. There have been many eras in history in which led to a rude awakening to new knowledge and wondrous new discoveries to be made. Knowledge and the search for truth is an ongoing process [41].

In most societies, people tend to believe they have "complete" or "near complete" knowledge of philosophical and religious beliefs and scientific knowledge and in some cases, they know all they need to know. There are always knowledge seekers who look beyond and search for the deeper meaning, interpretation and data gathering with its organization into theory. These seekers and listeners to the heart beat of nature will ever expand our view of us and the Universe.

A brief consideration should be made as to the manner in which scientific exploration expands and it is not all an orderly process. In 1899, the Commissioner of Patents suggested closing the Patent Office because almost everything had been invented! Yet many more inventions were made and patented, some by the authors of this text. After an illustrious career, Lord William Thomson Kelven retired in the 1880's. He announced that all the discoveries in physics had been made and that all that was to adjust the last decimal point in various measurements and hence students should not go into physics. He also pointed out that there were two blots on the horizon of physics. One was the interpretation of the Michelson-Morley experiment [42] and the other was the problem of the fit of the Rayleigh-Jean's law of black body radiation called the ultraviolet catastrophe [43]. The first, is said, to have led to the relativity theory and the second to the quantum theory through Max Planck's correct fit with the introduction of his, Planck's constant, $\hbar$. One should always be suspicious of statements such as "we almost know everything". Such a position is almost always a gateway to a new scientific revolution! The end of civilizations and the beginning of new ones most likely grow from such a hypothesis. Major changes in thinking such as the Copernican revolution as well as the advent of the quantum mechanics and relativity have vast philosophical as well as scientific effects so as to create a paradigm shift in thinking.

We believe we are heading for a crisis and hence a revolution I physics. For example the non understood hypothesis of dark matter and dark energy, Higgs particles and mini black holes, etc point to the need for a deeper more thorough reexamination. But through crisis comes new knowledge through the resolution of crisis and to the next deeper knowledge, ourselves, life existence and truth. We hope that our work is another step towards a new and viable approach that, as the uncertainty principle did, nonlocally does by discovering and formulating another truth and in some manner which incorporates the observer and observed in our ongoing glory of discovery [44,45].

Grand unified theories (GUTs) are an attempt to unify the mathematical description of the electromagnetic and weak force (electromagnetic force) with the strong force. Supersymmetry models, string theory, the "theory of everything" use GUT theories to form a unity with the gravitational force (a spin 2, tensor force). The GUT description finds its origin with James Clerk Maxwell's unification of electricity and magnetism. A further major step was taken with the development in the context of the quantum theory is the theory of quantum electrodynamics (QED). No GUT theory comprises a complete theory even an adequate unification and, of course, does not contain gravity, which is necessary for a so termed "theory of everything". String theories are notorious for not containing uniqueness and a huge number of such theories exist. No complete, comprehensive and unique supersymmetry, string theory or "theory of everything" (TOE) exists. The progress to a TOE is desirable but needs to proceed with caution. We would be most hesitant to embrace a TOE as it appears that knowledge is an ever expanding. Our work is a path to new knowledge, which leads us all to more new knowledge and truths. The fundamental question is can we find "the truth of everything" or only relative greater truths. What one sought and found as true yesterday, with new knowledge and new data, guides us in a new direction today, and tomorrow! Like traveling on towards a rainbow as it ever moves from us as we approach.

References

[1] Wheeler, J.A. (1978) Mathematical foundations of quantum mechanics, in A.R. Marlow (ed.) pp. 9-48, New York: Academic Press.
[2] Eccles, J.C. (1952) The Neurophysiological Basis of The Mind, Oxford: Oxford Univ. Press
[3] Rauscher, E.A. (2010) Quantum mechanics and the role of consciousness in the physical world, in R.L. Amoroso, (ed.) Complementarity of Mind and body: Realizing the Dream of Descartes, Einstein and Eccles, New York: Nova Science.
[4] Wells, H.G. (1920) Outline of History, New York: McMillian and Co.

[5] Adler, M.J. (ed.) (1952) The Great Books of the Western World, 54 volumes, Chicago: William Benton.
[6] Hofman, S., Bleicher, M., Garland, I., Hoffelder, S., Chwake, S. & Stockew, H. (2002) High Energy Physics-Phenomology, arxiv: Hep – Ph /0111002.
[7] Sirag, S.P. (1983) International J Theor. Phy. 22, 1067.
[8] Penrose, R. & Rindler, W. (2008) Spinor and Twistor Methods in Spinors and Spacetime, Cambridge: Cambridge Univ. Press.
[9] Hansen, R.O. & Newman, E.T. (1975) Gen. Rel. and Gravitation 6, 216.
[10] Ramon, C. & Rauscher, E.A. (1980) Found. Physics 10, 661.
[11] Rauscher, E.A. (1996) Some models potentially applicable to remote perception, A. Puharich, & B.D. Josephson, (eds.) The Iceland Papers, Select Papers on Experimental and Theoretical Research on the Physics of Consciousness, 2nd edition, pp. 50-93, Ottawa: PACE.
[12] Gerock, R. Heldand, A. & Penrose, R. (1973) J. Math. Phys. 14, 874.
[13] Georgi, H. & Glashow, S.L. (1974) Phys. Rev. Lett. 32, 438.
[14] Weinberg, S. (1995) The Quantum Theory of Fields, Cambridge: Cambridge Univ. Press.
[15] Georgi, H. (1982) Lie Algebras in Particle Physics, Redwood City: Benjamin Cummings.
[16] Green, M.B., Schwarz, J.H. & Witten, E. (1987) Superstring Theory, Cambridge: Cambridge Univ. Press.
[17] Muta, T. (1987) Foundations of Quantum Chromodynamics, Singapore: World Scientific.
[18] Rauscher, E.A. (1970) Bootstrap and a Uniform Formalism of the Four Force Fields, LBNL, UCRL – 20068.
[19] Chew, G. (1964) S-Matrix Theory of Strong Interactions, New York: Benjamin.
[20] Gross, F. (1993) Relativistic Quantum Mechanics and Field Theory, New York: John Wiley and Sons; and valuable private communication with Evyand Wichmann, UCB with EAR, 1964.
[21] Chew, G. & Pignotti, A. (1968) Multiperipheral bootstrap model, Phys. Rev. 176, 2112; and valuable private communication LBNL, 1964-2003.
[22] Landau, L.D. (1946) JETF 16, 574.
[23] Witten, E. (1985) Superconducting strings, Nucl. Phys. B249, 557.
[24] Witten, E. (2003) Perturbative gauge theory as a string theory in twistor space, arXiv:hep-th/0312155.
[25] Hall, G.M. (2005) A geometry for non geometric string backgrounds, arXiv hep-th/0406102, v.3.
[26] Duff, M.J. (1990) Recent results in extra dimensions, in T. Piran & S. Weinberg (eds.) Physics of Higher Dimensions, pp. 40-91, Singapore: World Scientific (1986) and Nucl. Phys. B235, 610.
[27] Duff, M.J. (1990) Nucl. Phys. B235, 610.
[28] Messiah, A. (1961) Quantum Mechanics, Vol. I, Amsterdam: North-Holland.
[29] Higgs, P.W. (1964) Broken Symmetries, Massless Particles and Gauge Fields, Phys. Lett. 12, 132.
[30] Higgs, P.W. (1964) Broken symmetries and the masses of gauge bosons, Phys. Rev. Lett. 13, 508.

[31] Rauscher, E.A. (2005) Cosmogenesis and quantum gravity, in R.L. Amoroso, B. Lehnert & J.P. Vigier, (eds.) Beyond the Statistical Model: Searching for Unity in Physics, pp. 43-72, Oakland: The Noetic Press.
[32] Lie, S. (1893) *Vorlesungen Uber Kontinulerliche Grupper mit Geometrischen und onderen Anwendungen*, G. Scheffer (ed.) Leipzig: Teubneir.
[33] Coxeter, H.S.M. (1991) Complex Regular Polytopes, 2nd ed., Cambridge: Cambridge U. Press.
[34] Gilmore, R. (1974) Lie Groups, Lie Algebras and Some of Their Applications, New York: Wiley Interscience.
[35] Boerner, H. (1963) Representation of Groups, New York: John Wiley and Sons.
[36] Mahanthappa, K.T. & Sundershan, E.C.G. (1965) Lorentz covariant SU_6, particle-antiparticle algebras, and supermultiple structure, Phys. Rev. Lett. 14, 458.
[37] Rauscher, E.A. (1971) A Unifying Theory of Fundamental Processes, LBNL / UCB book, USRL 20808.
[38] Hawking, S. (1975) Particle creation by black holes, Communications Math. Phys. 43, 199.
[39] Rauscher, E.A. (1972) Closed cosmological solutions to Einstein's field equations, Lett. Nuovo Cimento 3, 661.
[40] Smullyan, R.M. (1992) Gödel's Incompleteness Theorems, Oxford: Oxford Univ. Press.
[41] Amoroso, R.L. & Rauscher, E.A. (2009) The Holographic Anthropic Multiverse: Formulating the Complex Geometry of Reality, Singapore: World Scientific.
[42] Michelson, A.A. & Morley, F.W. (1887) Am. J. Science 34, 333.
[43] Planck, M. (1897) *Verlesungen Uber Thermodynamik*, Berlin: Veit and Co.
[44] Bohm, D. & Hiley, B.J. (1993) The Undivided Universe, New York: Routledge.
[45] Wigner, E. (1967) Symmetries and Reflections, Bloomington: University of Indiana; and private communication with EAR.

Chapter 14

Unity of Consciousness Experience, Nature of the Observer and Current Physical Theory

...the wise man looks into space, and does not regard the small as too little, nor the great as too big; for he knows that there is no limit to dimensions. - Lao-tse

...it is by logic that we prove, but by intuition that we discover. - Poincaré

He liked watching the glorious stars, thinking "there must be myriads of worlds out there". Then one night he shifted his awareness toward and into himself. "By God," he whispered, but only to himself, "there are myriads of worlds there, too!" [1]. We in the western scientific culture have just begun, *en masse*, to explore our inner cosmos; something most cultures in the East have been doing for thousands of years. Inner exploration has been an intellectual activity in the relatively recent past, and has been for the most part associated with psychotherapy. Now inner exploration is beginning to enter the domain of emotional and spiritual development as well. Certainly the nature of our inner being, the nature and structure of our consciousness, shapes and determines our concept of reality. This realization is having a vast impact on world societies and us, as individuals. One manner in which we internally organize the intellectual, emotional and spiritual information we receive is by a mental system of concepts or categories of information and their causal relations. We address the fundamental nature of conscious perception and how we comprehend existence. Techniques such as yoga, meditation, and processes of spiritual awakening have opened the horizons to the consideration of the attributes of the consciousness. There also appear to be clues as to the nature of consciousness in the structure of physical theory. In fact, the co-called internal journey and external validation system of science may be leading us onto a similar path of knowing. Until recently psychologists were not given tenure if they dared to look into the so called "black-box" of the mind. It was unheard of for physicists to consider the mind-body problem. But in recent times the issue of the observer in measurement theory has plagued physicists sufficiently to consider it.

14.1 Role of the Observer in Modern Physical Theories

This chapter is an attempt to make it easier for physicists to study the role of the observer in measurement theory and mankind's place in an anthropic multiverse. We explore the bases of the structure and representations of human thought and thought processes in order to better understand the scientific method and other modes of inquiry even to the extent of making a case to complete the tenets of epistemology. A better understanding of the relationship between our inner thinking, feeling modes on the external world and our universal connections, will allow us to better move to world peace, personal peace and freedom. What should start our discussion of the possible relationship between states of consciousness and modern physics? Let us start from a concept so well expressed Eddington: "Physics is the study of the structure of consciousness"[2]. It is the mind that is the ultimate instrument for "doing" physics. Not only do concepts in philosophy, psychology, and perhaps neurophysiology lead us to the conclusion that the structure and content of physics may depend deeply on the relationship of physical theory to the structure of consciousness. Also, recent discoveries in physics itself indicate a need to examine this relationship.

Discoveries, and/or creations of new concepts in physics lead to the observer/participant issue. Quantum mechanics, the theory of atomic microcosm, is a description that may imply that the state of the observer affects his interpretation of what is seen. In the context of quantum theory and relativity, we may be able to shed light on the relationship of discovery vs. creation and its validation of the properties of an external reality. In quantum physics, as well as in the structure of n-dimensional relativistic models, the implication for a fundamental remote connection of events is deeply implied by the structure of these models, and we can use this property for the experimental verification of these models.

The structure of physical theory, its very fabric, points towards a world view that speaks of concepts which have been considered external to the body of science but yet may be implied by modern physical theory itself. Can light be shed on the seeming paradox from research in the altered or unusual states of consciousness, paranormal and mystical phenomena, and states of meditation, etc.? One of the deepest, most profound discoveries for the western mind is the induced meditative state, that is, the discovery of an intrinsic variety of a state of consciousness common to most people who modify their behavior to experience it. Altered or mystical states of consciousness may hold great clues to resolving conceptual paradoxes in science and in life. We have an unprecedented opportunity in current world society to integrate the intuitive mystical and spiritual with the analytical pragmatic modes of scientific thinking and being.

14.2 States of Consciousness and Reality Perceptive Modes

We will define "altered states of consciousness" (ASC) for purposes of the present discussion as mental states other than awake, "beta wave," or sleeping, non-dreaming states of consciousness. Therefore, we are including dreaming as an altered state. Note that this is a matter of definition. Also, there is a great variety of subjective reports within the dreaming experience.

The methods for inducing such states may either be chemically or methodology produced, such as in the use of meditation, yoga, self-reflection, "magic mushrooms," *amanita muscaria*, or not require an external agent and resorting to meditation or sleep/dreaming. There appears to be a set of relatively well characterized states of consciousness [3,4] which can be induced and experienced by one's internal self that produce descriptions by individuals of very similar mental/emotional and spiritual experiences [4].

Visual imagery (audio, etc.) is well known in dreams but is also reported by meditators. One of the primary experiences induced by artificial means such as psilocybin, and hallucinogens is their visual pageantry which excites the imagined, stimulated sense. What is meant by image, hallucination and stimulus? A hallucination is usually defined in terms of an image in the mind which is not prompted by an external source of visual stimulus of light (photons) impinging on one's retina through the lens of one's eyes. We talk of "imagined," which is associated with not being real, i.e., not having a counterpart in the external physical world.

But then, how real is real? How real is an internal visual image? If it is an experience which is acted upon – it is seen, felt, heard – is it not real? The criterion for reality in the western world (and science) is one of a consensus reality. The usual test of an experience (for example, a visual image) is usually made against an assumed to be external physical reality. But we cannot develop a hard and fast proof of "It Exists", (cf. Rene Descartes'), "I think therefore I am! I did not create myself, therefore God did it."

Meditators also report a consensus reality (more on this later), but where is the image? It's in the mind! But what is and where is the mind? Is it a physical neumenon or a phenomenological neural artifact? Visual imagery that also appears not to be instigated by photons from external sources appears as the 'remote perception' of psychic information imagery. A subject-participant, in a laboratory experiment, describes a randomly chosen distant target location, under conditions in which an agent, termed a "blind judge", can match a target with a subject's description, blind to the correct match, out of many possibilities. We then bridge the gap between external information sources and internal mind imagery without external physical light input and yet a correlation of that perception is made to an external target and hence is an objective reality. This anomalous 'psi' artifact has been statistically measured in many forms during the past two centuries, but as of yet there is no comprehensive theory and repeatability is difficult and usually highly disputed by empiricists especially because of an inability to formally calibrate 'subjective reporting' and since there are essentially no 'virtuoso' subjects.

We have at least three sources of imagery:

- that prompted by an external stimuli;
- that prompted only by an internal stimuli;
- that which is prompted by an external stimuli but is not connected by any known physical means [5] to the process of perception.

Another example of a checkable reality match for a so-called external imagery is exemplified by Friedrich August von Stradonitz Kekulé, 19th century German chemist, who dreamed of six snakes biting each other's tails. He had been trying to work out the structure of the molecular form of benzene, C_6H_6, and based on his dream, demonstrated it was a ring structure and not a linear structure.

A resolution of the discussion of the validity of these various information sources which lead to mental imagery may lie in the concept of the existence of states of consciousness. If the model of consciousness is a set of states or its structure has "levels of awareness", this concept might resolve the disagreement about the criterion of internal vs. external validation. Both internal and external experiences and their source of origin may be equally valid, but may involve different distinct levels of consciousness in a vast possible array of states of consciousness being and perceiving.

Perhaps the concept of altered, non-ordinary or unusual states of consciousness or levels of perceptual awareness will shed some light on some of the current paradoxes in modern physics. The existence of a paradox implies lack of information or ignorance, as nature does not admit such a paradox. A paradox resolved leads to a new level of awareness and understanding. Perhaps current events are leading to a Kuhnian paradigm shift [6] which may involve a shift in understanding that some scientists and non-scientists are in a different state of consciousness as they examine the same event. The lesson of autogenic training and other processes which modify the central nervous system (CNS) may be that there is a host of new horizons, of new unexplored states of consciousness, to experience, and in which one can attempt to explore reality and discover a wider, vaster view of existence. Then, if we do deal with the framework in which consciousness is "all", or the seat or root of reality, then understanding more about states of consciousness is a vital key to comprehending the nature of reality.

14.3 Defining Science and Mysticism: Methods in the Search for Knowledge

There are two dominant methods for gathering information about reality and developing concepts about what one deems reality to be. They are: the scientific method, and the mystical, intuitive or experiential method. Science is defined as accumulated systematized knowledge, ascertained by observation and experiment, which is brought under general rules or laws. At the basis of the scientific method is experience, i.e., in the form of experimental tests. What is crucial to the method is the interplay of observation (experiment) and testable theoretical hypothesis. Reason and logic are fundamental to the construction of these theoretical models. Mysticism, on the other hand, relates to obtaining information by direct or intuitive experience. Mysticism is the belief that the most reliable source of knowledge or truth is intuition rather than reason or the scientific method.

Perhaps the most fundamental aspects of intuition are not at odds with science but are at the root of both science and mystical experience such as Kekule's dream. The relationship of the scientific and mystical view is rooted in the nature of consciousness. Forms and varieties of the experience of states of consciousness can give us clues as to the relationship and validity of the use of methods of science and mysticism in the search for knowledge. One of the ultimate desires of mankind is self-knowledge, i.e., understanding consciousness itself.

14.4 Some Basic Tenets of Science and the Western View: Duality, Causality and Object Grouping

The struggle between parts, separateness or duality and unity or whole is seen in many world religious and philosophies. Central to the struggle is an attempt to understand one's connection to the Universe. In physics we examine in more detail the possible relation of the observer and the observed. Often, when the dualistic view is set up, there is an attempt to overcome dualism and reunite that which has been conceptually divided. Socrates, via Plato's dialogues, suggests a model of reality consisting of physical objects and another world filled with a perfect "master plan" of images of those objects, such as a perfect chair or table. The perfect-image world also contains a representation of perfect love, or good, etc. Another model employing a multiplicity of objects, or noun concepts, or ideas is that of Immanuel Kant, who suggested that the mental attribute to categorize objects or concepts (for example, to identify and distinguish chairs from tables or love from hate) is an inborn characteristic of the human mind. These are, in essence, dualistic models (see Table 14.1), and the concept of categorization, or object identification and grouping, is also a key tenet of science.

William von Gottfried Leibniz attempted to reconcile the Greek mind/body dualism by hypothesizing the existence of monads. The monads, or "atoms of matter and mind", operate to link up, by a synchronistic mechanism at one's birth, in such a manner as to correlate one's intentions with one's actions. If one intends to raise one's arm, it will move, not because one directly wills it to do so but because at one's birth, mind/matter monads were set up to produce this coincidence! This is Leibniz's view. Enter the debate over free will vs. determinism. In this view, all events are now definable in terms of causal terms.

A predominant concept of western scientific truths is that of causality. The causality description is the way in which causes and effect are related and is certainly another dualistic model. Immanuel Kant suggests that causality, like categorization, is also an innate construct of the human mind.

Table 14.1 Western Philosophers and Their Concepts

Philosopher	Born/Died	Major Concepts
Descartes, Rene	(1569-1650)	1. Mechanistic view 2. Mind/body duality 3. Importance of pineal gland 4. Acausal 5. Space-time independence 6. "What you see is what there is"
Leibnitz (Wilhelm von Göttfried)	(1646-1695)	1. The monad as fundamental metaphysical unity 2. Synchronistic link of mind/body division 3. Space-time independence 4. Acausal
Kant, Immanuel	(1724-1804)	1. Innate categorization 2. Causality (cause-effect relationship)
Spinoza, Benedict	(1632-1677)	1. Contingency (like monads) 2. Causality 3. One-world unity 4. Process as primary
Hegel (George Wilhelm Friedrich)	(1770-1831)	1. Thesis, antithesis & synthesis: analogy to yin-yang concept
Jung, Carl	(1875-1961)	1. Archetype (as in Socrates "world of images") 2. Synchronistic (analogous to 6th century B.C. view of Lao Tze)

Besides Leibniz's theory, in which he attempts to reconcile the mind/body duality by introducing the system of monad synchronicity, is the synchronicity theory of Carl Jung. After Carl Jung's break with Freud, Jung suggested an acausal model in which events are correlated by meaningful coincidence of synchronicities. Not only do these coincidences occur "randomly," but also methods can be utilized to set up meaningful synchronicities which can yield useful information. Jung was particularly interested in the I Ching, in which information is accessed by a "random" process of a set of tosses of three coins of yarrow sticks. Jung describes a mechanism whereby the system of synchronicities are correlated by a universal or cosmic consciousness [7]. Again, we have thrust toward the holistic or universal model. It is difficult to summarize anyone's philosophical framework in a few words. In fact, one's philosophical concepts change over time and may explore different points of view, comparing and contrasting them to other's philosophies.

14.5 Physical Models of Interconnectedness

Much of the conceptual framework of western philosophical and scientific though treats the observer of world processes as an inert and passive agent with respect to the observed "reality". Reality is pictured as something external and untouched by the process of observation by human consciousness. How universal is this view? Are there philosophical systems which hold other tenets about the relationship of the observer of reality and the observed reality? Systems of thought such as physical science or mysticism are based on philosophical precepts. There are realms of physics and mysticism where the observer may not only be linked to the observed by affecting the observed, but may also be, to a degree, the creator of the observed reality.

Once one chooses the view of the observer and observed as separate, one is immediately confronted with a dualistic model (discussed in the previous section). Then the philosophical thrust is towards re-uniting the dualistic aspects. Although dualism is a predominant concept in western science and philosophy, there are also non-dualistic models or conceptual frameworks about reality, as we shall explore further. Starting with the observer/observed duality, let us explore some ways in which new discoveries in 20th century physics may imply a link between these two and a possible dissolution of this duality into a unity.

At the microscopic atomic level, the position of particles and information about their physical properties, which we are observing and measuring, appear to be altered by our probes. These probes are other particles which collide with the target particles. It appears that we cannot "look" at the world as though we didn't observe it. The apparent effect of the observer on the observed, in the micro domain of the atom, is termed the Heisenberg Uncertainty Principle.

Physicists are also interested in another observation about microscopic quantum phenomena which seems to imply a connection or correlation of systems of particles remote from each other in space. If particles which are juxtaposed in space are correlated with each other; that is, are related to each other by their respective physical properties, they remain correlated even when separated by large distances (several meters, which is indeed large on the atomic scale) [8]. This form of "interconnectedness" is not part of the conventional interpretations of physics but was proposed by Einstein and others in the 1930's [8]. The interconnectedness hypothesis relates to the formulation of completeness of the quantum mechanical description of matter and the universal applicability of the Heisenberg Uncertainty principle [9].

Bell [8] formulated a very detailed description of the behavior of a particle in the microdomain. Bell's work in 1964 is based on the earlier important questions raised by Einstein and others in 1935 about the nature of the completeness of the quantum description of the microscopic behavior of particles. In 1973 Clauser [10] and others tested Bell's hypothesis in the laboratory and found that the interconnectedness hypothesis formulated by Bell appeared to be valid. The experiment involves the production of a pair of photons (or particles or quanta of light) at a fixed source. These two photons move away from each other in opposite directions. (Photons move at the velocity of light since they are quanta of light.) The relative spins (photons rotate like little tops) are measured. It has been determined that this spinning is correlated even when the photons are separated by up to several meters. This correlation, although somewhat complex, seems to be a manifestation of a fundamental interconnectedness, at least on the microscopic level [5,10].

This nonlocal correlation, termed Bell's Theorem, may have implications similar to the Heisenberg Principle. It's as if a spider in its web feels each distant disturbance of the web's far reaches as it sets at the center of the web waiting for dinner. Bohm and Pribram, [12] as well as others, have suggested holographic models of events and consciousness [11,12]. This view may be consistent with some of the multi-dimensional models discussed below.

Perhaps the universe is constructed in such a way that what one does or thinks here on the planet Earth may indeed affect conditions in the far reaches of space. Discoveries in physics do relate to the philosophical precepts constructed by humankind. Volumes of philosophical discourses have been generated on the issue of observer effects, chance and free will, and the uncertainty condition proposed by Heisenberg. We anticipate that volumes will also be generated on Bell's "interconnectedness" theorem and its relationship to cognitive processes generating physics and mystical concepts, as the implications of the theorem seem to relate to the visions expressed by certain mystics. Do the implications of Bell's theorem (nonlocal interconnectedness) hold for large-scale processes? Are there other theories that could describe remote or nonlocal interconnectedness? There are a number of differing views by researcher on the first question and much work may precede its resolution. We shall examine some physical models which may answer the second question [13-16].

There are other models being examined which involve apparent macroscopic nonlocal correlations. One such model which we have been examining is a macroscopic interconnectedness model formulated in terms of multi-dimensional geometries [5,13]. Geometries involve more than the four dimensions of three spatial (dimensions) and one temporal dimension. The construction of these dimensions is such that there are conditions in which information can be accessed from remote space-time locations such that they appear juxtaposed and not remote from the perspective of the higher dimensional space. A set of these dimensions involve both real space-time and imaginary dimensions which are space-time dimensions multiplied by the imaginary number or $i = \sqrt{-1}$. We examine the possibility that physical effects can be transmitted over a distance, which does not violate our usual concept of the relationship of causes and effects [14,16]. The important point is that physicists are now examining the concept of remote interconnectedness (see [5] and [13] for further details).

The potential nonlocal connection of event may demonstrate certain types of unity which we will discuss as a mystical concept. Also, if there are systems where photons and other particles are nonlocally connected, this type of correlation may also involve the observer and the observed. Perhaps physical theories will describe certain systems in which the manner which one looks at something will determine what it appears to be.

The relativity of motion, as described by Einstein, implies that one's observational vantage point affects what one thinks one is observing. The relativity theory is, in general, a macroscopic description of causal connections of events and synchronization of time. It is observed in physics that the time recorded on a watch depends on the relative frame of reference of the observer and the observed (such as their watches'). If the timepiece moves very near the speed of light past the observer, the time reading appears to change depending on the velocity of the watch's motion. Although the principle of Lorentz invariance insures that the laws of physics remain unchanged by motion, translation, rotations, etc., one's observational perspective makes things appear different, depending on the way in which they are observed.

From the macroscopic domains it seems that we cannot make a physical observation of the world without affecting the patterns of the observed; that somehow, "the observer is a participator; it is a participatory universe," in the words of Wigner [17]. What is observed? What is reality? We assume (in western thought based on the logic system developed in Greece a few centuries prior to the birth of Christ) that there is an external, solid, "out there" reality; that, in essence this reality is immutable [18]. But is this the case? We discover, more and more, that this "reality" depends on our frame of mind, on our state of consciousness or, "on state specific science," Tart [3,4]. If one's mood can affect how one views life, can it also effect one's observation of the period of swing of a pendulum bob?

The results of the Clauser-Aspect experiment testing Bell's theorem appears to require giving up realistic, local models of reality, i.e. objectivity. The ability of the mind to transcend space and time demands giving up absolute "Newtonian" objectivity because the interaction of consciousness with physical matter changes and modifies matter and consciousness. The scientific method yields the valid results that it does because there is approximate objectivity, which is more applicable in the classical domain and begins to break down in the quantum domain.

Eddington goes further by suggesting that, "Physics is the study of the structure of our conscious minds!" [2]. In fact, it is certainly valid to say that our minds are the ultimate instruments for "doing" science; that is back of every telescope is an eye, and back of that is (hopefully) a mind! Eddington's thoughts may lead us to the Buddha Prince (Gautama, ca 550 B.C.) concept, that reality is in part, or completely, a construct of consciousness, and that not only is the universe "perturbed" by the observer consciousness, it is created by it! The ultimate question then is: Is what we believe not only creating how the world appears to us (state specific science), but determining how it really does work?

One of the major debates in science, and the so-called pure science of mathematics and science itself, is: does the human mind discover the workings of an "out there" external reality, or does mind create the reality? Do we create ideas, ideas about the workings of an external reality? In fact, do we create the reality through individual and/or collective consciousness? Certainly this hypothesis has been made. Let us term this model a "consciousness ontology": that being or existence is a creation of consciousness [19].

Much of western philosophy and science concerns itself with what we may term as theories of knowledge of epistemology. The main concern of such a system is how do we discuss the nature and structure of an external reality. Such a system, consisting of the "discoverer" and the "discovered" in the epistemological approach, necessarily leads to a dualistic model of the observer and the observed. If there is the mind (observer) and the physical reality (the observed), how then are the two (mind-body) linked once dualism is hypothesized? Where does one draw the line in the chain of the observer and observed. This is a major concern of quantum theorists since, as we discussed, it appears that the observer is not passively inert, but a participator. In defining a quantum system it is of major consideration where one decides the "eye" ends and the "universe" beings? See some constructs of unity we list in Table 14.1.

Another example in physics of the expression of a fundamental interconnectedness is a principle formulated by Mach (1838-1916) [19], we may find another "connectedness" clue and a possible relationship to the absolute "Achasic records". Mach's principle is not tightly formulated like Bell's theorem, but may relate to it. Although one of Einstein's axioms of relativity is that there is no fixed reference frame in space, the structure of his theory does not preclude this possibility. Mach's principle states that a local even, such as the rotation of a bucket of water, depends on the whole fixed star system, i.e., the rest of the universe. If one rotates the bucket of water fairly rapidly, the surface of the water forms a parabolic meniscus. The rotation of the bucket and the shape of the water's surface depend on the existence of the universe or some fixed reference frame defined by it [20]. Then it would appear that Mach's principle is also a statement of or aspect of a principle of interconnectedness which is more precisely and specifically formulated by Bell's theorem [5,17]. The application of Bell's theorem has been specifically made for quantum systems, although it is more general than the quantum theory. Mach's principle has been discussed primarily in terms of astrophysical applications. In the next section we will discuss the possible interpretation of the incompleteness theorem of Kurt Gödel.

Table 14.2

Objectivity, Subjectivity, and Reality: Contour Integral Model

1) What we perceive as reality depends on our assumptions and state of consciousness
2) Barriers are useful but limited and are artificial constructs of the mind
3) What is real must necessarily include that aspect called mind/reality as well as what is called external reality, to be complete or unified
4) It is as though we see only the isolated islands above the sea rather than the universal of the whole of the land beneath the sea
5) In a sense, when we include the mind's knowledge and structure in our description of reality, we find our perception appears to become dynamically "unbounded" with new aspects of possibilities. See Chap. 13.

14.6 Gödel's Incompleteness Theorem and the Search for Truth

"The way that can be described is not the way", from the Book of Tao by Lao Tze, China, ca 6th century B.C. Can we find an expression of this eastern thought in western culture? Prince Gautama Buddha expressed the concept that enlightenment comes to an individual but cannot be taught or explained. The only thing that can be taught or discussed are some of the possible conditions one can practice to make it possible for enlightenment to happen.

Gödel, developed a mathematical theorem in 1931 which states, in part, that all the truths of a mathematical system do not follow from its axioms. (There are more truths than axioms.) [21]. He first applied his theorem to algebraic systems, then to geometric systems, and demonstrated that every mathematical system (algebraic or geometric) was necessarily incomplete. Since all language systems, mathematics, English, Japanese, Chinese, Russian etc., are based on the logic axioms of arithmetic or algebraic systems, then Gödel's theorem implies that all language systems are necessarily incomplete. Since we communicate thought by the symbolic representation of language, it appears then that a complete thought system can neither be expressed verbally nor written in a language/logic system [21].

By definition, we believe that in order to construct an ultimate truth, this truth should necessarily be complete. Then it appears that in western logic we see a parallel to Buddha's contention about enlightenment. The ultimate truth then will "just come" when the right conditions are met and not reasoned; that is to say, one cannot write the ultimate truth as a mathematical equation or set of equations, nor can one even describe it to another. Can one at least suggest the path to take to obtain enlightenment? Or can one suggest the path, or Tao? Yes; this is what Buddha's teachings are about, how to set up the conditions to receive enlightenment.

So it appears that the facts of physics and the scientific method may not yield the ultimate answer to the riddle of the universe, but the scientific endeavor is enjoyable and may be a part of the Tao. Science involves maps or theories which are approximate to the territory or reality. So-called "occult truth" may be irrational, that is, "not of reason", or at least beyond the scope of reason, but maybe, via Gödel's theorem, all Truth is "irrational". One interpretation is that it is neither rational nor irrational but a-rational. That is, it is not one or the other but at the root of both. Reason and feeling may be derived from the ultimate truth and have their roots in it.

Gödel also entered the search for certainty in mathematics by demonstrating it is not absolute, just as Heisenberg had done for the physical sciences with his Uncertainty Principle developed five years earlier [21]. Whereas Heisenberg demonstrated that the observer is a participator, Gödel formulated the incompleteness of mathematical systems. It was the completeness concept that led to the formulation of Bell's theorem [8]; the theorem is a quantitative formulation of the Einstein Podolsky Rosen Paradox formulate [22] by these three authors in the 1930's as a description of the completeness (or lack of it) and connectedness (and perhaps unity) become inexorably intertwined. If we search for truth, particularly a universal truth, we certainly would consider these two concepts as central.

14.7 Observer / Participant and Schrödinger's Cat

We have a variety of thinking modes which we utilize. These bring into question objectivity vs. subjectivity. One of the major tenets of the scientific method is the assumption of objectivity, that is, a consensus reality about external events such that a scientist replicates his own and other scientists' results so that under the same conditions he gets the same results. This is also called Lorentz invariance. Subjectivity, on the other hand suggests that observations are dependent on and unique to a particular observer.

Some have thought that internal reality states, such as in meditation are therefore subjective, and yet we do find a consensus reality here also, even though the "objects" observed are in one's head. But then where is the object of "blueness" (qualia) of the sky in our head? and what is "out there" prompting our perception? Cognitive psychologists argue about our ultimate experience of internal and external perceptions. Let us discuss the observer/observed issue in terms of physics. What is the definition of an external/internal boundary? Is the skin (sense), retina (sight), eardrum, anvil and stirrup (hearing)? Is this definition in some sense arbitrary?

When we attempt to define the observer/observed link, we find that the definition affects the manner in which we define a so-called objective system. We have a test then that this boundary between observer and observed is not arbitrary. In the 1920's and 30's there was much discussion of the interpretation over the newly developing description of microcosmos with the quantum theory. Some said that the theory was just a pragmatic method of predicting the outcome of a specific experiment termed the Copenhagen view, after Neils Bohr's interpretation and one could not build a model of reality from this theory. Others said the purpose of doing physics is to comprehend reality by building testable models of it.

One *gedanken* (thought experiment) that pinpoints some of the problems of the interpretation of the quantum theory and the link of the observer with the observed is the "Schrödinger cat paradox", named after one of the developers of the quantum theory, Schrödinger [23]. A cat is locked in a room where it will eventually be killed by a poison gas pellet activated by a "random" quantum atomic decay process. Before an experimenter looks in the chamber, and after a given time interval, is the cat dead or alive? The theoretical physicist sits down with pen, paper, calculator and computer, and using the quantum theory figures out what will happen after a certain time interval. The solution to the Schrödinger wave equation, used to describe the process of radioactive decay, has two solutions. In one possible universe (solution), or eigenstate, the cat is still alive; and in the other equally possible universe, the cat is dead. The total solution is the sum of these two solutions and the wave function eigenstate only "collapses" to the alive or dead cat eigenstate, after the cat is observed. In this sense, the observer's consciousness participates, but then the cat knows if it is alive (and perhaps if it is dead, if there is life after death for the cat) [23]. The famous Schrödinger cat paradox brings many concepts into consideration. It basically asks whether our physical models describe the universe objectively or just define the limits of our own knowledge.

As Walker [5,24] suggests, the problem in quantum measurement is not a problem at the microscopic level (which the Schrödinger wave equation describes), but is at the connection of this level, hooked to a macroscopic (classical) measuring device. This is just another way of describing the problem of where and how the observer is connected to the observed. This may be ultimately expressed thus: that the paradox invokes our lack of understanding of our connection to the universe. Can we find unity? Is unity experienced in some state of awareness and not in others? What role do the constructs of space and time play in our confusion over this issue [25-2]. There may be a connection path for the observer and observed in a similar manner to that of the remote connection of the two photons in the experimental result of Bell's theorem as well as the single photon through one or two slits in the Young's double slit experiment where diffraction occurs when both slits are open, no diffraction when one slit is open when only one photon at a time is emitted from a source through the slits to a screen.

14. 8 Particle, Processes, Geometry and Spacetime Independence of Consciousness

We can ask ourselves what is the primary essence of the Universe? What is its fundamental building block? The Greeks called it "atom" (indivisible), the Hindus called it "processes". Are there sufficient clues in nature and/or in our minds to suggest an answer? What are space and time? These apparent restricting aspects of reality appear to be rigid physical constraints and yet are transcended by consciousness. These fundamental concepts appear to be "breaking down" as we probe the elementary particles at the micro-level, and perhaps even the macro level. Even the concepts of processes vs. particle are coming under close scrutiny in particle physics. Let us examine these ideas in the following subsections.

14.8.1 *Particle and Process, Logic and Reason*

The two main tenets of the scientific process are logic and reason. The origin of these words are *logos*, "the word" (from Greek) and *ratio* (from Latin). As Pribram [28] points out, words or noun concepts are the objects of scientific law (such as pressure, momentum, etc.) and ratios of these concepts are incorporated into scientific law. For example, in the ideal gas relation we have $P_1/P_2=V_2/V_1$, so that the volume becomes smaller, as the pressure of a fixed quantity of an ideal gas becomes larger [28].

In western culture the primary structures of the Universe are postulated to be objects, particles, noun terms, etc., which are inert and immutable. Suppose the Logos were verbs or other action concepts. Suppose, as Benedict de Spinoza suggests [26], the God is not Being (existing) but Becoming (or process) [29]. This concept is closer to the Hindu Vedic concept where process is primary, not inert objects [30]. He also suggested that the God concept ultimately refers to ultimate nature – nature as all, an infinite whole of possibilities (even in a closed universe). We have still the universal in the sense of the whole containing all the variances, dualities, change, flux and dynamism of the "Universe" or Multiverse, we observe, yet somehow cohere, even by the measure of our comprehensive states to even live in it and with it. It is dynamic, change whole as nature, as existence that is many faceted with dynamical forces interacting that may be seen in Nature as God.

If the ultimate absolute is a dynamical process with infinite possibilities; this is fortunate because it gives infinite possibilities for achieving a Socratic "good" from this world of abstract images of things of reality. Some associate Socratic good with the idea of God. We relate to these two conceptual frameworks as experiences – an ebb and flow of the tide, or the seemingly immutable commuter traffic jam. If we developed a physics around the concept of process and not object, we would describe the same universe but our theories might look quite different!

Finkelstein suggest a model of quantum phenomena in which process is primary. His theory of "spacetime code" suggests that quantum processes are not random, i.e., "God does not play with dice," quoting Einstein [31]. These processes may appear random, due not to some intrinsic Universal property, to our ignorance of it. Some experimental evidence supports this view to the degree that several experiments have found small (5% or so) deviation from randomness of radioactive decay. Wheeler develops a geometric model of the Universe in which geometrical constructs are utilized to express the fixed, immutable, symmetric aspects of reality. This is also the approach that was taken by the "geometrizing" Greeks [32]. As stated above the entrance to Plato's academy, "Geometry is the highest form of religion", and his "Noetic Insight" he considered the highest form of knowing. The nature of force and dynamics and the manner in which they originate, from constant constructs in a model, which assumes static geometry as primary, is not well understood. Starting from dynamics to express change or process in the manifold and deriving the constant construct of geometry may be a way to proceed, or putting dynamics and process on an equal footing may lead to a new conceptual framework for physics and for science in general [33,24]. Geometric models are useful for expressing constant constructs in the physical universe.

14.8.2 *Intrinsic and Extrinsic Geometry, Mind-Universe*

People have observed, under a variety of conditions, extended before them a detailed and often colorful regular geometric pattern or lattice-work of structure and color. Approximately one person in ten, according to Shepard, have such an experience upon awaking, in meditation, in a Lilly sensory-depravation tank, etc. Rauscher has noticed such a phenomenon in meditation, in a Lilly tank and while observing a variable-frequency stroboscopic flash and in Kundalini yoga. Many scientists and mathematician have developed models of the Universe based on geometric constructs. Perhaps there is an intrinsic structure in the mind which prompts us to describe our perceptions of reality by means of geometric structures or constraints [35]. Maybe Eddington has a valid point about the study of physics divulging an aspect of the nature of consciousness.

14.8.3 *Spacetime Independence in Physics, Psychic Phenomena and Mental Imagination*

Some of the concepts of cause and effect formulated by Immanuel Kant and utilized by the currently defined scientific method are based on space and time as primary constructs of the universe. Now, both recent multidimensional models in astrophysics by Rauscher [15,16,19,20] and earlier geometric models by Wheeler [35] as well as recent discoveries in the descriptions of elementary particle processes by Chew [33] and Stapp indicate that space and time have lost their central and inert place and are no longer primary!

But it is the subjective aspect of space and time of which mystics, psychics and others speak. Techniques such as yoga, meditation or simple contemplation take us out of the normal space time ego self [35] which perhaps it does, as do other self-referential experimental exploration. Our research in remote perception (clairvoyance and telepathy) seems to imply that conscious perception can access remote information in space and time and transcend space and time [4,36]. Also, in the words of Einstein, 1941, "time and space are modes by which we think and not conditions in which we live" [36], or in the words of Eddington, 1923, "time is a mental construct of our private consciousness. physicist construct the concept of a worldwide time from a string of subjective instances" [36,37].

We can only detect space-time transcendence of consciousness and we can formulate the relationship between ordinary, real space-time and the complex multi-dimensional space-time domain of consciousness as we have mentioned earlier [5,13]. Particle physics and psychic phenomena tell us something is wrong with our present formulation of space and time. Since the causality concept is an expression of even connections in space and time, we see that this fundamental concept may bear the brunt reformulation. Such modifications are afoot in particle physics Chew [33] and discussed in [5].

Another form of interconnectedness is that expressed in multidimensional geometries is Einstein linking space and time, and matter and energy [36]. There now appears to be evidence that a multi-dimensional Universe which relates to matter, energy, space, time, momentum and force having a fundamental link [19]. This concept is termed a Descartes geometry after Rene Descartes (who suggested such a geometry might be possible) [19,20]. This is an extension of the relativity theory. In the words of Einstein (1921), "It was formerly believed that if all material things (matter and energy) disappeared out of the Universe, time and space would be left. According to the relativity theory, however, time and space would disappear together with the things" [36].

There is experimental evidence that a so-called vacuum, supposedly devoid of all matter and energy, is not really devoid at all but seething with virtual (not directly physical) energy, which indeed can be observed as affecting observable physical (particle) processes and therefore, in that sense, has a physical reality. This virtual energy makes itself known, for example, in observable modifications of the conductivity and dielectric constant of hot fully ionized gases, called plasmas, consisting of ions and electrons. The energy of this system (of which the Sun is an example) excites and polarizes the vacuum "sea" of energy which in turn interacts with the plasma, affecting its electrical properties [38].

The type of geometrical picture of the Universe which is multidimensional leads us to the possible existence of a macroscopic remote connectedness which may extend over great distances: thousands of miles. This model may be consistent with the test of the space-time transcendence of consciousness tested in remote perception, at least over terrestrial distances [5]. This virtual sea of energy could possibly be accessed by consciousness to remotely manipulate matter in the so-called psychokinetic (mind-movement) phenomena which would truly be a measure of the connection of the observer to the observed!

14.9 Mystic Oneness: Unity

What is the experience of oneness from feeling, rather than thinking point of view? Meditators describe their experience as unity or oneness. In the words of Kriyananda, in 1967, we can see the description of this experience [39]. We read, "See how meditation is like a boundless sphere of light. The light has started to grow – light and joy fill the air of the room, the people, the objects nearby. All these in the peacefulness of that blue light of joy, are one with you – this light embraces your country, your continent, the world! – the limits of the solar system, to the distant stars, to the galactic fringe. You are boundless, Eternal!" [39]. This so well expresses the meditative experience. It is the reality of this experience that leads to the constructions of traditions such as those expressed in the Tantric Upanishads and Vedic literature.

The "ontological consciousness" concept is not at odds with the view of western science and it is becoming more like this tradition. Science deals with collective agreements about so-called external reality, called external validation realities; for example, in general, people agree the sky is blue. Meditators can agree on internal conscious states of reality such as the so-called "blue pearl" of mediation. Meditators often report seeing a blue-green light when their eyes are closed. Rauscher has discussed the color of this light with other meditators and we agree on its form, shape and color. It appears as a speckled pattern like laser light and the color of Chernov radiation in a reactor. (The normal field of vision for closed eyes in semi-darkness is reddish, since one "sees" the blood in the eyelid vessels.)

How different is the discussion of the color blue of the sky from the color of the "pearl" perceived in meditation? Some may "explain" the blue light of meditation as a neurophysical response to the meditative experience. What is the image? Where is it? What is its reality context? Some may call it a hallucination. No matter the explanation, the observation leads us to ask where is the perception of the color or awareness of any reality, which is the basic essence of consciousness?

The basic unity or oneness of the universe is central to the mystical experience, as well as to the present direction of physics, as we have discussed in the interpretation of Bell's theorem. "The world thus appears as a complicated tissue of events in which connections of different kinds alternate or overlap or combine and thereby determine the texture of the whole" [23] again, we have unity and wholeness expressed by another: "He on whom the sky, the earth, and the atmosphere are woven and the wind, together with all life-breath, Him alone known as the one Soul (unity and God)" [30]. (from the Mundaka Upanishad)

To paraphrase Jean Paul Satre the turning point in his own intellectual development with the thought: "Everything we experience is hallucination or illusion, *Maya*. Reality is a structural-mathematical-logical principle that we do not see. That is, each person creates his own universe out of his own imagination, biases and belief systems. Science is nothing else but the search for the unseen structural integrities that underline these appearances" [35], this again brings us to Eddington's concept [2]. We all struggle through this *Maya* to truth which we will ultimately find within each of us. *Maya* may not imply total falsehood, but just our limited view. The Vedic literature gives a detailed description that may involve moving beyond our present state of consciousness to understand.

In the system of the tantras and in the Vedic literature (Rigveda), all creation is the manifestation of a Supreme Consciousness, which is unbounded. His Consciousness "spreads" itself out into manifestation and this becoming does not exhaust It's being. Consciousness has two aspects: S'iva is the static and S'akti is the active or kinetic (motion) aspect. These terms are from the Tantra Sastra tradition and the parallel terms in the Vedantic tradition are Sat, as being and Cit, as action. The static state is one of Supreme Unitary experience wherein the "I" and the "this" are without distinction. The active state, the S'akti, negates itself, becoming the object of experience, leaving the S'iva consciousness to become mere "I". Here arises the beginning to dualism of being and doing (action). By the operation of consciousness, that limits itself, called *Maya*, the united consciousness is severed and from this separation follows the multiplicity of creation. By a series of Tattvas (or steps) the pure become impure; the entirety of creation becomes the inner and the outer.

It is said that the travesty of the Maya of inner and outer, observer and observed, can be overcome by the mantra power which is S'akti in the form of sound, or mandala in the form of light. The earthly striving is to re-unite one's S'akti consciousness to the S'iva of beingness or bliss consciousness and, in so doing, overcome the Maya of the divisional reality of the physical world. This model presents us with a possible prescription for overcoming dualism to obtain unity. The true essence of reality is Universal Consciousness. The essential feature of this model is an ontological consciousness one. Eddington alludes to such a model in his earlier quote that it is consciousness that can know only the structure of consciousness and hence dispel the duality of consciousness and matter [40].

Often the scientific and mystical are thought to be at odds. Perhaps they are not. The methods of science and mysticism or intuition, utilized by conscious minds in the search for truth stand as two paths to find truth, or, in part or in whole, to create truth. Einstein expressed the workings of these two paths together so well:

> The most beautiful and most profound emotion we can experience is the sensation of the mystical. It is the sower of all true science. He to whom this emotion is a stranger, who can no longer stand rapt in awe, is as good as dead. The view of the mystical is my idea of God [41].

Another quote by Einstein brings us to the whole from the scattered parts we often perceive in our daily lives.

> A human being is part of the whole called by us "universe", a part limited in time and space. He experiences himself, his thoughts and feelings, as something separated from the rest, a kind of optical delusion of his consciousness. This delusion is a kind of prison for us, restricting us to our personal desires and to affection for a few persons nearest to us. Our task must be to free ourselves from this prison by widening our circle of understanding and compassion to embrace all living creatures and the whole of nature and its beauty.

The experience of unity is a fundamental state of consciousness which can be experienced. It appears to have a universality beyond the manner in which it is achieved, such as by autogenic training, *amanita muscaria*, meditation, or Za Zen. The dynamic, flowing oneness with the Universe, in which the observer participates and creates reality, is a common experience reported by many throughout yogic and other practices. These descriptions are often taken to be subjective babblings of a few, yet the basis for western thought, the methodology of science, yields similar concepts deep within its structure[1].

Such concepts as the Heisenberg uncertainty principle and relativity lead us to concepts in which the observer is a participant and that the perspective from which he observes from determines what he sees. Bell and Mach have developed concepts which lead to a universal connectedness. Some recent work by us and others indicate that some multidimensional geometries consistent with astrophysical data also lead to the concept of interconnectedness. Bell's [7,5] theorem of interconnectedness originated

---

[1] Rauscher grew up on indigenous Witton Indian lands in Northern California and her intuition is that everything has a form of consciousness and is alive.

from the discussion of the completeness of physical theory (in this case the microcosmic quantum theory). Gödel also shows us that mathematics is open-ended and not complete, and therefore not absolute – shades of Buddha.

Unity and wholeness are major experience and concept creators in some eastern mystic traditions as well as some westerners who have some experience with a variety of states of consciousness. As more conceptual frameworks of thinking are exchanged throughout the world, we gain new insights in our personal experiences of the world. The crux of these experiences, and the new discoveries in physics, is the multi-level nature of consciousness and the multidimensional nature of reality. Space and time are no longer hard and fixed absolute constructs, but are themselves becoming fluid, becoming Einsteinian flowing clocks or fluid time. Particle physics also is telling us that time and space no longer are the totality of the arena for the dance of particles but, as particles dynamically interact, the space-time continuum participates and is itself dynamic too. In the words of Chew, "particles are no longer isolated entities separated in spacetime, but are created out of each other and spacetime, or are "bootstrapped" from and with spacetime itself. Perhaps spacetime is also bootstrappable".

Perhaps in a sense, all we know is "wrong" in the sense that it is incomplete (Gödel's theorem). Science is a dynamic process, in that we continuously add to and modify our body of scientific knowledge. An altered state of consciousness does affect how one feels and how one interprets what one sees or what one actually sees. Observer/participant physics seems to tell us that what we see is determined by how we look at it. Also altered unique states of consciousness may affect the structure of physical theories we create. Examining this question could be a test of Eddington's concept that the structure of the physical is a reflection of the structure of consciousness.

Perhaps with the infusion of the recognition of the role of states of consciousness in our observation of reality we will be in a better position to understand the relationship of parts to the whole. The "melting point" of the observer and the observed and the obvious connection of the two will surely lead us to a reformulation of models of physics and a new comprehension of reality. We have discussed the relationship of the multiplicity of the parts and the universal whole in the Tantric literature. There is the universality of the S'iva and the separateness of the S'akti. This is a model which attempts to dissolve the relational philosophy by methods (meditation – enlightenment) in moving from the Shakti to the Shiva. Until there is no more Shakti, there is still a very basic duality in this model; the duality of the universal and the particular. All universal models appear to contain within themselves relational aspects. Unified theories attempt to resolve the relational aspects by explaining the relation.

We can now ask: Does the duality of the mind/body or observer/observed, etc. come about as a fundamental property of reality? Is the relativity of the separate parts real or imagined? The resolution of the universal vs. particular and the relations of parts may lie in just that – the relational description. Parts and whole may both exist but the key is that they are related and there are methods to describe their relation. The basic concept of universality is not that there are no parts but that there are no truly isolated entities or parts! A useful model of reality involves understanding the connections of the parts to each other and to the whole. The Chinese expressed this as Yin-Yang.

The observer does not stand aloof and isolated from the world, unconnected from the observed. There is mounting evidence, both physical and mystical that there is this interconnection. The observer, we, is a participant and is connected to our Universe. Perception of this depends on our state of awareness. On our growing living planet we must embrace existence globally.

14.10 Nature of the Physical Observer and the New Noetic Paradigm

The nature of the observer has long plagued physical science. Here we review the current status of cognitive science in the context of a cosmology of mind in an Anthropic Multiverse. The concept of an *élan vital* or life force has long been considered the elementary action principle driving the evolution of living-systems by theologically minded scientists and individuals. Sufficiently extending Einstein's original model of a Static Universe, to a Holographic Anthropic Multiverse (HAM) cosmology, provides a context for solving this centuries old problem for introducing this type of teleological principle into Physics, Biology, Medicine and Psychology [42,43]. This means the contemporary framework of *biological mechanism* should no longer be considered the formal philosophical basis for describing living systems and contemporary allopathic (scientific) medicine. The new noetic action principle has far reaching implications for medicine and transpersonal psychology.

We introduce a quantum-informational-unitary field noetic model of brain-consciousness-universe interactions based on the holonomic neural networks of Pribram, the holographic quantum theory of David Bohm, the nonlocality properties of the quantum theory and the unified field (tantamount to the *élan vital*, *prana, Qi, chi* or spirit of God). We consider this model an extension of the interactive dualism of Sir John Eccles. His ideas (stemming from Descartes) of an interconnection between brain and spirit by means of quantum microsite (dendrons and psychons), has deeply influenced the development of our conception of consciousness.

We propose a dynamic concept of consciousness, a new teleological action principle driving self-organization, that generates a flux of the noetic field interconnecting quantum brain dynamics with the unitary nature of the universe. This scale-invariant self-organizing flux is embedded in the holographic mode of neuronal information can be optimized through practices of deep meditation, prayer, and others states of higher consciousness underling the coherence of cerebral waves to improve health generally and totally prevent conditions like influenza by blocking viral adhesion.

Brain mapping studies performed during the occurrence of these harmonic states have shown a spectral array of brain waves highly synchronized and perfectly ordered like a unique harmonic wave, as if all frequencies of all neurons from all cerebral centers played the same symphony. This highly coherent brain state generates the nonlocal holographic informational cortical field of consciousness interconnecting the brain and the holographic cosmos. Comprehending this holonomic quantum informational nature of brain-mind-universe interconnectedness allows us to solve the old mind-matter Cartesian "Hard Problem", unifying science, philosophy, and spiritual traditions in a trans-disciplinary, holistic, integrated paradigm.

The discovery of the interactive Noetic Theory represents a Copernican class discovery; one that comes along only once in several hundred years. In sharing this occasion, it seems fitting to enjoy an ancient verse by Lucretius:

> I am blazing a trail through pathless tracts of the Muses' Pierian realm, where no foot has ever trod before. What joy it is to light upon virgin springs and drink their waters. What joy to pluck new flowers and gather for my brow a glorious garland from fields whose blossoms were never yet wreathed by the Muses round any head. This is my reward for teaching on these lofty topics, for struggling to loose men's minds from the tight knots of superstition and shedding on dark corners the bright beam of my song that irradiates everything with the sparkle of the Muses [44].

What would it take to make psychology a hard science like physics or chemistry? Hipparchus, a Greek mathematician 2,000 years before Copernicus was first to make calculations revealing a heliocentric cosmology in conflict with Aristotle's principle of perfect circles or spheres. After some intellectual struggling Hipparchus discarded his calculations as false because elliptical planetary orbits were considered unphysical theologically. Hipparchus' influence was so strong that his bias suppressed the truth for 2,000 years! A similar problem exists today. Scientists insist that consciousness is a product of brain only. Noetic Science is in radical opposition to current thinking in six main fields of scientific endeavor: Psychology, Philosophy, Biology, Physics, Cosmology and Computer science. Progress in medicine is driven by advances in these disciplines.

Progress in the evolution of human consciousness most often takes place in a constant series of tiny steps; however on rare occasions like that of Galileo, Newton, Copernicus, or Pasteur for example, a radical transforming event occurs. You dear readers are witness to such a historical moment. The purpose of this chapter is to introduce the revolutionary concepts of noetic science precipitating a revolution where mankind leaves the 'modern Age' enters the Age of Consciousness.

The current vogue – *Biological Mechanism* states that: 'The laws of chemistry and physics are sufficient to explain all life; no other principles are required'. Providing a physical basis for the action of the '*life force*' or *élan vital* would finally change this myopic naturalistic perspective. The empirical formalization of such an action principle leads to a whole new class of consciousness based medical conditions and associated '*spirit*' or transpersonal based treatment modalities. When psychology is recast as a physical science 'Moral Psychology' will also have a pragmatic basis because one will be able to experimentally measure which types of behaviors or mental conditions promote life and health or disease and death.

This immense task is accomplished by first extending the standard model of cosmology from the current Darwinian naturalistic (,mechanistic, atheistic) 'Big Bang' theory to one that contains an inherent teleology or purpose. Making this change creates a domino effect that runs through all the other standard models of science. Evolution remains in the new model, not as a random Darwinian progression; but one considered to be 'guided' by the teleological action inherent in the Conscious Multiverse [42,43]. Such a Noetic cosmology called the Holographic Anthropic Multiverse (HAM) has now been developed in general form [45-47]. The associated comprehensive theory of mind is now sufficiently mature; and is not only empirically testable, but also able to rigorously define qualia[2] and begin categorizing the associated fundamental conscious elements in a manner similar to that performed in developing the periodic table of the chemical elements in past centuries. This will lead immediately to new 'conscious technologies' allowing dissolution of the 1st person 3rd person barrier. Because of the teleology inherent in Noetic Cosmology, the HAM represents philosophically what is called a form of Cartesian substance dualism / interactionism. This means that the brain is not of paramount importance to consciousness; the brain is not the seat of awareness as cognitive psychologists currently define it [48]. The brain plays only a secondary role with three main biological functions related to the operation of the complex self-organized living system:

---

[2] Qualia – short for 'quality of the feel', the 'what it feels like' sensation of awareness.

- The brain couples awareness to temporal reality.
- The brain acts as a transducer for processing sensory data and intentional action.
- The brain represents a form of naturally occurring 'conscious quantum computer' that data processes and operates the moment-to-moment homeostatic and metabolic functions of the body.

This important discovery has not been feasible earlier because the currently dominant model for consciousness research (cognitive psychology) has rejected it by definition by asking myopically what processes in the brain give rise to awareness instead of posing the more scientifically proper query – What processes give rise to awareness?

14.11 Allopathic Medicine – The Demise of Vitalism

Contemporary Western Medicine is comprised of Traditional and Alternative treatment forms. Traditional scientific medicine is the orthodox style also known as Allopathic; a term derived from the Greek *allo* – reversal and *pathos* – to suffer. Traditional medicine is characterized by four treatment modes:

- Pharmaceutical drugs,
- Surgery,
- Radiation and Chemotherapy,
- Psychotherapy,

which sadly are all applied only when a person's life is threatened.

The development of organic chemistry began in the middle of the eighteenth century when alchemy began to evolve into modern scientific chemistry. There were unexplained differences between substances in minerals and those observed in living systems; compounds from living systems were difficult to isolate and tended to decompose more readily than compounds from minerals. Swedish chemist Torbern Bergman [49] was first to express this difference between *organic* and *inorganic* substances in 1770. Many chemists at that time believed this difference was the result of a *vital force* which they believed precluded the ability to prepare organic compounds in the laboratory. But in 1816 when French chemist Michel Chevreul discovered that soap made from alkali and animal fat could be separated into glycerin and a number of pure organic compounds he called *fatty acids;* Vitalism was dealt a severe blow. This was the first time organic substances were converted into other substances without the influence of a vital force.

About a decade later in 1828 German chemist Friedrich Wöhler converted the inorganic salt ammonium cyanate into the organic compound urea. By 1850 the scale had tipped heavily against Vitalism [49].

Not until the beginning of the twentieth century did standard scientific (allopathic) medicine become totally dominant. Before that allopathic physicians prescribed harsh and distasteful cures based on mercury, purgatives, emetics and blood-letting which were not considered more effective than popular alternatives such as phrenology, homeopathy, botanics, eclecticism or folk remedies. Allopathic theory was based on the mechanical or material laws of physics and chemistry. The adherents of alternative medicine generally believed that health was based on a vital force related to the soul or spirit. A combination of adherence to the educational standards of state and local medical boards, the complete adoption of science (which history had shown meant progress) and development of a strong professional identity by the class of allopathic physicians led to the inevitable demise of Vitalism which became considered old fashioned by an increasingly progressive science and technology based society [50].

Is it time for a rebirth of Energy Medicine? First to clear up any nomenclature conundrums, Although there may be a loose association with contemporary discussions of Energy Medicine and Mind-body Medicine; any of these modalities would be considered primitive in terms of the advances Noetic Medicine will introduce. The standard models associated with the current state of medicine are

- Darwinian naturalism,
- Biological mechanism and
- The cognitive brain model of psychology.

To summarize briefly this implies:

- Evolution by natural selection,
- The laws of physics and chemistry are sufficient to explain life, and
- The mind is state of brain processes.

Obviously Noetic medicine would be considered a radical heresy by these standards. Noetic medicine redefines the basis for living-systems based on a new cosmology that is an advanced form of Einstein's static universe model. This model includes what Bergson [51] and others called the *élan vital* or vital force. Currently use of Energy Medicine and Mind-body Medicine uses the vital force in only a superficial manner like the early history of electricity with only 'amber and fur' not the highly advanced transistor based devices of modern technology.

14.12 Status of Cognitive Theory

The study of awareness has been recently classified as a 'Hard Problem'; with the easy problems of awareness being ones that are nearly impossible to research by scientists [48]. The nature of mind has been called the oldest and most difficult problem facing human epistemology [52-54]. While people of faith have always been complacent with theological doctrines stating that individuals have an immortal soul created by God; it is only recently that a framework for posing the question of the nature of mind has reached sufficient maturity that any real scientific progress has been able to be made [55]. Chalmers' initial premise that 'awareness is the fundamental principle from which to formulate a theory of mind' [48] is a reasonable assumption for studying consciousness; but he mistakenly goes on to ask: 'what processes in the brain give rise to awareness?', which creates the very 'hard problem' he wants to solve because this manner of posing the question represents a category error for philosophy of mind. While it is true that the brain is the most complex structure known in the universe it is not the seat of awareness. If the mind is instead a whole cosmology; then trying to save the problem in terms of the brain alone will be forever impossible. Historically whenever there has been a 'hard problem' in science, it has turned out to be because the underlying principles have been poorly understood. Although it has been postulated that the mind/body is a naturally occurring form of conscious quantum computer; mind is more than brain or algorithm [56-58] and it is impossible to formulate a correct or sufficient theory of awareness from the point of view of AI, computer science or neurobiology alone. Mind, to be adequately described, must be represented by a complete cosmology with mankind imbedded in it [45-47,55,56,59-61]. Currently about 93% of scientists mistakenly believe the brain is sufficient to model the mind.

14.13 Philosophy of Mind - Vitalism / Teleology

The noetic model of cosmology called The Continuous State Holographic Anthropic Multiverse (HAM) requires reintroducing concepts like Vitalism and teleology that have been historically disdained by science. Mechanistic models of the universe have allowed no place for these 'philosophical constructs' considered non-scientific and non-physical even by their major proponents. In the HAM they finally become physicalized and thus subject to falsification or study by empirical scientific methods. So in one sense we cannot blame science, because by definition it only allows concepts that have been empirically tested even if it is obvious to many that they exist.

Teleology is the philosophy based on the supposition that the universe has intelligent design and inherent purpose beyond the mechanics of a Newtonian Big Bang universe driven acausally by a Darwinian type of natural or random evolution. Evolution obviously exists, but it is not a random series of accidental events. It is guided by a teleological quantum of action (God defined in the coldest scientific terms) inherent in the higher dimensions of the HAM. In perennial philosophies, teleology represents a basic argument for the existence of God, that the order and self-organization of the natural world are not accidental. If mind is fundamental to existence, an ultimate designer or teleological principle exhibiting a quantum of action must exist. Since God is unseen in the usual sense, he does not exist in our temporal locale but resides in a higher dimensional realm that acts nonlocally on our three dimensional domain.

Modern teleologists like H. Driesch or H. Bergson proposed a principle of *vitalism* - the processes of life result from a self-determining fundamental rule not explicable by currently observed physiochemical laws. Bergson, proposed an *élan vital* or vital force [51] as the spontaneous energy of the evolutionary process and defined mind as pure energy responsible for all organic evolution denying sciences claim to explain the universe on purely mechanical principles. This *vis vitae* is shown to be physical when discussed in terms of the noetic field [59-61].

14.14 Current Theories of Mind

Consciousness is not a brain state but a complex multi-factor cosmology. Often individual researchers consider their component theory to be a complete fundamental model. The time has come when it is possible to go beyond this "elephantness" consciousness to formulate a true comprehensive theory. The elephant metaphor relates to six blind men trying to feel an elephant, each attempting to describe it: One thinks of the tail as a rope, another the leg as a tree, the elephant's trunk as a hose, the body as a wall, the tusks as swords and the sixth thinks the ear is a large fan. Until now this has been a major problem for consciousness researchers.

A number of partial theories of merit at the forefront of consciousness research are outlined below and then integrated into one comprehensive theory called Noetic Field Theory (NFT) representing The Quantization of Mind in a Holographic Anthropic Multiverse (HAM). NFT is the first complete and empirically testable theory of mind. We have proposed a number of empirical tests and await funding so that our group can perform the experiments or hope other experimentalists might be inspired to make tests in their laboratories. Historically it took one hundred and fifty years for Copernicus' views to be accepted, fifteen years to perform the simple test of

Einstein's trivial photoelectric effect because even after his *annum mirabelis* his contemporaries thought he was an idiot. Can we dare hope for only a one and a half year delay for fruition in our case?

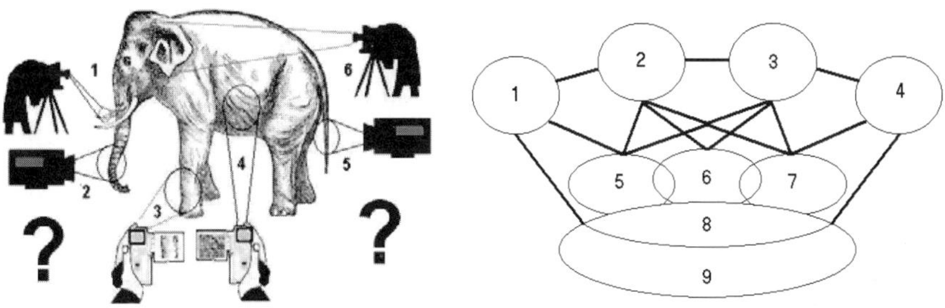

Figure 14.1. A) The Elephant of consciousness – see text. B) Another metaphor for current theories of mind illustrating how they are integrated into one comprehensive model by noetic field theory. 1. AI and Computational models, 2. Neural Networks, 3. Synaptic Tunneling, 4. Quantum Brain Dynamics, 5. Holonomic Brain Theory, 6. Orchestrated Reduction (Orch-OR), 7. Dualist / Interactionist Mind-Body Theory, 8. New Physics, 9. Noetic Field Theory: The cosmology for Quantization of Mind.

14.14.1 *Computational and AI Models*

The artificial Intelligence (AI) model states that the mind is merely a computer; and if the correct algorithm was known it would be able to completely describe all the functions of human consciousness. This view stretches from considering a thermostat as a conscious entity because it has two bi-stable states on one hand to an advanced autonomous android on the other.

Current classical computers are much less complex than the human brain and do not have enough degrees of freedom to handle consciousness. Furthermore they are pre-programmed and unable to change or escape from this condition as a sentient being is [58,59] able to make choices based on spontaneous volition.

14.14.2 *Neural Networks and Cellular Automata*

The neural network model of mind states that the subjective process of awareness is a result of computational information dynamics in various biophysical networks such as neural, quantum and sub-cellular systems. Cognition is a problem of both processing and representation [62]. There is a wide spectrum of belief among researchers regarding the nature of mental data processing in networks. Some think the linear electrical pulses along neurons are sufficient. Others believe quantum processing with nonlocal effects is required [63,64]. Or is it nonlinear, parallel processing or a more complex form of distributed processing throughout the whole neural network as in the holonomic model of Pribram [65]? A similar conundrum occurs among researchers of the neural model regarding representation. This is called the problem of *qualia* - the nature and origin of qualitative subjective experience. Do qualia emerge into a mind from the neural substrate? How does thought bind to the conscious system? This is called the binding problem. These questions have been called the hard problem of consciousness [5]. For example, a 6D hypercube with 64 nodes and 6 connections per node representing connectivity for computation in a neural net or cellular automata. This is a form of computer modeling used to study the possible neural network structure of the brain [66].

14.14.3 *Synaptic Tunneling*

The linear action potential along a nerve fiber is electrical; and is converted to nonlinear chemical transmission at the synapse (see Fig. 14.4) which are 'boutons' at the end of nerve fibers that release various neurotransmitters. There is always a low level continuous release of neurotransmitter acting as the baseline of activity. Quantum tunneling is the charge transport of electrons through an otherwise impenetrable barrier or insulator at the synapse. Acts of volition or other neurosensory inputs are believed to be the phase regulators that trigger, through quantum tunneling, the release of neurotransmitter vesicles which is called exocytosis at the synaptic grid [67,68]. The most a neural impulse can evoke is a single exocytosis, probably because of the paracrystaline nature of the material the vesicles are imbedded in. Exocytosis is the most fundamental action of the cerebral cortex; and is an all-or-nothing response each of which results in a brief excitatory postsynaptic depolarization [67].

The trigger model itself is still incomplete because it has only been developed at the classical level of the electron transmission. What is still needed is a description of the coherent process that couples a mental event by quantum probability selection to the actual biochemistry associated with action. At the synapse and synaptic grid neurotransmitter vesicles are released by quantum tunneling of electron transmission. The tunneling mechanism is believed to be the trigger action of intentional mental activity or the site of the mind-body connection [67].

14.14.4 *Quantum Brain Dynamics*

Quantum Field Theory has several branches, Quantum Electro Dynamics (QED) for electromagnetic interactions, and Quantum Chromo Dynamics (QCD) for strong interactions. Quantum Brain Dynamics (QBD) is a quantum field theory describing biological systems and the fundamental mechanics of the brain [69]. QBD is mediated by an exchange field called the corticon [69], a quantum of the water rotational field which interacts with electric dipole oscillations along neural proteins. When synchronization of the water corticon and electromagnetic field occurs, nonlocal coherence is manifest giving rise to long-range order and collective phenomena. Nonlocal coherence provides a much stronger correlation than a classical collective mode could describe.

QBD of the water rotational field and interacting electromagnetic field although providing an excellent model of neuromolecular computation is not sufficient to describe consciousness because freewill or intentionality is still left out of the picture and the founding fathers of quantum mechanics said it was not capable of describing biological systems. The Schrödinger equation describes the evolution of a particle on a manifold; so just because QBD describes action on a brain manifold it is not a sufficient extension of the theory. For this we need an extension not only of the orthodox Copenhagen interpretation but also are required to go beyond the quantum ontology of Bohm into a higher dimensional extension of Cramer's theory [70,71]. Bohm described the quantum potential as a nonlocal pilot wave effecting the probability matrix of the Schrödinger equation.

As we will show Noetic Field Theory: The Quantization of Mind completes Bohm's work by introducing a noetic effect [72] that mediates equalibrium set points in both mind and body. Neurocomputing models of the brain are linear closed systems; Once a Turing computer is programmed there are no remaining degrees of freedom for rational intentional input.

In summary water has been theorized to play two important roles in consciousness:

- To provide a storage buffer to amplify or attenuate the corticon field,
- To allow switching between neurosensory quantum computation and mental intentionality.

Although the role of ordered water in the dynamics of consciousness remains a qualitative model at this point in time; a growing body of literature from both experimental and theoretical areas are converging to suggest an important role of water in the quantum physics and molecular biology of consciousness.

14.14.5 *Pribram's Holonomic Brain Theory*

The holonomic brain theory relying heavily on the Fourier relation and the holographic application of Fourier's theorem by Gabor in 1946; has been expanded by Pribram [73] to include a phase space of interaction in the brain; called the holoscape [65]. Integration of holonomic theory and the Bose-Einstein model provide a substrate for explaining recent work on quantum information processing represented as conformational changes of alpha and. beta tubulin dimers in microtubule protein structure [74] providing a stage for the first application of these concepts to tangible brain material However the brain, a Fermi apparatus with Einstein-Bose interactions; is viewed here as only one of three key aspects of consciousness.

Integral to Pribram's holonomic brain theory is the concept of the holoscape, a neuronal manifold which embodies the polarization occurring in dendritic networks [65]. The holoscape is the active manifold of entrained neural processing that couples phenomenal information to the phase space of what Pribram calls the Heisenberg matrix which includes the raster of consciousness (of an analog TV screen) below it. Gabor and Fourier relationships describe the activity of information processed in the neural ensembles as a raster of mental functioning.

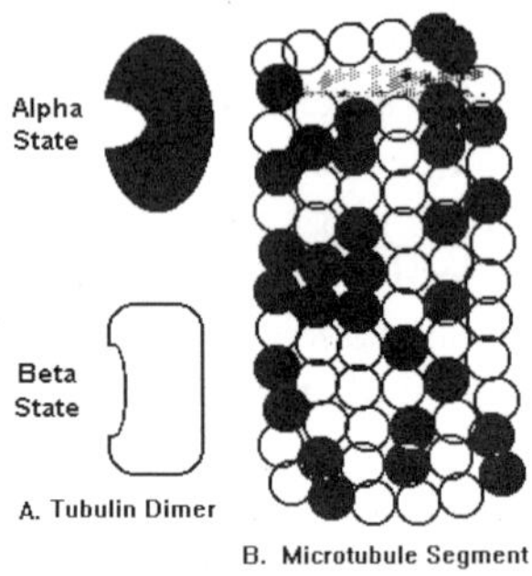

Figure 14.2. Alpha and Beta configurations (Two quantum states) of tubulin protein dimers provide a bit state model for information processing at the quantum level in cell structures that contain microtubules.

Freeman [75] relates that chaotic dynamics can create information in the Shannon-Weaver sense of information. This is the relationship with the Gabor logon utilized by Pribram [65] in the holoscape. Pribram has skillfully integrated his holographic model with quantum activity associated with QBD in what might be thought of as a dualistic picture of consciousness. The question remains, does consciousness originate from qualia at the level of holoscape dendritic microprocess or is it underpinned by the quantum domain? There has been general skepticism of quantum effects having any relevance to such a hot entropic matrix as the brain. However there is a distinct difference in coherence at these levels. In the quantum realm there is essentially thermodynamic equilibrium. Much could be written about the holoscape, where the major philosophical issues are information coding and processing, the binding problem.

In Fig. 14.2 is a segment of a microtubule composed of tubulin rings. Shading represents conformational ordering as data I/O patterns in active quantum states as a basis for brain level dynamics of consciousness. A tubulin dimer is about 8 nanometers long. The quantum dipole shifts or conformational shape changes occur with transitions of $10^{-9} - 10^{-11}$ seconds.

14.14.6 *Orchestrated Reduction (Orch-OR) Model*

The Hameroff-Penrose theory states that quantum events at the microtubule (MT) (Fig. 14.2) and other nanoscale objects are sufficient to process the necessary amount of information to satisfy the needs for consciousness. The conformational states of the tubulin dimer are coupled to Van der Wall dipole moments. Each conformational state could represent a bit for information exchange [76].

There are several types of microtubules in the cytoskeleton that seem to have complementary features such as acidic and basic tubulin subunits. The tubulin polypeptide dimer has been found to have seven alpha and over ten beta species. Other differences include dynamic or stable, more or less curly, and variance in turnover rate. MT's are involved in a wide variety of cellular functions. They form the spindles during mitosis and meiosis, the cytoskeleton plays a major role in cell morphology, MT's aid transport, and maintain cell surface sites like receptor caps [77-79].

Microtubules do not handle all the information processing of mental states as Hameroff and Penrose have proposed. There is an integrated system of data processing that includes DNA, cell topology, microtubules, cAMP, a variety of synaptic neurotranmitters and water [80], not only in the brain, but also coupling the noetic field throughout the entire body modulated by baseline neural firing in muscle dynamics, thought, and other psychosphere processes related to the noetic unified field [60,61]. One problem with the Orch-OR model is that it attempts to utilize a conservative model of gravitation asking: 'what is the minimum gravitational mass required to collapse the wave function' which Orch-OR considers the process of mental action. This limits conscious systems to creatures only as small as a planaria; but we know from noetic theory that even the prion responsible for mad cow disease is a conscious system albeit a purely mechanical one [72,81,82].

14.14.7 *Dualist / Interactionist Theory of Mind-Body*

Over 400 years ago Rene Descartes claimed to receive a revelation from God that consciousness was divided into mind stuff - *res cogitans* and body stuff - *res extensa* [83]. This dualism of mind and body has endured until today because intuition dictates self as separate from world and until now there has never been a comprehensive mind/body theory. The nature of rationality or free will, as opposed to Newtonian determinism evidenced in a computer program or robot suggests that the basic theory of quantum mechanics also is not equipped to describe consciousness. The violation of the 2nd law of thermodynamics and entropy flow by living systems, and the smoothness of our perception of reality versus the discreetness or discontinuity of its origins at the microscopic brain level all show the inadequacy of our current thinking on the nature of cognition. Dualism states that although the mind has an independent eternal existence from the temporal body, it acts in concert with it [84]. Traditionally considered beyond physics because by definition only measurable quantities are deemed to exist; the dualistic view has funneled most scientists into the erroneous belief that brain equals mind. Since the brain is a physical object, scientists have believed this is the only basis for developing a physical theory of mind.

The complaint against the current thinking of Cognitive Psychologists regards the limits of inquiry bounded by its myopic metaphysical foundation of considering the brain as equal to mind. Science fits the basic definition of a theology by its rigid adherence to its principles. This heresy is not a call for science to embrace an *a priori* philosophy. Since Galileo the profound value of empiricism has been well learned. But the finite limitations surrounding the measurement problem in quantum theory and the need for a more advanced approach strongly suggests that we have come full circle to the time for mandating another evolutionary step to improve:

- The ability to pose foundational and empirical questions, and
- Data gathering and evaluation techniques that accept input in ontological terms, allow subjectively or both.

There may be no alternative to integrating a noetic based science for progress to occur. The Perennial Philosophy, attributed to Kant and others, states:

- Deity exists
- Is knowable
- Provides a path to be found [85].

Benefits to utilizing the perennial philosophy include, insight into the nature of absolute truth [86], which promises a more efficient compass for reality testing; and insight into the utility of subjectivity by developing an acceptable methodology for instituting the radical empiricism of James [87].

Aspects of the following premises are based on noetic insight[3] using elements of the Cartesian modality (institution and verification by revelation or meditative insight), and presented axiomatically as a bold call for testing this hypothesis. It must be stressed that utilizing the 'Cartesian modality' does not interfere with the pragmatism of the empirical method. It is a time saver; if the correct model is 'divined' it may save hundreds of years in finding it, but it must then still be experimentally verified. Descartes distinction between *res extensa* and *res cogitans* has not been tested. If this turns out to be the correct model as is presented here; is it any wonder little progress has been made - if no one has been looking where the answer lies.

14.14.8 *Beyond the Brain - Elemental Intelligence*

While the brain services the temporal aspects of our Earthly existence; current thinking has ignored the eternal aspects of mind and body. Elemental Intelligence is the fundamental eternal condition of individuality and exists outside of time and the bounds of the phenomenological reality we observe in our 3D world view. This bound, although currently an ineffable domain not yet having an empirical foundation waits for vacuum quantization and a deeper understanding of nonlocality to open the avenue to a more empirical explanation, and is currently only known to exist by noetic insight. Simply stated if individual intelligence has no domain, i.e. is not bounded in some manner; it cannot exist with any connotation of individuality. Apparently there is as much to us behind the curtain of reality as we see in front of us. So at this time only transcendent or philosophical arguments can be given for Elemental Intelligence as follows:

Firstly individuality must be separated from ‘The One’ at some level for absolute unity is again nothingness, and nothing has no boundaries and cannot exist by its very definition. For even the demarcation of nothing as such demands its qualification by something extant which gives it existence. This idea of nothingness is not meant also as in the abstract sense of redness for example. For though redness is not assigned "thingness", it still has existence in sentient apprehension and is therefore not nothing. This is the abstract content of consciousness often deemed immaterial. However,

[3] Noetic Insight: Plato said Noetic Insight was the highest form of knowing (epistemology) because it was transcendent – beyond ones intelligence and knowledge.

according to the tenets of Noetic Field Theory [60,61] thought is deemed a physically real unitary noetic field that is encoded with information; thus a typical case of abstractness in this sense is now relegated to tangibility.

Secondly without some form of separation from absolute unity there can be no self-identity. Without this identity or boundary “it” would disappear into the ‘one’ or nothing as stated. Absolute unity is nothingness, cannot exist and cannot be comprehended. Further this complement of elemental intelligence is fixed nonlocally and promotes the separateness mandatory for individuality to exist.

14.14.9 *Consciousness is a Universal Cosmological Principle*

The second compliment of consciousness is a cosmological principle that fills and orders the immensity of space. It could be said to be equivalent to the life principle, *élan vital, Qi, chi, prana*, or Holy Spirit. In contrast to elemental intelligence above this aspect is not fixed but represents flux and promotes the unity of mind and body. This is the root of the mind - brain problem – cognitive brain science versus Cartesian dualism. We are complementary aspects of both unity and separation; monism by itself cannot be an absolute.

The cosmological aspect of consciousness exists in all matter and is itself a pure material with the properties of light. However as ordinary photons originate in atomic geometries coupled with properties terminating in space, Photons of mind (psychons as termed by Eccles) originate in complex higher dimensional geometries. Noeons is the term given to the unitary field in Noetic Field Theory. They are confined to the spacetime backcloth like quarks. This is why they haven't been measurable by standard methods of Physics and why an extension of QT is required.

Consciousness pervades every point in spacetime and enters all atoms, is related to the teleological anthropic organizing power deeper than gravitation that controls or orders the large scale structure of the universe, causes gravitation, and the noetic flux of which gives life. Plant life does not appear to make direct use of the component of elemental intelligence, only the cosmological ordering principle and the ‘body state’ of matter. Sentience is caused by the autopoietic (self-organized) integration of elemental (eternal) and cosmological (spiritual) intelligence. This basic holistic framework incorporates ‘the implicate and explicate order’ described by Bohm is his discussions of the holographic principle which seems to be conformally scale invariant and was empirically found by Pribram to operate in the brain.

14.15 Origin of Complexity in Biological Systems: A New Model for the Origin of Life

Generally unicellular prokaryotes are considered the most fundamental form of living system. Many researchers include viruses since they commandeer cellular machinery in their replication; while others insist viruses are merely complex infective proteins. New biological principles are introduced suggesting that even the prion, the infectious proteinacios compound responsible for transmissible spongiform encephalopathies by merely a change in the conformational state of its molecule, qualifies as the most fundamental form of life; and remains in general concordance with the six-point definition of living systems put forth by Humberto Maturana and his colleagues in their original characterization of living organisms as a class of complex self-organized autopoietic systems [88].

"What is the necessary and sufficient organization for a given system to be a *living unity*?" [88]. Maturana and his collaborators posed this question in their effort to formalize the general definition of a living system. They further stated that all other functions are secondary to the task of establishing and maintaining this unitary organization; defining this process as *autopoiesis* [88]. For review, the description of an autopoietic living system is as follows: *Autopoiesis* from the Greek 'self-production' is a fundamental expression of the basic complementarity of structure and phenomenology [89-91]. An autopoietic system is self-organized, complex, open, dissipative, self-referential, auto-catalytic, hierarchical, far from equilibrium, incursive, recursive, anticipatory and autonomous. A system is autopoietic when its primary function is self-renewal through self-referential activity. This contrasts an allopoietic system like a robot deriving function from an external source. Stated another way autopoiesis is a network of production components participating recursively as a globally stable structure operationally separable from the background in which the system exists [88,89].

These properties operate in an ascending hierarchy:

- An autopoietic system is an open non-equilibrium system. If closed in equilibrium all processes eventually stop and a state of maximum entropy will prevail.
- The processes are cyclical.
- As a complex self-organized system, operations occur within multi-levels where higher levels contain all lower levels.
- Function – the primary function of the system is self-organized autopoiesis as defined above [88].

14.15.1 *Summary of Maturana's Six-Point Key for Determination of Life*

1. Does the entity have identifiable boundaries?
2. Does the entity have unique constitutive elements?
3. Is the entity a *mechanistic system* possessing properties satisfying certain relations for its interactions and transformations?
4. Do the components constituting the boundaries of the entity act through preferential relations and interactions between the components?
5. Are the components constituting the boundaries of the entity produced by interactions of the components either by transformation of previously produced components, or by transformations and/or coupling of non-component elements that enter the entity through its boundaries.
6. If all the other components of the entity are produced by the interaction of the components as in 5 above, the entity is an autopoietic entity in the space in which it exists [89].

14.15.2 *Non-Autopoietic Entities That Seem to Satisfy Maturana's Conditions*

- **Automata** - Superficially automata [92] seem to obey Maturana's six points for autopoiesis, especially in terms of self-reproduction and autonomy; but they are readily disqualified for two salient reasons: Automata are generally nonphysical and cannot naturally escape or exist outside of the computer system they are programmed in.
- **Crystals** - Crystalline structures conform to many of Maturana's six key requirements. The symmetry of the *unit cell* contains the geometric framework of the whole periodic structure, which is repeated in translations of the unit cell. So although a crystal has open self-organized boundary conditions, appears to be recursive and can reproduce; a crystals main failing is that it remains mainly a chemical reaction because its 'unique constitutive elements' can only be reproduced and remain structure preserving under precise conditions of chemical reactivity.
- **Ribosomes** - Although partially comprised of components produced by the ribosome, as entities they are produced by processes beyond those comprising their operation and their function is not completely self-referential. Ribosomes have high level metabolic properties but they are organelles not unique unities.

- **Belousov-Zhabotinsky Reaction** - A key aspect of a self-organized autopoietic system is its globally stable structure over an extended time. These are called *dissipative structures* because they maintain a continuous production of entropy, which is then continually dissipated. The best known dissipative structure is the Belousov-Zhabotinsky Reaction produced by the oxidation of malonic acid by bromate where rotating concentric or spiral waves create interference patterns oscillating with a periodicity maintaining itself for many hours [89,93]. Although self-organized with environmental interplay, can this be more than a recursive chemical reaction?

Jantsch and Maturana both state that dissipative chemical reactions like the Belousov-Zhabotinsky reaction and the glycolytic cycle qualify as primitive autopoietic systems [88,89]. Should these or any of the entities above be accepted as living systems? Maturana's six-point key is not experimental; but a set of logical premises, and in that sense arbitrary philosophical deduction. Even if these systems are considered autopoietic by the claim of definition, the thesis developed here is to not accept these types of entities as living-systems but to make a case for requiring additional physical principles added to Maturana's key to complete the requirements for properly defining a unique class of autopoietic systems qualifying as true living-systems. Our conclusion is that Maturana's autopoiesis at best only defines the mechanistic components of self-organization.

14.15.3 *Mechanism In Biology as a Semiclassical Limit*

Autopoietic systems as defined by Maturana are a special class of *mechanistic system*. This is a challenging philosophical issue. It is generally considered an open question whether all biological process can be described completely in terms of the 'mechanisms' of physics and chemistry. In the philosophy of biology *mechanism* is defined as the view that every event described as a biological event is the same as those exemplified in non-biological physical chemistry [94,95]. Beckner in a discussion of *mechanism* states:

> It is plausible to suppose that biology contains terms that could not be defined by reference to physics and chemistry, particularly if we count psychological phenomena as special cases of the biological, but perhaps even if we do not. Biological theory takes account of the circumstances of an event's occurrence in a way that the physical sciences do not. For example, it is a biological fact that lions hunt zebras. The biological

> mechanist ought to insist merely that everything that happens in a given case of zebra hunting is identical with a sequence of physicochemical events, not that the concept of hunting can be defined in physicochemical terms. It may be the case that *hunting* can be defined only in intentional language [95].

This has left the final sense of reduction for the standard model of biology an open question; and until recently this is where conceptual development had to remain. The philosophy of biological mechanism reviewed here is akin to philosophical naturalism that states that 'the natural world represents the whole of reality without requiring any additional teleological parameters'. This suggests that the current limits of scientific pragmatism provide sufficient explanation for all universal phenomena. Arguments on mechanism and naturalism have probably not been quite beaten to death but let it suffice here to postulate that additional scientific laws are yet to be discovered because 'lion hunting' as intentional action is not describable by the laws of physics and chemistry.

One cannot in good conscience label the Belousov-Zhabotinsky reaction [89,93] as a living system any more than one can logically allocate consciousness with reasonable definition to the bi-level state of a thermostat as is often done in Artificial Intelligence (AI) circles. The sophistication of self-organization in autopoietic systems cannot be discounted. While this inherent complex order provides a highly efficient substrate for living systems to be built on, like a little finger applied to the helm of a megaton ship, mechanism alone provides an insufficient basis for describing living systems. A teleological principle, inherent in a conscious universe [2-4], acting in concert with mechanism is required for life; providing components of what cosmologists have recently called the holographic cosmological principle.

14.16 New Cosmology Leads to Redefinition of the Observer

Until the advent of the Noetic cosmology [45-47] physical cosmologists generally believed that the universe could not be ordered enough to have a symmetric spacetime with an inherent periodicity where events are structured such that the future-past prepares the 'nows' evolution into the future [45-47]. These spacetimes were considered non-physical and appeared to violate the causal principles of quantum theory [96]. The semi-classical limit in physics refers to the boundary between classical mechanics and quantum mechanics where an incomplete understanding of the dynamics of a system allows only statistical predictions to be made on the behavior of a system rather than a precise determination [97]. Consciousness is able to violate quantum causality. Quantum mechanics is known to be both incomplete and not able to describe biological systems; therefore how can biological mechanism offer a complete framework for living-systems!

Self-organization produces *freedom* and the degree of autonomy a system achieves in relation to its environment provides one way to loosely define *consciousness*. Jantsch says "this autonomy appears as an expression of the fundamental interdependence of structure and function which is one of the most profound laws of dissipative self-organization" [89]. Drãgãnescu further adds "If a virus is alive it has a phenomenological subconscious, if not, it cannot have any form of consciousness, because there is no structural organization with sufficient complexity to process structural information significantly" [91]. This is similar to Maturana's idea that the autonomy obtained by autopoiesis relegates a primitive form of consciousness, even to chemical dissipative structures, which he calls a *cognitive domain* in relation to the systems environment [90]. This is where we will draw a line in the sand giving a definitive description of the term *cognitive domain* that goes beyond mechanism.

Consciousness, and not necessarily that with self-awareness, requires a sufficient number of degrees of freedom beyond those of an allopoietic mechanistic automaton. While one might reluctantly concede that the Belousov-Zhabotinsky reaction [89,93] is autopoietic by Maturana's original definition [88]; one cannot proscribe a cognitive domain with the structural-phenomenology of intentional awareness to such an autocatalytic pattern-producing chemical reaction. How is this ultimately different than programmed automata? We believe that embracing biological mechanism leads one into the trap of 'conscious thermostats'. The autocatalytic chemistry of the Belousov-Zhabotinsky reaction has a cyclical self-organization that keeps the cycle in motion recursively by a chaotic component in the symmetry of the boundary conditions leading generally to a global stabilization of the reaction until a chance occurrence of an ordered ground state occurs. One could argue the reaction is the result of the inherent activity in the reactions so-called *cognitive domain* because it includes a self-referential multilevel hierarchy that maintains the cycle of the reactions self-production. One could carry this argument further to lend correspondence with Prigogine's symmetry breaking factors in the thermodynamics of evolution [98,99]. But the driving force described by these arguments is not an intrinsic *intentional awareness*; it is more like the incongruent geometric symmetries driving the chain of unstable intermediaries in a radioactive decay series, an automatic unraveling continues as long as a stable ground state with boundary conditions that preserve the unity of the intermediate atom cannot be reached.

14.17 Living-Systems and the Physical Observer

Recall Jantsch's claim that the Beluzov-Zhabotinski reaction, as a result of its classification as an autopoietic system by Maturana's definition [90], has rudimentary consciousness [89,100]. For decades researchers have believed that consciousness is merely a computer program, "a special software in the hardware of the brain or just a matter of information processing" [101]. This isn't seem acceptable; and is more a reflection of the current state of bias in the field of consciousness studies where the dominant cognitive model is aligned with the standard model of biology. This *philosophy of biological mechanism* provides only half the story of mind. Our aim is to show that an addition to and clarification of Maturana's key allows classification of the prion [81,82] as the fundamental living system.

The *cognitive domain* [90] of a prion[4] does not create and dissipate entropy in its own right like higher life forms. The prion is not even at the same level as the virus where this critical factor of far from equilibrium complex processing is satisfied by proxy when the virus protein commanders the existing cellular machinery of the host. The prion, as the zeroth case of a living system, does not 'live' at the viral level. The factor that separates the prion from the non-autopoietic entities listed in section 14.17 (which utilize only the mechanistic half of the complementarity required for a complex self-organized living system) is the prions utilization of the coherent energy of the *élan vital* in its propagation. This is a prediction of the noetic theory we intend to demonstrate empirically [72].

14.18 Is There More to Biology Than Mechanism?

Returning to the analysis of the fundamental philosophy of biology we summarize Brillouin's [102] categorization of the issues of mechanism versus teleology into three general positions:

- Knowledge of physics and chemistry is essentially complete and life could be explained without introduction of any additional *life principle*.
- Considerable physics and chemistry is known, but not everything. A new law or principle needs to be discovered to explain life; but this concept will not be outside the laws of physics and chemistry already

[4] The prion propagates through conformational changes in the geometry of its protein structure [72,81,82].

known. Whether or not this is considered a *life principle* or not is irrelevant.

- A *life principle* is mandatory for an understanding of life because living systems are much different and more complex than inert matter. The laws of thermodynamics describe only inert and *dead* matter to which life is an exception requiring a new principle to explain.

Theories of mind abound with great disparity between them [91]. It could be said to be like the early days of electromagnetism when 'for every 100 theorists there were 101 theories'. Simply stated, and reducing from the top down, mind theory can be generally categorized as follows:

A. Classical Reductionism – Newtonian mechanics deemed sufficient to describe mental activity
- Neural action – Consciousness can be completely explained by brain processes
- Information processing in Neural Networks / Cellular Automata / Physics and Chemistry

B. Heisenberg Cut – Additional degrees of freedom, possibility of nonlinear & nonlocal interactions
- Quantum computation in brain microstructures like synapses, microtubules or ordered water
- Copenhagen phenomenology – collapse of wave function essential for mental activity

C. Cartesian Cut – Requires additional 'life' and/or physical principles beyond mechanistic theory
- Dualism / Interactionism – ontological extension of quantum theory, collapse not required for evolution
- Monism – all is mind, consciousness is ineffable

The first four types above fall under the domain called the philosophy of biological mechanism. Theories in the Classical and Heisenberg arenas have defined consciousness as a hard problem too difficult to research [103]. This provides significant motivation to explore below the Cartesian divide where additional physical laws are anticipated. What evidence exists to justify such a search?

Continuing with the premise that quantum theory is incomplete, Schrödinger in relating the 2nd law of thermodynamics and life says: "We cannot expect that the 'laws of physics' derived from it to suffice straightway to explain the behavior of living matter…We must be prepared to find a new type of physical law prevailing in it. Or are we to term it a non-physical, not to say a super-physical law [104]?" But what can this new physical law be?

14.19 Complex Systems Theory: A New Model for the Origin of Life

It appears unanimous that unicellular prokaryotes are considered the most fundamental form of living system with the inclusion of viruses controversial. By defining awareness as a fundamental physical quantity like the concept of *charge* in electrodynamics [72,105-109], it is possible to show how the prion recapitulates, in the sense of its organization, the propagation of its infective state by maintaining the 'charged' form of its conformation by merely being coupled to the Noetic Field. Prion propagation therefore represents the most fundamental form of biological mechanism and provides the root of its redefinition. Although slightly more complex, the self-organization pertinent to viral replication also falls under this new definition of biological mechanism. Something else happens at the level of bacteria or perhaps any motive unicellular life form. The cognitive domain has sufficient capacity for activity based on an *interactive computational model* [110]; the evolution of the content (qualia) is driven by more than the mere presence of teleology as in the case of the prion or virus, i.e. more degrees of freedom are available.

The *continuous state* of this new action principle, as already suggested, is a 'force for coherence' like the well-known radiation pressure in the QED of light propagation. This symmetry enhancing force acts not only on the topological states of prion conformation by constructive interference as the base state of biological mechanism, but also by higher order conditions of self-organization. The structural-phenomenology of the new noetic action principle [45-47] is a complementarity of mechanism and the noetic field, together forming a teleology that is the general driving principle governing all aspects of complex self-organized living systems [60,61]. Applying the concept of a *unit cell* from the nomenclature of crystal structure to this fundamental teleology in the topology of spacetime, forms the scale-invariant hierarchical basis of living-systems from the microscopic origins of mechanism to macroscopic intentional systems. The complementarity of mechanism and teleology is a structural-phenomenology that is the primary cosmological principle of the conscious universe; the fundamental least unit of which is defined as awareness [72].

Defining awareness as a fundamental principle like charge in Electromagnetic Theory [100,103,105] provides two paths to formulate a theory of life and consciousness. 1. The currently popular cognitive avenue poses the question '*what processes in the brain give rise to awareness*?'

Unfortunately this creates a *hard problem*, which at present is deemed impossible to study empirically [100,103] - an investigative dead end! Charge has been considered fundamental physically and indivisible; but this definition appears to hold only to the semi-classical limit. Physicists are finding out that the so-called unit of elementary charge arises from a deeper wormhole structure in the higher dimensional topology of spacetime [77]. This is also true in defining the fundamental unit of awareness. Charge, or in this case awareness, does not arise as a brain process. 2. Only looking beyond the brain leads to a model of awareness (consciousness) that is both definable and empirically testable. In brief, the fundamental basis for the least unit of awareness has three complementary components [105-109]:

- Elemental Intelligence – A nonlocal atemporal HD domain or set of boundary conditions co-eternal with God that define an individual entity.
- Noetic Ordering Principle – A new action principle synonymous with aspects of the unified field and mediated by an exchange particle called the noeon that is synonymous with spirit or an *élan vital.*
- Local Fermi and Bose brain/body States – Classical, semi-classical and quantum modes associated with neural activity and other aspects of simpler autopoietic or complex self-organized living systems.

Remaining problems center around the fundamental nature of space; suffice it to say that Einstein's superceding of Newton's 3D absolute space with a 3(4)D or (+++-) signature relativistic space was a significant milestone, but not a final answer. The triune complementarity above provides a sufficient structural-phenomenology of the 11(12) noetic space to define the psychosphere of an individual's mind and body.

14.20 Action of the Unified Noetic Field

Fröhlich [109,111] proposed a new energy that produces coherent long-range order in biological systems. Some authors have suggested this coherence is a type of Bose condensate. Einstein and Hagelin [110,112] further postulate this coherent principle arises from the unified field, which is also proposed here by Noetic Field Theory. The action of the unified field is the basis for a life principle governing the evolution of complex self-organized living systems.

We will show generally how the continuous transformation of the topology of the 12D superspace of the noetic least unit introduces by periodic holophote action evanescence of a life force from the HD energy covering of each moment of the present [45-47,72,100,103,105]. First we illustrate one of a number of possible models of how at the semi-classical limit from the stochastic background of the vacuum zero-point field, this energy of the *élan vital* is harmonically injected into every point and atom in spacetime by a mechanism like a 'chaotic gun' [113,114].

14.21 Physical Self-Organized Basis of Qualia

Qualia, plural of *quale*, is 'the subjective quality of experience; a *qual-itative feel* associated with an experience' [115,116]. The physical HAM cosmology of *élan vital* leads to a rigorous model for representing qualia [117,118] allowing immediate application on the mind-side to psychology and on the body-side to medicine. In 'What's it like to be a bat?' Nagel [116] states that current reductionist attempts fail by filtering out any basis for consciousness; becoming meaningless since they are logically compatible with its absence. He assumes if an organism has conscious experience, "there is something it is like to *be* that organism". This is the subjective character of experience for any conscious entity whether bat or Martian. Every experience has a specific subjective nature [116].

To Nagel "there are facts which could not ever be represented or comprehended by human beings, simply because our structure does not permit us to operate with concepts of the requisite type"; because "to even form a *conception* of what it is like to be a bat one must take up the bat's point of view". If one removed the viewpoint of the subjective observer; what would be left? Nagel suggests the remaining properties might be those detectable by other beings, the physical processes themselves or states intrinsic to the experience of awareness. This changes the perspective of qualia to the form "there is something it is like to undergo certain physical processes". "If our idea of the physical ever expands to include mental phenomena, it will have to assign them an objective character". Nagel recognizes that:

> Very little work has been done on the basic question (from which mention of the brain can be entirely omitted) whether any sense can be made of experiences having an objective character at all. Does it make sense ... to ask what our experiences are *really* like, as opposed to how they appear to me?...This question also lies at the heart of the problem of other minds ... If one understood how subjective experience could have an objective nature, one would understand the existence of subjects other than oneself [116].

These are questions an integrative Noetic Science can answer. Standard definitions of qualia are an inadequate philosophical construct describing only subjective character. In the physical sense of Noetic Field Theory (NFT) components describing qualia from the objective sense are introduced - i.e. distinguishing the phenomenology of qualia from the noumenon or physical existence of the *thing in itself*.

A comprehensive definition of qualia includes three forms considered physically real by NFT because the noetic fields of HAM cosmology on which the noetic model is based are all physically real. See [117,118].

Type I. The Subjective - The *what it feels like* basis of awareness. Phenomenological states of the qualia experience. (The current definition of qualia Q-1)

Type II. The Objective - Physical basis of qualia independent of the *subjective feel* that could be stored or transferred to another entity breaking the 1st person 3rd person barrier. The noumenal elements of qualia upon which the phenomenology is based.

Type III. The Universal - Living systems represent a Qualia substrate of the conscious universe, acting as a 'blank slate' carrier from within which Q-II are modulated into the Q-I of experience by a form of superradiance or hyper-holographic evanescence.

A standard image requires a screen or other reflective surface to be resolved; but if the foci of two parabolic mirrors (Casimir-like plates in our model) are made to coincide, the two images superpose into a real 3D image that does not need a screen. A science toy called the 'magic mirage' is used to demonstrate this effect of parabolic mirrors. Objects placed in the bottom appear like solid objects at the top of the device.

The holophote action of *élan vital* energetics arises from the harmonic oscillation of least unit boundary conditions tiles the spacetime backcloth and pervades all self-organized living systems. The inherent beat frequency of this continuous action produces the Q-III carrier wave that is an *empty slate* modulating cognitive data of Q-II physical parameters into Q-I awareness states as a superposition of the two (Q-III and Q-II). This modulation of qualia occurs in the HD QED cavities of the cognitive domain. The QED cavities are a close-packed tiling of least unit noetic hyperspheres; the Casimir surfaces of which are able to reflect *quaneme* subelements. While the best reflectors of EM waves are polished metal mirrors, charged boundary conditions also reflect EM waves in the same way radio signals bounce off the ionized gases of the Kennelly-Heaviside layers in the Earth's ionosphere. This reflective 'sheath' enclosing the cognitive domain is charged by the Noeon radiation (exchange particle of the noetic field) [59] of the *élan vital*, the phases of which are 'regulated' in the complex HD space of the least unit HAM cosmology.

Cosmological Origin and Production of the Three Types of Qualia

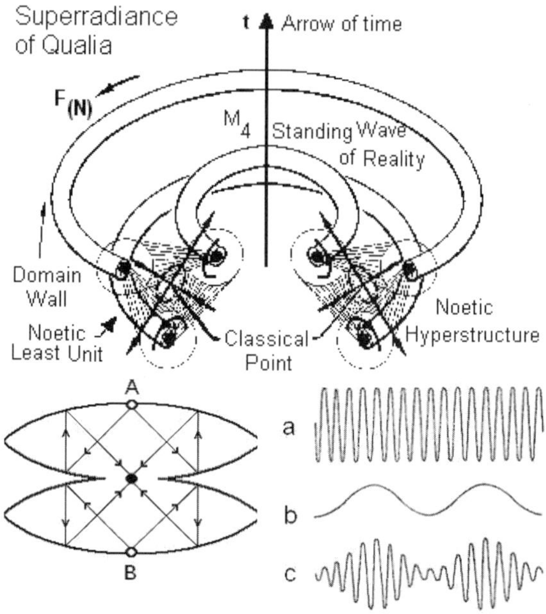

Figure 14.3. Metaphor for the emergence of qualia from the continuous action of the noetic least unit (1a), a microcosm of the HAM where past oriented compactification periodically produces a classical spacetime point. The standing-wave domain walls represent the lightcone singularities of Q-III propagation, the surfaces of which act structurally as Casimir-like plates, and phenomenologically as a carrier wave base for Q-I qualia evanescence by Q-II modulation. 1b represents two pairs of parabolic mirrors (the Q-III Casimir domain walls) whose foci overlap; this is the high frequency wave in 1c denoted as *a*. The longer wave *b* represents Q-II qualia which is modulated by the Q-III wave into the usual Q-I qualia *c*. Thus *a, b*, and *c* in 1c represents the three forms of qualia and how they work together to form Q-I by superradiance of the noetic field.

How does noetic theory describe more complex qualia than the simple qualia of a light pencil? (The qualia-II of a light pencil is assumed to be *the* pencil of light [117,118] Light quanta are microscopic in contrast to the macroscopic sphere of awareness. It thus seems reasonable to assume that scale invariant properties of the HAM least unit of awareness would apply. Like phonemes as fundamental sound elements for audible language there are qualia-nemes or *quanemes* for awareness all based on the physical modulation of Q-II states by the geometric structural-phenomenology of the Q-III carrier base of living systems [117,119].

14.22 Cosmology of Noetic Medicine – An Introduction to Catastrophe Theory

Regarding homeostasis - living-systems and every component subsystem, especially those related to health and well-being, biophysically are forms of *dynamical systems* that generally operate in a framework of stability and equilibrium – the maintenance of which is the charge of medical practice. Technically these systems have a restrictive class called gradient systems which contain singularities or points of *extrema*. Some causal action can institute a bifurcation of an extrema that can initiate a qualitative change in the physical state of the system. Catastrophe theory[5] describes the breakdown of stability of any equilibrium system causing the system to jump to another state as the control parameters change. The changes in the singularities associated with the bifurcation of extrema are called elementary catastrophes [119-121] and can be described by real mathematical functions

$$f : R^N \to R. \qquad (14.1)$$

The equation describing an elementary catastrophe utilizes variables representing *Control* and *State* parameters of the system and is a smooth real function of r and n where R represents the resultant singularity or catastrophe

$$f : R^r \times R^n \to R. \qquad (14.2)$$

The r variables are the control parameters of the state variables n. The function f is therefore an r-parameter family of functions of n variables. If we let

$$f\left(a_i,...a_r;\ x_j,..x_n\right) \qquad (14.3)$$

be a smooth real-valued function of $r + n$ real variables we get equation (2).

The number of elementary catastrophes depends only on r and is finite for $r \geq 5$ totalling eleven (Table 14.1) and infinite for $r \geq 6$.

---

[5] The groundwork for Catastrophe Theory began with the Poincairé work in 1880 on the qualitative properties of solutions to differential equations; and became formalized in the 1950's by R. Thom's work on mapping singularities in structural stability, which he called catastrophes [119-121].

14.23 Catastrophe Theory and Anticipatory Effects of the Noetic Formalism

The structural-phenomenology of Double-Cusp Catastrophe (DCC) Theory in $\geq 9D$ appears homeomorphic to the Riemannian manifold of both 10(11) dimensional M-Theory and the topological geometry of the continuous state dimensional reduction spin exchange compactification process inherent in the action of the corresponding scale invariant least unit of noetic superspace which because it is a complex self-organized system has inherent anticipatory properties mediating the catastrophes. In this general framework the double-cusp *equilibrium surface* is analyzed in terms of a hierarchy of *jumps in state* providing a framework for expanding the basis of allopathic medicine and psychology. One can say FAPP that the noetic least-unit tiling [109] of the Planck backcloth is a complex HD catastrophe manifold mediated by the unitary noetic field.

r (Control Factors)	Number of Catastrophes	Name		Dimensions
r = 1	1	A_2	Fold Catastrophe	2D
r = 2	1	$A_{\pm 3}$	Cusp Catastrophe	3D
r = 3	3	A_4	Swallowtail	4D
r = 4	2	$A_{\pm 5}$	Butterfly	5D
r = 5	4	A_6	Wigwam	6D
r = 3	-	D_{-4}	Elliptic Umbilic	5D
r = 3	-	D_{+4}	Hyperbolic Umbilic	5D
r = 4	-	D_5	Parabolic Umbilic	6D
r = 5	-	D_{-6}	2nd Elliptic Umbilic	7D
r = 5	-	D_{+6}	2nd Hyperbolic Umbilic	7D
r = 5	-	$E_{\pm 6}$	Symbolic Umbilic	7D
r = 6	∞	X_9	Double Cusp	9-11D

Table 14.3. The general forms of catastrophes showing how the dimensions increase as the number of control factors increase. The names bear some resemblance to the geometric pattern of the catastrophe. The double cusp catastrophe is utilized in development of Noetic Theory because it models most closely noetic superspace transitions and is compatible with the fundamental equation of consciousness.

The noetic action of consciousness $F_{(N)}$ is not a 5th fundamental force but an integration of the electromagnetic and gravitational force at the unitary level where it is confined to the Universal sea of consciousness embodying a 12D Noetic spacetime metric $S_{(N)}$ [88-90]. The well-known Schrödinger equations central to quantum theory make correspondence to Newton's second law of motion $F = ma$ which is also the starting point for deriving the noetic formalism. Newton's law of gravitation $F = Gm_1m_2 / r^2$ is not chosen because it is not the fundamental form of gravitation and also contains an undesirable constant of dimensionality. Whereas $F = ma$ is dimensionless. Likewise Einstein's gravity is also not chosen.

Substituting Einstein's mass-energy relation $E = mc^2$ into Newton's 2nd law we obtain: $F_{(n)} = E / c^2 a$ where $F_{(n)}$ is the noetic force and E becomes the self-organized autopoietic energy [88,89] related to ψ_e of the cosmology of mind defined in the fundamental dualistic interactionist relationship of noetic theory:

$$\left|\Psi_M\right\rangle = \left|B\right|\psi_b\rangle + \left(\left|\psi_e\right\rangle + \left|\psi_c\right\rangle\right), \tag{14.4}$$

i.e. the mind Ψ_M is not merely quantum brain dynamics $B\psi_b$, but a classical $\leftrightarrow$ quantum $\leftrightarrow$ unitary continuum of brain, élan vital ψ_e and HD elemental intelligence ψ_c. E is scale invariant through all levels of the HAM beginning at the highest level in the supralocal Multiverse as a hyperdimensional Wheeler Geon - a *ball* of photons of sufficient size to self-cohere through gravity [123]. At the micro level the Geon becomes synonymous with the de Broglie wave-like mental energy of a conscious entity. The Prion [115,123-126], the infectious protein responsible for spongiform encephalopathies (mad cow disease) is designated the simplest known life form, if correct that the prion protein is 'animated' by the self-organizing properties of the *élan vital* of the noetic field [119]. The E unit is comprised of a factor of *Einstein's,* the fundamental physical quantity defined as a 'mole - Avogadro number (10^{23}) of photons'.

Next the derivation of the noetic equation is generalized for the conscious universe by taking an axiomatic approach to cosmological scaling from the work of Kafatos et al, [127] suggesting that all lengths in the universe are scale invariant. Beginning with the heuristic relation $c \equiv \dot{R}$ or $\dot{R} = L / t = c$ where $\dot{R}$ represents the rate of change of scale in the universe. This corresponds to the Hubble relation for perceived Doppler expansion of the universe where $H_0 = \dot{R}/R$ and $a = \dot{R} \times H_o$. By substituting $\dot{R}^2/R$ for a in the original $F_{(n)} = E/c^2 a$, for final substitution we have $F_{(n)} = E/c^2 \times \dot{R}^2/R$. Since $c = \dot{R}$ the c^2 & $\dot{R}$ terms cancel and we are left with the simple equation

$$F_{(N)} = E/R \qquad (14.5)$$

which is the unexpanded fundamental formalism for noetic action within a conscious entity in the HAM cosmological model. It should be noted that *R* is a complex rotational length with standing wave properties and could be derived in terms of angular momentum or spacetime spinors at HD levels in domains described by future developments in M-Theory using Calabi-Yau dual mirror symmetry.

When applied in concert with the fundamental noetic equation of consciousness [106] and the model of interactive computing [110,128] double-cusp catastrophe theory provides a mathematical basis for the noetic action principle called the "Noetic Effect" that applies to medicine and psychology where new medical technologies are under development to ameliorate autoimmune etiologies and balance mental disparities in Transpersonal Psychology. The processes of metabolic homeostasis and intentional action are modulated by the ubiquitous flux of the unitary noetic field as described by the anticipatory effects of the $F_{(N)}$ formalism.

Equation (14.4) is a standard equation for the equilibrium surface of the DCC [120-122] as modeled in (Fig. 14.8); where $B \pm Q$ is the state variable and μ_d and υ_d are the control parameters.

$$(B+Q)^3 + (B+Q)\mu_d + \upsilon_d = 0 \qquad (14.6)$$

The position of the two cusps is found at $\mu_d = 0$ and $\upsilon_d = 0$.

If Fig. 14.5a is considered as a present moment; 14.5b represents a flag of temporal permutations as the noetic catastrophe cycle evolves through its equilibrium cycle in time from future to past and higher to lower dimensions in the same manner as the 'eternal present' undergoes the continuous-state spin-exchange dimensional reduction compactification process in the dual mirror symmetry of the Calabi-Yau background spacetime manifold of HAM cosmology for the spaces: $R^{12} \supseteq \ldots R^4 \supseteq R^3 \supseteq R^2 \supseteq R^1 \supseteq R^0$.

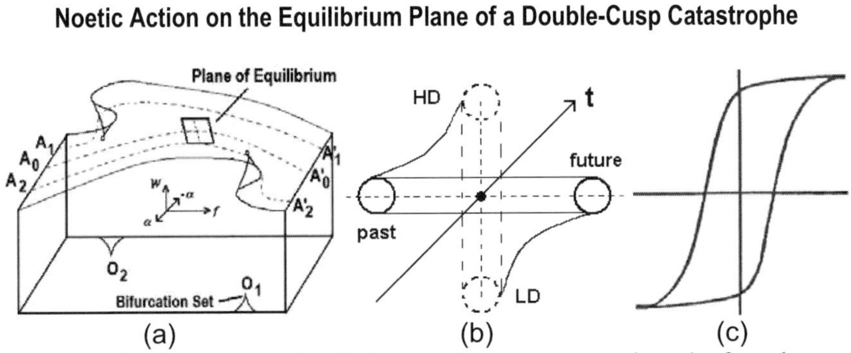

Figure 14.4. a) illustrates the DCC showing cusps at each end of a plane of equilibrium. The DCC occurs in ≥ 9 D and the catastrophe most compatible with NFT symmetry. The plane of equilibrium is a topological manifold tiled of noetic *least units*. The equilibrium manifold undergoes a 'conscious' quantum computation best described by interactive computation [110,128]. b) graphically illustrates the fundamental scale invariant noetic equation $F_{(N)} = E / R$ of conscious action. Any internal or external stress or change in E is a nonlinear dynamic process producing stability or instability in the boundary conditions of R; an instability in $E \rightarrow$ stress $\rightarrow$ displacement $\rightarrow$ catastrophe $\rightarrow$ jump…whereas stable flux is homeostatic. 14.4b like noetic HAM cosmology is also a form of hysteresis loop generalized in c).

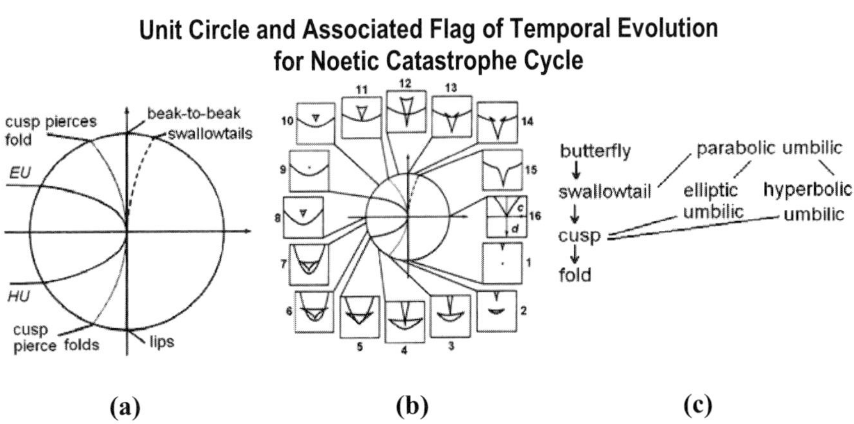

Figure 14.5. [120,121] a) represents a plane of the unit circle with corresponding cross sections in 14.5b Section 16 for example shows a cusp. A single point in 1 grows to the 'lips' in 2. In 3 to 4 the original cusp 16 penetrates the mouth becoming a hyperbolic umbilic point at 5, turning into an elliptic umbilic at 6, shrinking to a point in 9. Growing again in 10 to pierce the fold line in 11 and through it in 12. A 'beak-to-beak singularity in 13 breaks in 14, collapsing to a swallowtail 15. The 7 fundamental catastrophes contain 'subcatastrophes' according to the diagram in c).

TABLE 14.4. Geometry of 0D to 12D showing points and lines contained

N space	Point	Lines	Souar	Cubes	Tesseracts	5T	6T	7T	8T	9T	10T	11T	12T
0	1												
1	2	1											
2	4	4	1										
3	8	12	6	1									
4	16	32	24	8	1								
5	32	80	80	40	10	1							
6	64	192	240	160	60	12	1						
7	128	448	672	560	280	84	14	1					
8	256	1,024	1,792	1,792	1,120	448	112	16	1				
9	512	2,304	4,608	5,376	4,032	2,016	672	144	18	1			
10	1,024	5,120	11,520	15,360	13,440	8,064	3,360	960	180	20	1		
11	2,048	11,264	28,160	42,240	42,240	29,568	14,784	5,280	1,320	220	22	1	
12	4,096	24,576	67,584	112,640	126,720	10,137	59,136	25,344	7,920	1,760	264	24	1

The putative significance of Table 14.4 for the application of double-cusp catastrophe theory to the noetic HAM formalism is that the structure of possible boundary conditions and the number of control points is revealed. For example, in this simplistic view, a 3D point in real spacetime might have 16 control photon-gravitons (noeons) covering it. Carrying the analogy all the way up to the 12D holoscape of the Multiverse, the same 3D point might be controlled or guided by a total of 8,176 photons. The number arrived at by summing the points of D4 to D12. No point in the universe is isolated; so this metaphor does not include the possible power factor by associated points in both the HD and LD HAM backcloth. Within the inherent continuous-state dimensional reduction compactification process, the LD domain (dimensions less than 3) might be coupled to orders of magnitude more photon-gravitons. This detail of Noetic Theory has not been completely worked out yet.

14.24 Example of Noetic Medicine: The Mechanism of Protein Conformation in Prion Propagation

Fatal neurodegenerative disorders known as transmissible spongiform encephalopathies (TSE'S) have been shown to spread by a proteinaceous infectious particle or prion [124-126]. According to Prusiner's definition these prion elements propagate conformational variation leading to replication by a mechanism not well understood until now [124]. Two conversion hypotheses have been proposed:

- The *template-assisted conversion model* [186] where a putative cellular chaperone called protein X assists conformational transition by altering the thermodynamic equilibrium of a kinetic barrier in favor of transition state protein formation.
- The *nucleation-polymerization model* where highly ordered aggregates of the infectious element form. This shifts thermodynamic equilibrium allowing this nucleus to act as a seed for further prion propagation. Protein folding thus appears in both cases to be the primary autocatalytic mechanism propagating prion diseases.

According to Prusiner [126]:

> Nascent prions are created either spontaneously by mutation of a host protein or by exposure to an exogenous source. Prions are composed largely, if not entirely, of a modified form of the prion protein (PrP) designated PrPSc. Like other infectious pathogens, they multiply but prions do not have a nucleic acid genome to direct the synthesis of their

> progeny. A post-translational, conformational change features in the conversion of cellular PrP (PrPC) into PrPSc during which alpha-helices are transformed into beta-sheets. Since this structural transition in PrP underlies both the replication of prions and the pathogenesis of the CNS degeneration, much of the effort in the laboratory is devoted to elucidating the molecular events responsible for this process. Indeed, prion diseases seem to be disorders of protein conformation.

And further relative to the theory of propagation proposed here:

> During prion replication, an as yet to be identified factor that we have provisionally designated protein X binds to PrPC. The PrPC/protein X complex then binds PrPSc; by an unknown process, PrPC is transformed into a second molecule of PrPSc [126].

A Postulated 3D X-bundle structure of the PrPC was chosen by Prusiner from four penultimate PrPC models reduced from ~300,000 possible configurations by both theoretical and experimental constraints. These four choices correlated best with human prion mutations. A Conceptual model of the orientation of the four helices of the X-bundle model looks like two X's nearly superimposed on each other. Since prions have no nucleic acid based genome to direct their propagation. Noetic theory proposes that prion replication is directed by fundamental mechanisms of complexity theory and that the action principles driving this complexity are a more fundamental form of mechanism than that perceived currently by the philosophical basis of mechanism in biology.

14.25 Implications for Transpersonal Psychology and Autoimmunity

The immune system is comparable in the complexity of its functions to the nervous system. Both systems are diffuse organs that are dispersed through most of the tissues of the body. In man the immune system weighs about two pounds. It consists of about a trillion (10^{12}) cells called lymphocytes and about 100 million trillion (10^{20}) molecules called antibodies that are produced and secreted by the lymphocytes. The special capability of the immune system is pattern recognition and its assignment is to patrol the body and guard its identity…The immune system is subject to continuous decay and renewal. During the few moments it takes to read this chapter one's body produced 10 million new lymphocytes and a million billion new antibody molecules. This might not be so astonishing if all these antibody molecules were identical. They are not. Millions of different molecules are required to cope with the task of pattern recognition, just as millions of different keys are required to fit millions of different locks. - *Niels K. Jerne,* 1973, Scientific American

Much biochemistry, biophysics and quantum dynamics has been developed in the study of the immune response relating to the principles of allopathic medicine. But if one recalls that the founders of quantum theory emphatically stated that quantum theory and the Schrödinger equation did not describe biological systems; something must be missing in all this work. This is of course the 'life principle' that is introduced by the Holoinformational principles of interactive dualism. First let us outline all the brilliant models that are insufficient:

- F. Popp's biophotons – Sure all of human physiology is a thermodynamic heat-bath and radiates a spectrum of radiation. But this is after the fact of any immune response and has nothing to do with consciousness.

- K. Pribram's holographic brain model utilizing Fourier and Gabor principles for a spectral holographic domain throughout the brain – A fabulous advance in quantal brain function, but again devoid of principles of consciousness.

- W. Schempp's quantum holography and his discovery that all information about objects, including their three-dimensional shape is carried in the quantum fluctuations of the Zero Point Field, the vast memory store predicted by Puthoff. With this information Schempp calculated, recovered and reassembled three-dimensional MRI images through Fourier transformation. – Again no conscious principles.

- Marcer and Schempp developed a mathematical map of how information is processed in the brain that is a mathematical demonstration of Pribram's theory. – Still no consciousness.

- S. Hameroff's showed microtubules are an exceptional conductor of quantum pulses and these pulses are transmitted through pockets of protein. Hameroff discovered a coherence among neighboring tubulin dimers; calling them "light pipes" or "waveguides" for photons sending these waves from cell to cell throughout the brain with Fröhlich coherence. – A model of consciousness, but insufficient.

- K. Yasue and M. Jibu also theorized that the quantum messaging must take place through vibrational fields and quantum coherence along the microtubules.

- K. Pribram, K. Yasue, S. Hameroff and S. Hagan assembled a collective theory about human consciousness in cytoskeletal microtubules in which microtubules and the membranes of dendrites represent the Internet of the body. Theoretically every neuron could log on at the same time and speak to every other neuron simultaneously via this quantum processes by quantum superposition, 'Superradiance' and 'Self-Induced Transparency'. – A logical step forward but still missing the conscious action principle.

- Similarly physicists Del Giudice and Preparata suggested Hameroff's 'pipes' contained coherent energy fields. They also suggested that water molecules in the brain extend coherence effects as far as 3 nanometers outside the cell's cytoskeleton suggesting water inside the microtubules is also ordered. They showed that this focusing of waves would produce beams 15 nanometers in diameter - precisely the size of the microtubule's inner core.

- All this led to the heretical thought already occurring to F. Popp that consciousness was a global phenomenon occurring everywhere in the body, not simply in the brain. Perhaps consciousness is fundamentally a coherent light within a unified theory of mind and matter like Bohm's "unbroken wholeness", where the universe is a vast dynamic and intelligent Holoinformational web of information exchange containing all possible versions of all possible forms of energy and matter. – Yes, but what kind of light?

This is the crux of the problem because any place where there are atoms and molecules and spacetime there are quantum fluctuations. This is all that is addressed by Copenhagen interpretation whether in microtubule, synapse or neuron - there is no consciousness described here even though this is the currently dominant cognitive model accepted by 90% of consciousness researchers today.

A holoinformational cosmology of consciousness is required because consciousness is deeper than the 3D brain. The sphere of action must be taken to where the unified field - spirit of God lies, hidden behind this virtual barrier, the regime described by the Copenhagen interpretation as outlined in the bullets above. We must follow Einstein's view that quantum theory is incomplete: "God does not play dice".

Several hypotheses are considered by immunologists:

- Clonal Deletion theory, proposed by Burnet, according to which self-reactive lymphoid cells are destroyed during the development of the immune system in an individual.

- Clonal Anergy theory, proposed by Nossal, in which self-reactive T- or B-cells become inactivated in the normal individual and cannot amplify the immune response.
- Idiotype Network theory, proposed by Jerne, wherein a network of antibodies capable of neutralizing self-reactive antibodies exists naturally within the body.
- The so-called "Clonal Ignorance" theory, according to which host immune responses are directed to ignore self-antigens.
- The "Suppressor population" or "Regulatory T cell" theories, wherein regulatory T-lymphocytes (commonly $CD4^{+}FoxP3^{+}$ cells, among others) function to prevent, downregulate, or limit autoaggressive immune responses.

Autoimmune diseases can be broadly divided into 1) systemic and 2) organ-specific or localized autoimmune disorders, depending on the principal clinico-pathologic features of each disease. An example of a systemic syndrome is Rheumatoid Arthritis; and Local syndromes include virtually any circumscribed mass of body tissue like Addison's disease or multiple sclerosis. A comprehensive list of autoimmune conditions can be found at [120]. Current treatments for autoimmune disease are usually immunosuppressive, anti-inflammatory, or palliative. Non-immune therapies, such as hormone replacement in Hashimoto's thyroiditis or DM Type 1 treat outcomes of the autoaggressive response. Dietary manipulation limits the severity of celiac disease. Steroidal or NSAID treatment limits inflammatory symptoms of many diseases. More specific immunomodulatory therapies have been shown to be useful in treating rheumatoid arthritis. These immunotherapies may be associated with increased risk of adverse effects, such as susceptibility to infection [121-126]. These therapies treat the 'branch' but not the 'root' of the problem; and therefore provide no cure.

14.26 A Noetic Effect on Autoimmune Systems

Holoinformational-noetic-conscious medicine is far different than scientific-allopathic medicine; it is a spiritual medicine relying on the *élan vital*. This is the gap we wish to bridge. We postulate that all autoimmune etiologies are diseases on consciousness and therefore require a Holoinformational cosmological perspective, rather than the current limitations to brain or biochemistry. Our starting point for correspondence to current theory is the network immune theory of N.K. Jerne [127,128]. This means that medical treatments must be found to regulate the flow of the unified Noetic Field. Complex self-organized systems like living systems are driven by an action principle. This action principle is the teleological life principle equated with the cosmology of unitary field. This is the missing component from Scientific or allopathic medicine; this is the principle that must be added.

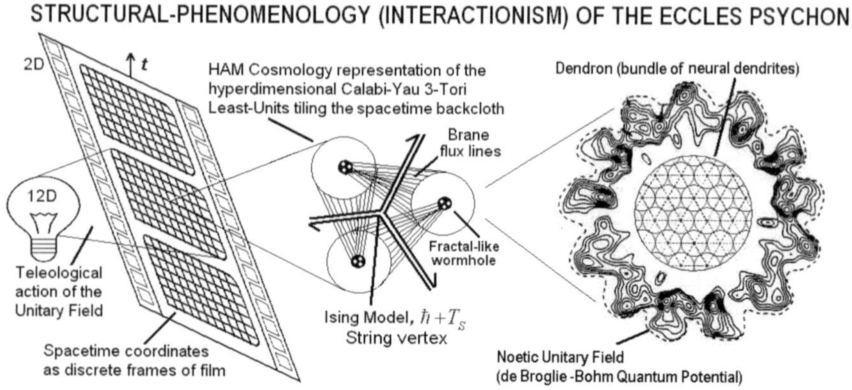

Figure 14.6. Conceptualization of Interactionist cosmology, a) showing injection of the noetic field or *élan vital* into spacetime points, b) Planck scale least-units mediating the noetic field, c) an Eccles Psychon field coupled to a brain dendron where autoimmune interactions may interrupt normal homeostasis.

In allopathic-scientific medicine if a tumor exists we surgically remove it or give chemo-radiation. If a deficiency we give a vitamin or any of the tens of thousands of pharmaceuticals discovered to control the human biochemistry and its myriad etiologies. But this is the branch only and rarely the root of the problem except for microbial causation. We want to address the root of the problem, the ~150 autoimmune conditions [120] that allopathic medicine has little insight into especially at the root. These are the noetic etiologies. Imagine a child with a magnifying glass focused by the sun on an ant that actually can catch them on fire. Imagine those foci are not on an ant but 'catastrophes' focused on many energetic microsites of the biochemistry or organ systems. Also imagine the converse – a cover is put over the lens blocking the flux of life energy. This life energy arises from a deep regime in spacetime backcloth itself in a Holographic Anthropic Multiverse [129]. The entry point of the life force where deep catastrophes may occur in the hysteresis (energy) loop of the propagation of the noetic unitary field with a simple relation that coupled with all the transpersonal-personality pre-disposition which forms a complex pattern specific to the ~150 different ways that lead to these conditions. I call it the noetic effect which is governed by the noetic field equation. The noetic effect is the switch for all the quantum/classical places discussed – neural, synapse, MT, maps etc. This deeper understanding is the fundamental key to all conscious medicine.

In Fig. 14.6 we see a conceptual view of the noetic cosmology of interactive dualism. To clarify our critique of the cognitive approach and allopathic medicine, we point out that current medical treatment modalities related to the work on consciousness and quantum theory are only related to quantum parameters associated with brain or physiological elements associated with Fig. 14.6c and are silent about 14.6a and 14.6b that relate to consciousness.

From Figs. 14.4 and 14.5 above we can understand that a sustained noetic effect resulting in catastrophes in the unified noetic field is responsible for all autoimmune disorders. The hysteresis loop in the center of the Double Cusp Catastrophe (DCC) is the energy available for this action. This can be applied to a Jungian type collective unconscious related to Transpersonal Psychology if we metaphorically consider the absorption-emission spectra of complex molecules to be like personality structure: We know already people are made of a web of molecules. We know a little of personality types and wish to apply this as a model to the absorption emission spectra of atomic structure. So in scale invariance people are like huge particles or atoms. The hysteresis loop in the middle of the DCC represents energy, bioenergy and mental energy depending on the relation. We know in general about extrovert, introvert, dependent personality, dominant, submissive, we know depression is like black hole and from Selye - stress causes foci of negative energy that is destructive. From the relaxation response and meditation we know of energy balance. For physics the future-past spacetime hysteresis loop can be applied directly to the formula for absorption-emission spectra. What is next is to formally systemize this model to generalize each foci of these noetic catastrophes. For Alzheimer's negative foci are on the brain, for colitis on the intestine Excessive anger in a certain way can cause a brain a tumor, another kind of dominance causes necropsy in the knees or a weak spine for back problems. But this initial idea must be systemized for the whole complex system. What are the precise entry points of the field; Which autoimmune molecules as Jerne suggests are related?

This is a global view generalized for personality structure to focused deficits in energy. The model is empirically testable and medical devices are being designed to implement conscious medicine for both diagnosis and treatment. In the near future new medical technologies will be developed that are currently considered science fiction. The popular US sci-fi TV-movie series Star Trek includes advanced medical techniques utilizing a device called the *Medical Tricorder*[6] able to diagnose and heal. Most notable at the

---

[6] Amoroso has a US patent pending for such a tricorder in hopes that such a device will arise in next 10-15 years.

fundamental level of triage is the tricorder's ability to heal a variety of injuries in a few seconds that today might take several months to heal. At the advanced level using combined features of replicator and transporter[7] technology as illustrated in the Hollywood movie "the Voyage Home" when Dr. McCoy utilizes the tricorder to repair a ruptured cerebral artery inside Chekov's head. The operation is accomplished by *beaming out* the damaged section of the artery and *beaming in* a new section.

We may not see the advent of the advanced form of the tricorder until our children's or grandchildren's day; but sufficient theory exists to construct the basic form of the tricorder now. In the interim noetic Holoinformational theory can be used to greatly expand the versatility of Transpersonal Psychology.

14.27 Completing Epistemology: The Utility of Transcendence as a Tool in Transpersonal Psychology

I want to know God's thoughts ... the rest are details - Albert Einstein

Human epistemology has steadily evolved from dark ages of superstition through enlightened periods of logical reason to the current pragmatic age of empiricism. Now another Galilean class revolution completing epistemology by integrating Science and Theology (S&T) utilizing transcendence seems immanent. S&T represent opposite ends of a long continuum of schools of thought rather than mutually exclusive disciplines as often believed. To implement the required paradigm shift an integrative noetic science must include an adequate understanding of Transcendence. Over 2,000 years ago the Greek philosopher Plato considered this type of noetic insight, paraphrased here as a corollary:

§ Noetic Insight: *No matter how great ones intelligence or how vast ones wisdom, noetic insight is cosmic insight transcending the capacity of the self* [130].

Human epistemology has come full circle to a time not only for another evolutionary step, but the final one completing the tools of epistemology through the use of transcendence.

For the first time since the Dark Ages, physicists Paul Ginsparg and

[7] The Star Trek Transporter disassembles the individuals atomic and life energy information and broadcasts it to a remote site for reassembly. The replicator is a similar technology for objects and tissue assembled from templates stored in a computer.

Sheldon L. Glashow wrote 12 years ago, we can see how our noble search may end, with faith replacing science once again [131].

This condition is not what is advocated here because it seems that no matter how advanced tools of transcendence may become, empiricism leads directly to engineering which is an integral part of temporal existence. In some arenas current science has already reached, at least in terms of experimental design, the limits of empiricism; for example some experiments in particle physics require an accelerator the size of the universe and some calculations require a computation cycle with a duration the age of the universe. Only about 70 years ago Cosmology was not considered science. The universe was believed to be clock-like as described by Newtonian mechanics. Since the advent of Quantum Theory the majority of scientists have considered the universe to be quantum.

But recent studies extending the standard models have allowed a growing number of scientists to embrace forms of an Anthropic Conscious Multiverse. The form utilized here in Noetic Theory has continuous-state properties with temporal reality cast as a virtual subspace of a higher dimensional eternity [45-47]. This new cosmology yields key elements pertinent to premises here (especially the periodic properties enabling introduction of an inherent spirit-based action or life principle); some of which are:

- The fabric of reality continuously cycles between classical, quantum and unitarity (continuous-state).
- Phenomenological reality is virtual; because of the arrow of time much of the underlying noumenon is 'filtered' out of perception.
- Dimensionality cycles continuously from spatial to temporal to energy. This "energy" is synonymous with the quantum potential or pilot wave in 4D and the unified field or spirit of God in HD.
- Matter by Einstein's $E = mc^2$ is continuously created, annihilated and recreated (the well-known wave- particle duality) forming the holographic backcloth of perceptual reality.

Inherent in the periodic properties this noetic model is the unified field or spirit of God, acting in governance as a higher dimensional de Broglie-Bohm super quantum potential [132,133]. Periodicity allows for the pervasive ubiquity of this supernumerary action principle. Since a conscious universe is implied the field is one of information. This is key to our idea of transcendence. In an Anthropic Holographic Conscious Multiverse (HAM) human beings are spiritual beings and a path to enlightenment is possible by following certain laws related to this condition. Because of the nonlocal (and because of the additional dimensionality – supralocal) character of the

Holographic Principle individuals perceive themselves as separate entities in 3-space. But in Higher Dimensionality (HD) we are unitarily imbedded in the holographic backcloth, which because of its spiritual nature –

- Transcendence may occur and
- Information received in the process

The coming paradigm shift does not merely represent a significant intellectual breakthrough like Copernicus' transformation of egocentricity into heliocentricity, the advent of quantum theory or Einstein's theories of relativity; but a profound paradigm shift where Humanity will leave the so-called Modern Age behind an enter an Age of Consciousness.

An 'empirical metaphysics' [134] is under development that will violate the uncertainty principle and allow actualization of Plato's noetic insights in a manner useful for scientific exploration. Noetics, the study of the cosmology of mind, comes from the Greek word *nous* meaning intellect. Noetic insight when used scientifically to complete epistemology is the highest form of knowing because it utilizes and integrates the pure logic of philosophical reason, the rigors of scientific empiricism and the absolute truth of theology.

All scientific theory formation has at least low level metaphysical components. Without entering into a technical discussion of the nature of creativity, we assume here that this is what any creative process entails. The latter-day Mormon prophet Brigham Young went so far as to say "All scientific discovery comes as a revelation from God" [135]; while this may indeed be considered true in an Anthropic Universe, one would suspect the vast majority of scientists are currently neither generally interested in, consciously aware of its occurrence, nor even consider this possibility, especially since statistics have demonstrated that only about 20 to 30% of scientist believe in some form of god in contrast to 95% of the general population.

Does this mean that only few might be initially prepared to take advantage of the premises of noetic transcendence? An informal survey of my colleagues has revealed that some have already begun using transcendent abilities in various ways in scientific endeavor and daily routine with reasonable success. We know of no team efforts yet at this writing; although a five year budget for implementing bulk quantum computing approved in March 2010 will be used to test the premises presented here. As principle investigator Amoroso has chosen a question, because of infinite possibilities, that can only be answered by transcendence. Should we be successful certainly "the game will be afoot" as Sherlock Holmes would say at the beginning of a case. Hopefully preliminary results will be available before this volume appears in 2011. We think a result like this is required to create sufficient pause for engendering a Galilean class paradigm shift in epistemology. Since inception in the 1980's all attempts have failed at implementing bulk quantum computing. By utilizing transcendence we have been provided a unique approach by the *Zeitgeist*; that approach demands the development of what Amoroso calls "the Noetic Transform" [136].

The value of the high level addition of Transcendence as a tool of science in theory formation would be to accelerate progress by saving considerable time, energy and funds by optimizing both avenues for empirical research and efficiency in contemplating and defining fundamental new tenets of a model. For example, early in my career I sat in on a round table discussion by an august body of great thinkers of the age. They divided up a challenging problem into every logical possibility like spokes of a wheel. Each agreed to take a spoke or two, intending to spend the remainder of their careers working on their arena of interest of the problem. The utility of transcendence in cases like this would be to narrow the field to a spoke or two.

Currently all the standard models of science are Darwinian or naturalistic excluding any place for God or Spirit. For example Biological Mechanism, the basis for allopathic or scientific medicine and psychology states: The laws of chemistry and physics are sufficient to describe all life; no additional life principle is required [93-95]. The founding fathers of quantum theory stated it could not describe biological systems, Big Bang cosmology is also naturalistic; therefore something most assuredly must be missing in physical theory.

The noetic model for the integration of S&T is based on three premises:

§1. That transcendence is a universal Anthropic Principle able to provide an interface or *common ground* between S&T.
§2. Rigorous application of The Golden Rule (see below) spontaneously leads to transcendent abilities under certain optimal conditions because
§3. Man is inherently a spiritual being (*The spirit and the body are the soul of man* [137]) imbedded in a conscious universe guided by a unitary field tantamount to this spirit.

A common ground [119,138] uniting S&T is required because traditionally scientific principles are not accepted by faith based theology; and religious dogma is generally considered an unacceptable anti-intellectual mode of epistemological inquiry by the common definition of scientific pragmatism in place since Galileo showed that reason, in the case of heavier objects falling faster, failed. Similarly today Hubble discovered redshift, not a Doppler expansion of the universe. Other interpretations are available [45-47] supporting HAM cosmology.

To achieve this integrative result a model of an Anthropic Continuous-state Holographic Conscious Multiverse (HAM) [45-47] (and this volume) is utilized that includes an inherent basis for defining complex self-organized living systems in a manner that includes the physical basis of spirituality and therefore transcendence [60].

According to the Perennial Philosophy: God exists and has revealed a path to find him [139]. This perennial philosophy is not only universal to all theology but ultimately to all truth whether theological or scientific as we make the case for here. The HAM [45-47] an extension of Einstein's Static Universe model, is shown to naturally include a new action principle governing complex self-organized living systems. This HAM elucidates the physical basis of spirituality. All legitimate religions or life paths in principle provide avenues to transcendence. Achieving transcendence is not based on the superficial icons of the world's theologies. Superficial artifacts like phylacteries, crosses, rosaries or rituals like bowing east or genuflecting are not relevant.

We believe that because human beings are inherently spiritual [60,137], transcendence can be achieved universally by practicing principles of love, service and charity; or adhering 'perfectly' to what is called the Golden Rule - Do unto others, as you would have them do unto you [140-147]. The Golden Rule is the most fundamental moral or ethical principle; it is the basis for the theology of virtually all world religions, the basis of social order, interpersonal relations, sound business practices and international diplomacy. The Golden Rule has many similarities to the Hindu belief in karma and is the basis for all the world's major religions.

14.28 The Golden Rule Subsidiary to Love for God the Great Commandment

For simplicity we will argue our case only form the point of view of Judeo-Christianity but the reader is asked to keep in mind that as illustrated in Fig. 14.11 the premises here are postulated to apply to all legitimate theologies. The monotheistic religions Judaism and Christianity teach that the Golden Rule and other moral commands for human relations are subsidiary to the Great Commandment relating to God, e.g., Jehovah, Emmanuel or Jesus The Christ explicitly identified the Great Commandment as supreme love for God, as affirmed in the Hebrew Torah and Christian Bible [148]. In contrast to the ancient 'an eye for an eye', Jesus gave a new command - "Love one another as I have loved you" [149]. We hypothesize that love is a real physical force of nature and will develop this in future work. By categorizing 'Love your neighbor as yourself' as the Second command like unto the first, Jesus placed the Golden Rule and human relationships as not subsidiary but tantamount to one's ideal relationship with God the father.

The paramount statement relating to our purpose here is Christ's teaching regarding the two great commands, specifically as stated in the last sentence: Master, which is the great commandment in the law? Jesus said unto him, Thou shalt love the Lord thy God with all thy heart, and with all thy soul, and with all thy mind. This is the first and great commandment. And the second is like unto it, Thou shalt love thy neighbor as thyself. On these two commandments hang all the law and the prophets [150].

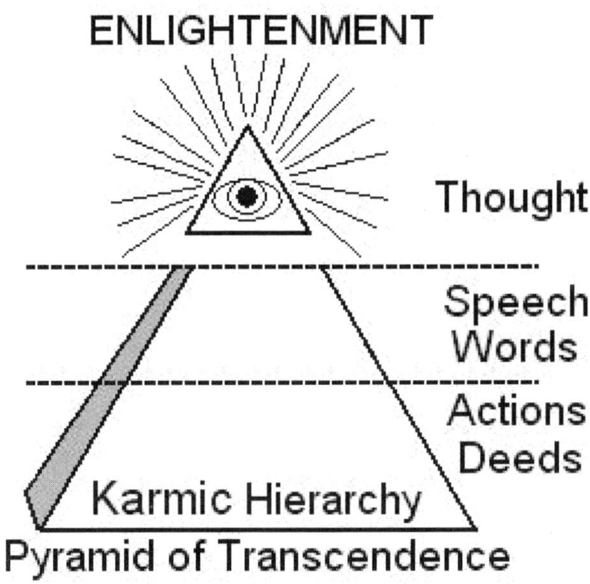

Figure 14.7. Pyramid of Transcendence / enlightenment. Individuals 'Living' operationally at the top of this 'karmic pyramid' spontaneously develop transcendent abilities. 'Sin' or violation of ethical principles, commandments or Karmic law can be classified into three weighted categories of decreasing severity: 1. Actions, 2. Speech, 3. Thoughts. The pyramid of transcendence is an ascension to refinement.

Prophets are seers and revelators – users of transcendent abilities. The requirements for transcendence may be further clarified in terms of a three-level pyramid (Fig. 14.7). The base represents crimes or sins of action like murder, theft or adultery for example. The middle of the hierarchy is represented by sins of word like lies or insults, which under extreme conditions could lead to another's harm or death. Goethe's 1774 classic *Sorrows of Young Werther* [151] is purported to have produced a rash of suicides on its publication; whereas a statement like 'where'd you get that stupid shirt' may or may not only hurt ones feelings. The top of the pyramid represents sins of thought. Thoughts by nature are fleeting, but are real and have real consequences. As long as an evil thought is not dwelt on; it can be forgiven as quickly as contemplated. At this level of living the limitations of being human come into play. Deity can expect no more of a mortal being than trying to manage ones thoughts.

According to metaphysical law of the perennial philosophy as applied to HAM cosmology, one is virtually guaranteed attainment of a degree of transcendence when ones 'moral crimes' hover at the apex of the pyramid (Fig. 14.7); provided one has sufficiently good karma or repaired any karmic debt or made restitution for negative conditions of the past.

Noetic Cosmology suggests that by routinely living at this apex a universal Anthropic Principle of Transcendence comes into play whereby anyone maintaining this mode will spontaneously achieve a state of transcendence. If the premise for this noetic Principle of Transcendence is correct, any team of scientists whether comprised of any combination of Jew, Christian or Shinto for example will be able to utilize Transcendence as a tool in scientific theory formation (Fig. 14.7). Likewise any dialogue between scientists and theologians could achieve similar fruition. Based on the fundamental premise that Men are spiritual beings [60,137] living in an Anthropic Multiverse; the following postulate is said to hold true:

> **Postulate 1:** Any individual or group of individuals living by the Golden Rule, to the extent where those individual's moral offenses[8] occur generally only at the level of thought, will spontaneously develop transcendent abilities.

Two conditions apply. The past history of the individual must be relatively free of serious offense. The postulate may not apply to those guilty of unpardonable offenses like murder or blasphemy against God[9] or the absolute truth of the Multiverse. The activity of thought is at the limit of human control. Human beings cannot be expected to have perfect control of their thoughts. The karmic rule is satisfied if one does not dwell on negative thoughts.

---

[8] Moral offense – We wish to skip for the most part a detailed delineation of what constitutes moral offense. For our purpose here we chose to simply state that good has a tendency to bring people together and moral offense has a tendency to separate or harm.

[9] Unpardonable Blasphemy – This is not a condition of swearing or cursing of the general kind; but a rare occurrence of a fully transfigured person who has beheld God like a Moses who then turn against God.

A power factor exists. Christian doctrine states: *Charity covereth a multitude of sins* [152,153]. This charity or good works, (of time or substance) provides a power factor for eliminating residual or negative Karma enabling the time to be shortened in reaching the apex of the pyramid or the transcendent state.

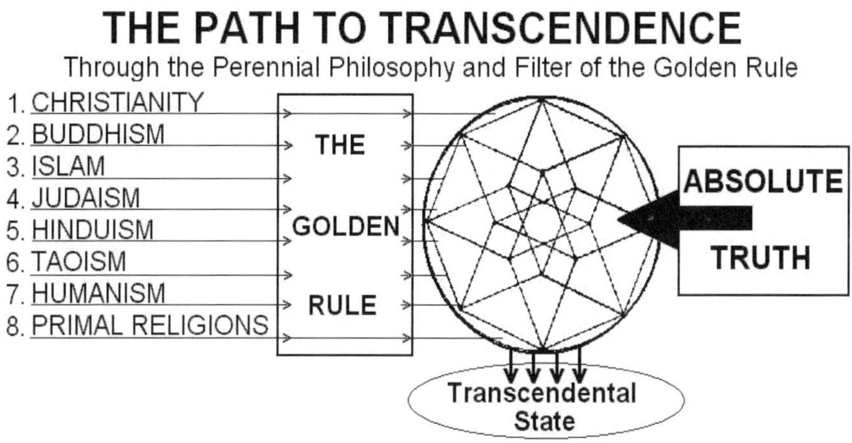

Figure 14.8. Because of the inherent spiritual nature of mankind as part of an Anthropic Cosmology with an inherent teleological life principle and the concomitant existence of 'Absolute Truth' in regard to spiritual matters, the Golden Rule, as a universal principle of the Perennial Philosophy provides a path to both find God and spontaneously develop transcendence.

The basic needs of all life on Earth is optimized by 'The Golden Rule'- treating other entities and the environment holistically in the same manner as we would like to be treated. This perennial philosophy is an absolute truth that relates to all sentient consciousness universally throughout the Holographic Multiverse where intelligent life is the rule not the exception. Transcendence can be achieved by a high level adherence to the universal tenets of the Golden Rule. Empiricism has been an impossible challenge for theology; and scientists have historically denigrated any dialogue utilizing religious dogma based on faith-based logic put forth by theologians as merely a product of pre-Galilean imagination. Therefore only by developing a common basis for utilizing transcendence as a universal epistemological tool can S&T be united pragmatically. Producing a universal framework for transcendence seems of grave import because such a completion of human epistemology could have broad impact ultimately leading to world peace, higher quality of life and amelioration of environmental concerns.

14.29 Transcendence as a Tool in Scientific Theory Formation

Since there are about 10,000 religious sects or spiritual paths in the world today, most of which have conflicting teachings or dogmas; how could developing an empirical metaphysics be possible? Whatever ones spiritual path - the dance of a twirling Dervish, fasting, meditation, charity, chanting, prayer or peyote, it is achieving the resulting pure transcendental state that is of paramount importance.

As performed here for Noetic Cosmology a team of investigators or an individual therapist attempting to utilize transcendence might also utilize historical tracts or scriptures pertinent to their individual path as a starting point to help guide the questions posed to the universe or for the therapy. We realize the extent of this challenge; one must have sufficient faith in the veracity of a scriptural tract to use it as a starting point. The spirit of truth gained from entering the state of transcendence is then used in the Platonic sense [130] for verification. Noetic insight is received through diligent study after sufficiently following ones individual path to perfection (Fig. 14.7) in conjunction with prayerful meditation. Alternatively when one comfortably 'hovers' at the apex of the pyramid if ones is studying a physics manuscript a passage on the Bessel function may leap out while reading or later while pondering as a transcendent suggestion that the Bessel function is pertinent to the engineering or other theory at hand.

We wish to make it clear that receiving such 'revelations from God' need not interfere with experimental verification; because as we mentioned earlier all theory formation has a metaphysical element when initially formulated in the creative mind(s) of its inventor(s). Therefore the metaphysical act of theory formation is independent of the pragmatic demands of hard science which is the second step or companion step in theory testing. There is already a growing movement for integrating science and theology.

Examples of noetic insight from history are Friedrich Kekule's dream of a snake joining head to tail in the discovery of the benzene ring, or perhaps more pertinent to our interest in the nature of consciousness here, Descartes claim of receiving a revelation from God designating to the distinction between mind and body [154]. Descartes 'vision' has remained controversial for over 400 years and is only now about to be tested by the methods of noetic theory. Science, if my work here has been successful, has finally progressed to the point where this is possible.

The great value of developing an integrative discipline of science and spirituality is that potentially 10's, 100's, or even 1,000's of years could be saved, along with the resources expended on spurious research paths that could alternatively be used to alleviate human suffering or maintain the environment etc. The timing in the Zeitgeist seems on target as history already suggests an asymptotic increase of technical information; so to keep pace transcendence seems timely. We have the potential now to integrate science and spirituality and complete the tools of human epistemology.

As an example of a test question, for example Amoroso's noetic theory considers the Big Bang an erroneous interpretation of astrophysical data (Hubble merely discovered redshift not a Doppler expansion); more horrific at the time of this writing when a Nobel Prize has just been given out for the Big Bang's discovery touted as one of the most profound discoveries of the last century. If this noetic premise is proven true, some of the best minds in astrophysics could have more efficiently expended hundreds of thousands of man-hours over the last 75 years. Science by definition is satisfied only by empirical evidence and theology by quietly submitting to faith. Francis Crick believes that the concept of a soul is a myth and that modern neurobiologists (except notably the late sir John Eccles [155] see no need for a religious concept to explain the interaction of nerve cells [156]. He calls this an astonishing hypothesis since over ninety percent of the earth's population believes in the soul.

14.30 Absolute Truth in Theology and Science

The philosophical or theological concept of absolute truth is something has been argued for centuries; whether there is such a thing, what form it takes, can it be proven, and what are the implications if any. A very simple perspective is taken here: Absolute truth indeed exists, it is independent of opinion or even what some kinds of empirical tests might show; because sometimes interpretation can be ambiguous. Absolute Truth can only be verified through transcendence. For example in near history the Earth was considered flat (as can be seen from any mountain top or the seashore) and the center of the universe.

Although we might be interested in forms of theological Absolute Truth like 'the Gods organized the Earth and gave life to man' [157]; some theological elements will not easily lend themselves to standard experiential-experimental forms of 'empirical metaphysics' and will have to be 'confirmed' by mutual verification by teams of noeticists experiencing the same transcendent "facts" or remain faith-based until a viable experimental protocol can be designed. Critics might consider the "divinations" of a particular group a form of group hysteria, which might be dispelled if disparate groups are causally separated.

If we consider God to be the Great Physicist, it is physical truths that science would be most interested in and also most readily verified by standard empiricism after transcendent discovery. It is difficult to predict what the world might do when it realizes that the path to transcendence is formulaic and while not necessarily easy or mundane but certainly no more difficult than learning to play the piano proficiently. And the earlier one began the easier the journey. This is not unreasonable considering that most scientists undergo an average of 22 years of study in preparing for an academic life, plus the lifelong study to keep abreast of developments in one's field(s).

Interestingly there also exists a concept of absolute or immutable truth in science:

> A truth that represents a permanent and final grasp of some limited aspect of nature. Most people would say this is incompatible with the expectation that our theories will be falsified. I adhere to the expectation that our theories will be falsified, and look for the immutable truth only in those theories that have already been falsified. Newtonian mechanics...is an example of the most certain and permanent truth man has ever achieved. Its only failing is its scope; it does not cover everything [158].

Now that it has been falsified it is an 'absolute truth' in the domain it describes.

14.31 The Path to Transcendence

Consciousness is an ubiquitous cosmological principle of the universe; and the human mind is a complex system imbedded in this universe. Inherent in the nature of the human mind is a fundamental spiritual component; that allows absolute truth to be perceived from any valid perennial path. Transcendent abilities seem to derive from three main avenues:

- A specific type of innate personality structure, which comprises our psychological makeup, level of intelligence, knowledge and wisdom, all of which occupies the spacetime structure of an individual psychosphere [59].
- Special gifts that the universe bestows upon us for its own purposes, or more likely through modification of number 1 above or that we have developed by certain forms of psychological stress or earned as in 3 below.
- Personal preparedness; which seems to equate in direct proportion to living life by the golden rule and any other ethical principles.

There are exceptions to the ascension of the basic karmic pyramid and more details beyond the scope of the discussion here, but as a simple generalization as one climbs the Karmic pyramid of perfection the threshold of spiritual enlightenment or reaching the transcendent state occurs when our imperfections become limited to misdeeds of thought only. There is a motivational factor also, and wisdom can also relate to mastery of the principles related to the chosen path that might enhance or vary this threshold. It helps to be actively engaged in a worthy cause or service to humanity. Idleness would be a detractor to spiritual awareness. One must at least be involved in meditation or prayer.

One must also choose a viable spiritual path. It does not seem reasonable that one could pay singular homage to a stone, currency, psychotropic pharmacopoeia, or ‘legal’ forms of passion and expect a significant degree of success while mentally occupying the top of the pyramid. Of the 10,000 spiritual paths existing on Earth today, one must use one that works. Ones stage of personal growth limits the choice of perceived path. Some paths are significantly better; and it seems that there are relatively few that enable true enlightenment in a reasonable length of time. The path must therefore be chosen carefully. "It takes nearly a quarter century to become a great physician. Why, oh, why do people think they can fathom the most spiritual depths without the necessary experimental and laboratory work accompanied by compliance with the laws that govern it?" [157]. Kimball further states this expertise comes from personal righteousness followed by revelatory experience precept upon precept.

14.32 The Law of Hierarchies and Noetic Epistemology

In applying Noetic Field Theory [55,108,159,160] to the quantization of the soul [161,162] and “The spirit and the body is the soul of man” [60,137]; how does the metaphor of the Karmic pyramid relate physically as a law of hierarchies as the means for reaching the transcendental state? Following the work of Plato we have defined noetic insight [130] as the highest form of knowing; and stated that transcendent communion operates because ‘the spirit and the body is the soul of man’ [60,137] and ‘all spirit is matter’ [137].

All matter is not spirit but can become so by perfection. But in the meantime in our temporal existence the human soul is comprised of earthy matter and spiritual matter in a complementarity of temporality and eternity. Our consciousness is imbedded in temporality and this is where our sensory apparatus is coupled to. It is a misconception that there is a ‘sixth sense’. What actually happens is that the senses couple to higher dimensionality instead which is in closer proximity to the flux of the vital noetic field. This is what occurs when one achieves the transcendental state. In a crude metaphor this could be likened to an electron going to a higher orbit in an atom when it is energized. The confinement of the electron to the higher orbit is similar to the senses being coupled to a higher plane of spacetime.

Newtonian mechanics was cast in 3 dimensions. Einstein showed us that we live in 4D, which is the limit of our normal perceptual phenomenology; but God dwells in the complete hyperstructure of at least 12D because this is the minimum number to describe eternity – meaning being causally free of temporal reality. Sins of deed and thought maintain a gulf from the 12D of perfection separating us from the full unity of the spirit and confining our matter to the 4D subspace.

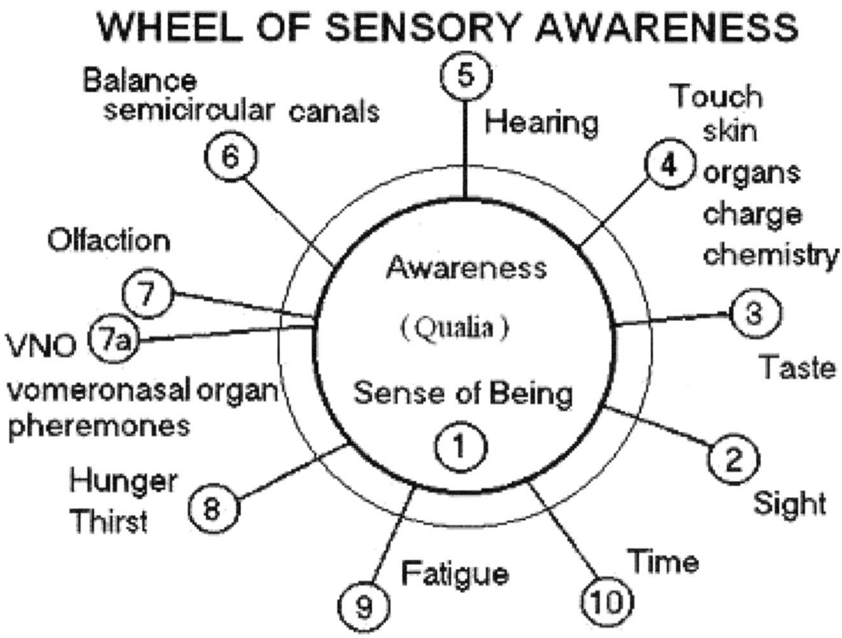

Figure 14.9. There are many more than the 5 common senses; all of which are connected to awareness. The term '6th sense' is a misconception. All of the senses are normally coupled to receive input from external sources but through a different orientation the mind can be coupled to higher dimensional spacetime to receive nonlocal input which is how 'paranormal' effects occur.

Our goal should be to separate our being from the dross matter of imperfection and *precept-by-precept* climb the ladder of dimensions to the full 12D complement of light. Like the light in a laser reverberating between the mirrors of coherent reflection, a light explosion in all the 12 directions, not attenuated by any darkness that stops the light or makes it tarry into dissipation.

References

[1] Swann, I. (1975) To Kiss the Earth Goodbye, New York: Northern Publications.
[2] Eddington, A.S. (1946) Fundamental Theory, Cambridge: Cambridge Univ Press.
[3] Tart, C.T. (1972) States of consciousness and state specific sciences, Science, 176, 1203.
[4] Tart, C.T. (1975) States of Consciousness, New York: E.P. Dutton; and private communications.
[5] Rauscher, E.A. (1979) Some physical models applicable to remote perception, in A. Puharich (ed.) The Iceland Papers: Frontiers of Physics, Select papers on experimental and theoretical research on the physics of consciousness; pp. 50-93, Amherst: Assentia Research Associates.
[6] Kuhn, T.S. (1970) The Structure of Scientific Revolutions, Chicago: University of Chicago Press.
[7] Campbell, J. (1971) The Portable Jung, New York: Viking Press.
[8] Bell, J.S. (1964) On the Einstein Podolsky Rosen paradox, Physics 1, 195.
[9] Einstein, A., Podolski, B. & Rosen,, N. (1935) Can quantum-mechanical descriptions of physical reality be considered complete? Physical Review 47, 777.
[10] Stapp, H.P. (1977) A Whiteheadian approach to quantum theory and the generalized Bell's theorem, Lectures at the University of Texas, Austin, April and May; and private communication.
[11] Bohm, D. (1977) Quantum theory an indication of a new order in physics: Implicate/explicate order in physical law, Found. Phys. 3, 139; and private communication.
[12] Pribram, K. (1976) Problems concerning the structure of consciousness, in G. Globus (ed.) Consciousness and the Brain, New York: Plenum Press,; and private communication.
[13] Ramon, C. and Rauscher, E.A. (1980) Superluminal transformations in complex Minkowski space, Lawrence Berkeley National Laboratory report LBL-9752, Berkeley, CA (1979), and Found of Phys. 10, 661.
[14] Rauscher, E.A.Einstein's field equations and the quantal force, Lawrence Livermore National Laboratory report UCRL-71435, Livermore, CA (1968).
[15] Rauscher, E.A. Closed cosmological solutions to Einstein's field equations, Nuovo Cimento Lett. 3, 661-665 (1972).
[16] Rauscher, E.A. The Minkowski metric for a multidimensional geometry, Nuovo Cimento Lett. 7, 361 (1973).
[17] Wigner, E. (1972) The place of consciousness in modern physics, in Consciousness and Reality, A. Yound, (ed.) New York: Outerbridge & Lazard; and private communication.
[18] DeWitt, B.S. (1970) Quantum mechanics and reality, Phys. Today, p. 30, September; and private communication.
[19] Rauscher, E.A.A Unifying Theory of Fundamental Processes, Bull. of Amer. Phys. Soc. 13, 1643 (1968); and UCRL-20808, University of California, Berkeley, Lawrence Berkeley National Laboratory (1971).
[20] Rauscher, E.A. A group-theoretical representation of the generalized Heisenberg relations, Nuovo Cimento Lett. 5, 925 (1972).

[21] Kleene, S.C. (1950) Introduction to Metamathematics, New York:Van Nostrand.
[22] Clauser, J.F. & Horme, W.A. (1974) Experimental consequences of objective local theories, Phys. Rev. 10D, 526; and private communication with J. Clauser in 1971, 1974-1976.
[23] Heisenberg, W. (1972) Physics and Beyond: Encounters and Conversations, New York: Harper; and private communication.
[24] Walker, E.H. (1977) Quantum mechanical tunnelling in synaptic and ephaptic transmission, Intl. J. Quantum Chem. 11, 103; and private communication, 1977-8.
[25] Rauscher, E.A. Conceptual changes in reality models from new discoveries in physics, PSRL-1076 (September 1981); Proceedings of the First International Symposium on Non-conventional Energy Technology, pages 114-140, October 23-24, 1981, University of Toronto, Ontario, Canada, and the ICF Press, New York.
[26] Rauscher, E.A. Some physical interpretations of multidimensional geometries, PSRL-7628, presented at the Mind/Being Research Conference, Los Altos, California, March 1982.
[27] Rauscher, E.A.Bull. Amer. Phys. Soc. 27, 35 (1982).
[28] Pribram, K.H. (1980) private communication, January.
[29] Spinoza, B. (1883) Ethic: Demonstrated in Geometrical Order, New York: Macmillan,
[30] The Upanishads (1896) Vol. 2, London: Theosophical Society.
[31] Finkelstein, D. (1972) Space-time code, Phys. Rev. 5, 320; and private communication.
[32] Wheeler, J.A. (1962) Geometrodynamics, New York: Academic Press; and private communication.
[33] Chew, G. (1968) Bootstrap: a scientific idea? Science 161, 762; and private communication.
[34] Stapp, H.P. (1964) Space, time and elementary particles, Lawrence Berkeley National Laboratory report UCRL-11688.
[35] Dune, D. (1967) Yoga, New York: Funk and Wagnalls.
[36] Clark, R.W. (1971) Einstein: The Life and Times, New York: World.
[37] Eddington, A.S. (1922) The Mathematical Theory of Relativity, New York: Chelsea.
[38] Rauscher, E.A. (1968) Electron interactions and quantum plasma physics, J. Plasma Phys. 2, 217.
[39] Kriyananda (1967) Yours the Universe, San Francisco: Hansa.
[40] Rauscher, E.A. (1990) Observer/Participator in quantum mechanics and life , Proceedings of the 12th International Conference on Human Functioning, Witchita.
[41] Einstein, A. (1949) Einstein Philosopher Scientist, P.A. Schilpp (ed.) The Library of Living Philosophers, Evanston, Illinois, pp. 85-683.
[42] Amoroso, RL (2002) Developing the cosmology of a continuous state universe, in RL Amoroso, G Hunter, M Kafatos & J-P Vigier (eds.), Gravitation & Cosmology: From the Hubble Radius to the Planck Scale, Dordrecht: Kluwer.
[43] Amoroso, R.L. (2005) Paradigm for a continuous-state holographic conscious Multiverse, in R.L. Amoroso & B. Lehnert (eds.) Extending the Standard Model: Searching for Unity in Physics, Oakland: Noetic Press.
[44] Lucretius, 55 BC (1957) On The Nature of the Universe, R.E. Latham (trans.)

Baltimore: Penguin.
[45] Amoroso, R.L. (2010) Defining a Context for the Cosmology of Awareness, in R.L. Amoroso (ed.) The Complementarity of Mind and Body: Realizing the Dream of Descartes, Einstein and Eccles, NY: Nova Science.
[46] Chalmers, D.J. (1996) The Conscious Mind, Oxford: Oxford University Press; Chalmers, D.J., 2002, The puzzle of conscious experience, Scientific American special edition, 12:1, 90-100.
[47] McMurry, J. (1992) Organic Chemistry, 3rd edition, Pacific Grove: Brooks/Cole Publishers.
[48] Haller, J.S. (1997) Kindly Medicine: Physio-Medicalism in America, 1836-1911. Kent: Kent State Univ. Press.
[49] Bergson, H. (1977) The two Sources of Morality and Religion, Notre Dame, Univ. Notre Dame Press.
[50] Flanagan, O. (1992) Consciousness Reconsidered. Cambridge: MIT Univ. Press.
[51] Freedman, D.H. (1994) Quantum consciousness. Discover, June, pp. 89-98.
[52] Horgan, J. (1994) Can science explain consciousness? Scientific American, July, pp. 88-94.
[53] Amoroso, R.L. (1997) Consciousness a radical definition: The hard problem made easy, Noetic J 1:1 pp. 19-27.
[54] Amoroso, R.L. (1997) The theoretical foundations for engineering a conscious quantum computer, in M. Gams and M. Paprzycki, (eds.) Mind <> Computer, Amsterdam: IOS Press.
[55] Amoroso, R.L. (1995) The extracellular containment of natural intelligence: A new direction for strong AI. Informatica, 19, pp. 585-590.
[56] Amoroso, R.L. (1996) Engineering a conscious computer, in T. Toffoli & M. Biafore (eds.) Proc. Fourth Workshop on Physics & Computation, Physcomp 96, pp. 12-16, New England Complex Systems Institute..
[57] Amoroso, R.L. & Martin, B. (1995) Modeling the Heisenberg matrix: quantum coherence and thought at the holoscape manifold and deeper complementarity, in K.H. Pribram & J. King (eds.) Scale in Conscious Experience: Is the Brain too Important to be Left to Biologists to Study, Hillsdale: Lawrence Earlbaum.
[58] Amoroso, R.L. & Amoroso, P.J. (2004) The Fundamental Limit and Origin of Complexity in Biological Systems: A New Model for the Origin of Life, in D.M. Dubois (ed.) CP718, Computing Anticipatory Systems: CASYS03-6th Intl. Conference, Liege, Belgium August 11-16 2003, New York: American Institute of Physics.
[59] Amoroso, R.L. (2010) The Physical Origin of the Principle of Self-Organization Driving Living Systems, in R.L. Amoroso (ed.) The Complementarity of Mind and Body: Realizing the Dream of Descartes, Einstein and Eccles, New York: Nova Science Publishers.
[60] Perus, M. (1997) Consciousness: network-dynamics, informational and phenomenal aspects, Noetic J, 1:2:183-197.
[61] Stapp, H.P. (2000) Why classical mechanics cannot naturally accommodate consciousness but quantum mechanics can, in R. Amoroso et al (eds.) Science and The Primacy of Consciousness, pp. 134-149, Oakland: The Noetic Press.

[62] Amoroso, R. L. (2000) Call for a model of deep ontology – A commentary on Stapp: Why classical mechanics cannot naturally accommodate consciousness but quantum mechanics can , in R. Amoroso et al (eds.) Science and The Primacy of Consciousness, pp. 150-153, Oakland: The Noetic Press.
[63] Pribram, K.H. (1991) Brain and Perception, Hillsdale: Lawrence Erlbaum.
[64] Hameroff, S.R., Rasmussen, S., Karampurwala, H., Vaidyanath, R., and Jensen, K.S. (1990) Computational connectionism within neurons: A model of cytoskeletal automata. Physica D 42:428- 449.
[65] Beck. F., and Eccles, J.C. (1992) Quantum aspects of brain activity and the role Consciousness, Proc. Natl. Acad. Sci. USA 89: 11357- 11361.
[66] Walker, E.H. (1997) The quantum theory of consciousness, Noetic Journal, 1:1, pp. 100-107.
[67] Jibu, M. & Yasue, K. (1995) Quantum Brain Dynamics & Consciousness, Amsterdam: Benjamins.
[68] Bohm, D. (1952) A suggested interpretation of the quantum theory in terms of hidden variables, I & II. Physical Review 85, pp. 166-179 and 180-193.
[69] Cramer, J.G. (1986) The transactional interpretation of quantum mechanics, Revs. Mod Physics 58, pp. 647-87.
[70] Sun, Y., Rauscher, E.A., Giandinoto, S., Chu, J. & Amoroso, R.L. (2008) Empirical Mediation of the Primary Mechanism Initiating Protein Conformation in Prion Propagation, in D. Dubois (ed.) Proc. CASYS07, Liege, Belgium.
[71] Pribram, K. H., Nuwer, M. & Baron, R. (1974) The holographic hypothesis of memory structure in brain function and perception, in R.C. Atkinson, D.G. Krantz, R. C. Luce & P. Suppes (eds.) Contemporary Developments in Mathematical Psychology, San Francisco: W. H. Freeman.
[72] Hameroff, S.R. & Watt, R.C. (1982) Information processing in microtubules, Theor. Biol. 98:4;549-61.
[73] Freeman, W. (1993) The emergence of chaotic dynamics as a basis for comprehending intentionality in experimental subjects, in K.H. Pribram (ed.) Origins: Brain and Self Organization, Hillsdale: Lawrence Erlbaum.
[74] Hameroff, S.R. (1990) Computational connectionism within neurons: a model of cytoskeletal automata, Physica D. 42: 428-449.
[75] Schutz, E. (1987) Posttranslational modification and microtubule stability, Journal of Cell Biology, Vol 105, pp. 2167-2177
[76] Webster, D.R., Gundersen, G.G., Bulinski, J.C. & Borisy, G.G. (1987) Differential Turnover of Tyrosinated and Detyrosinated Microtubules, Proc Nat. Acad. Sci, USA, 84:24; 9040-9044.
[77] Clevland, D.W. & Sullivan, K.F. (1985) Molecular Biology & Genetics of Tubulin, Ann. Rev. Biochem. 54: 331-365.
[78] Koruga, D. (1992) Neuromolecular computing, Nanobiology 1:5-24.
[79] Prusiner, S.B. (1982) Science, 216, p.136-144.
[80] Prusiner, S.B. (1998) Proc Nat. Acad. Sci, USA, 95, p. 13363-13383.
[81] DesCartes, R. (1960) Discourse on Method and Meditations, L.J. Lafleur (trans.) Indianapolis: Bobbs-Merrill.
[82] Eccles, J.C. (1986) Do mental events cause neural events analogously to the probability fields of quantum mechanics?, Proc. Roy. Soc. Lon. B227, pp. 411-428.

[83] Smith, H. (1991) The World's Religions, San Francisco: Harper Collins.
[84] Kimball, S.W. (1978) Absolute Truth, Ensign, September, pp. 3-8, Salt Lake City: The Church of Jesus Christ of Latter-day Saints.
[85] James, W. (1912) Essays in Radical Empiricism, F.H. Burkhardt et al (eds.) 1976, Cambridge: Harvard Univ. Press.
[86] Varela, F.G., Maturana, H.R. & Uribe, R. (1974) Autopoiesis: The organization of living systems, its characterization and a model, BioSystems, 5, 187-196.
[87] Jantsch, E. (1984) The Self-Organizing Universe, New York: Pergamon.
[88] Maturana, H. R. (1970) Biology of cognition, Report BCL 9.0 Urbana: Biological Computer Lab, Univ. of Il.
[89] Drãgãnescu, M. (1997) On the structural phenomenological theories of consciousness, Noetic J., 1:1, 28-33.
[90] von Neumann, J. (1966) The theory of self-reproducing automata, in A. Burks (ed.) Urbana: Univ. of Il. Press.
[91] Zhabotinsky, A.M. (1974) Self-oscillating Concentrations, Moscow: Nauka
[92] Haldane, J.S. (1923) Mechanism, Life and Personality, New York: Dutton.
[93] Beckner, M.O. (1972) Mechanism in biology, in P. Edwards (ed.) The Encyclopedia of Philosophy, Vol. 5, pp. 250-2, New York: Collier Macmillan.
[94] Goldberg, J.N. (1981) Spacetime, in Encyc. of Physics, R.L. Lerner & G.L. Trigg (eds.) Reading: Addison-Wesley.
[95] Messiah, A. (1999) Quantum Mechanics, Mineola: Dover.
[96] Prigogine, I. (1973) Irreveresability as a symmetry breaking factor, Nature, 248: 67-71.
[97] Prigogine, I., Nicolis, G. & Babloyantz, A. (1972) Thermodynamics of evolution, Physics Today, 25: 23-28; 38-44.
[98] Chalmers, D.J. (2002) The puzzle of conscious experience, Scientific American Special, 12:1, 90-100.
[99] Searle, J. R. (2002) Consciousness, Review Roumaine de Philosophie, Tome 46:1-2, pp.87-108.
[100] Brillouin, L. (1949) Life, thermodynamics and cybernetics, American Scientist, 37: 554-568.
[101] Chalmers, D. (1996) The Conscious Mind, Oxford: Oxford University Press.
[102] Schrödinger, E. (1945) What is Life? London: Cambridge Univ. Press.
[103] Amoroso, R.L. (2002) The Physical Basis of Consciousness: A Fundamental Formalism, Part 1 Noesis, XXVI, Romanian Academy.
[104] Amoroso, R.L. (2000) Derivation of the fundamental equation of consciousness, Part I, Boundary conditions, Noetic Journal 3:1, pp. 91-99.
[105] Amoroso, R.L. (2000) Consciousness, a radical definition: Substance dualism solves the hard problem, in Amoroso, R.L., Antunes, R., Coelho, C., Farias, M., Leite, A., & Soares, P. (eds.) Science and the Primacy of Consciousness, Orinda: The Noetic Press.
[106] Amoroso, R.L. (1999) An introduction to noetic field theory: The quantization of mind, Noetic J 2:1, pp. 28-37.
[107] Amoroso R.L. (2003) Awareness: physical cosmology of the fundamental least unit, Noetic Journal 4:1, 1-15.

[108] Wegner, P. (1998) Interactive foundations of computing, Theoretical Computer Science, 192, 315-351.
[109] Fröhlich, H. (1968) Long-range coherence and energy storage in biological systems, Int. J. Quant. Chem. 2:641-649.
[110] Hagelin, J.S. (1988) Is consciousness the unified field? A field theorist's perspective. Preprint.
[111] Ciubotariu, C & Ciubotariu, C. (2002) A chaotic-stochastic model of an atom, in R.L. Amoroso, G. Hunter, M. Kafatos & J-P Vigier (eds.), Gravitation and Cosmology: From the Hubble Radius to the Planck Scale, Dordrecht: Kluwer Academic.
[112] Argyris, J. & Ciubotariu, C. (1999) A new physical effect modeled by an Ikeda map depending on a monotonically time-varying parameter, Int. J. Bif. Chaos, 9:1111-1120.
[113] Chalmers, D. (1996) The Conscious Mind, Oxford: Oxford University Press.
[114] Nagel, T. (1974) What's it like to be a bat?, Philos. Rev., 83, pp. 435-450.
[115] Amoroso, R.L. (2003c) The physical basis of qualia: Overcoming the 1st person 3rd person barrier, Noetic Journal 4:3, pp. 212-230.
[116] Amoroso, R.L. (2007) Ce Este Constiinta? Trepte Intru Cosmologia Mintii (What is Consciousness: Introducing the Cosmology of Being, N. Bulz et al (trans.) Bucharesti: Editura Academiei Romane, in press.
[117] R.L. Amoroso (ed.) (2008) The Complementarity of Mind and Body: Realizing the Dream of Descartes, Einstein and Eccles, New York: Nova Science Publishers.
[118] Poston T. & Stewart, I. (1978) Catastrophe Theory & Its Applications, New York: Dover.
[119] Gilmore, R. (1981) Catastrophe Theory for Scientists & Engineers, New York: Dover.
[120] Qin, S. et al. (2001) International J of Solids & Structures, 38, pp. 8093-8109.
[121] Wheeler, J.A. (1955) Geons, Physical Review, 97:2, 511-536.
[122] Pruisiner, S.B. (2002) Research Summary, www.ucsf.edu/neurosc/faculty/neuro-prusiner.html.
[123] Huang, Z., Gabriel, J-M, Baldwin, M.A., Fletterick, R.J., Prusiner, S.B., & Cohen, F.E. (1994) Proposed three-dimensional structure for the cellular prion protein, Proc. Nat. Acad. Sci, USA, 91, pp. 7139-7143.
[124] Kurschner, C. & Morgan, J.I. (1996) Mol. Brain Res. 37, pp. 249-258.
[125] Kafatos, M., Roy, S. & Amoroso, R. (2000) Scaling in Cosmology & the Arrow of Time, in Buccheri, di Gesu & Saniga, (eds.) Studies on Time, Dordrecht: Kluwer Academic.
[126] Milner, R. (1993) Elements of interaction, Comm. of the ACM, 36:1, 78-89.
[127] Prusiner, S. (1982) Science, 216, pp. 136-144.
[128] Mitchell, E.D. (1976) Psychic Exploration: A Challenge for Science, J. W. White (ed.) New York: Putnam.
[129] Musser, G. (1998) String Instruments, Scientific American, V.10, pp. 17-19.
[130] Holland, P.R. (2000) The Quantum Theory of Motion: An Account of the de Broglie- Bohm Causal Interpretation of Quantum Mechanics, Cambridge: Cambridge Univ. Press.

[131] Bohm, D. & Hiley, B.J. (1993) The Undivided Universe: An Ontological Interpretation of Quantum Theory, London: Routledge.
[132] Meehl, P.E. (1966) The compleat autocerebroscopist: A thought-experiment on Professor Feigl's mind-body identity thesis, in P.K. Feyerabend & G. Maxwell (eds.) Mind, Matter and Method: Essays in Philosophy and Science in Honor of Herbert Feigl, Minneapolis: Univ. of Minnesota Press.
[133] Teachings of the Presidents of the Church - Brigham Young, (2002) Salt Lake City: The Church of Jesus Christ of Latter-day Saints.
[134] R.L. Amoroso, S. Giandinoto, G. Hunter, G. Nibart & E. A. Rauscher (2008) Universal Quantum Computing: Anticipatory Parameters Predicting Bulk Implementation, Part I – Philosophical Foundations of the Formalism, in D. Dubois (ed.) Proceedings of CASYS07, Liege, Belgium.
[135] Smith, J. (1989) Doctrine & Covenants, 88:11, Salt Lake City: LDS Church.
[136] Burns, J.E. (1994) Spaciousness: The common ground between science and spirituality, in R.I. Heinze (ed.) Proceedings of the 11th Intl. Conf. On the Study of Shamanism and Alternative Modes of Healing, Berkeley: Indep. Scholars of Asia.
[137] Smith, H. (1991) The World's Religions, New York: Harper Collins.
[138] Wattles, J. (1996) The Golden Rule, Oxford: Oxford University Press.
[139] Terry, Q.C. (2005) Golden Rules and Silver Rules of Humanity, Bloomington: Authorhouse.
[140] Holoviak, S.J. (1993) Golden Rule Management, Reading: Addison-Wesley.
[141] Bigelow, J. (1927) Toleration, and other essays and studies, New Church Board of Publication; or http://newearth.org/frontier/grmain.html.
[142] Hare, R.M. (1963) Freedom and Reason, Oxford: Oxford University Press.
[143] Gensler, H.J. (1996) Formal Ethics, New York: Routledge.
[144] Gensler, H.J. (1998) Ethics: A Contemporary Introduction, New York: Routledge.
[145] Citations for The Golden Rule: The Holy Bible: King James Version (1989) Leviticus 19:18; Mathew 7:12, 19:19, 22:39; Mark 12:31; Luke 6:31, 10:27; Romans 13:9; Galations 5:4, Salt Lake City: The Church of Jesus Christ of Latter-day Saints; Doctrine & Covenants of the Church of Jesus Christ of Latter-day Saints, 59:6, Salt Lake City: The Church of Jesus Christ of Latter-day Saints; Confucius (500 BC) Analects of Confucius (1998) D.C. Lau (trans.) Ch. 15, Verse 3, New York: Penguin Classics; also http://classics.mit.edu/confucius/analects.html.
[146] The Holy Bible: King James Version (1989) Mark 12:30; Deuteronomy 6:5, Salt Lake City: The Church of Jesus Christ of Latter-day Saints.
[147] The Holy Bible: King James Version (1989) John 13:34-35 Salt Lake City: The Church of Jesus Christ of Latter-day Saints.
[148] The Holy Bible: King James Version (1989) Matthew 22:36-40, Salt Lake City: The Church of Latter-day Saints.
[149] Goethe, J.W. von (2005) The Sorrows of Young Werther (Die Lieden des Jungen Werther) B. Pike (trans.) Mre York: Random House.
[150] The Holy Bible: King James Version (1989) James 1:26, 3:5-6,8; 1 Peter 3:10, Salt Lake City: The Church of Jesus Christ of Latter-day Saints.
[151] The Holy Bible: King James Version (1989), New Testament, 1 Peter 4:8, Salt Lake City: The Church of Jesus Christ of Latter-day Saints.

[152] Descartes, R. (1641) Meditations on First Philosophy, in The Philosophical Writings of René Descartes (1984) J. Cottingham, R. Stoothoff & D. Murdoch (trans.) vol. 2, 1-62., Cambridge: Cambridge University Press.
[153] Eccles, J.C. (1992) Evolution of Consciousness, Proc. Nat. Acad. Sci., 89:7320-7324.
[154] Crick, F. (1994) The Astonishing Hypothesis: The Scientific Search for the Soul, New York: Scribner's Sons.
[155] Kimball, S.W. (1978) Absolute Truth, The Ensign, Sept, pp. 3-8, Salt Lake City: The Church of Latter-day Saints.
[156] Misner, C.W. (1974) Cosmology and theology, in W. Yourgrau & A.D. Breck (eds.) Cosmology, History Theology, New York: Plenum.
[157] Amoroso, R.L. (1996) The production of Fröhlich and Bose-Einstein coherent states in in vitro paracrystaline oligomers using phase control laser Interferometry, Bioelectrochemistry and Bioenergetics 41, 39-42.
[158] Amoroso, R.L. (1999) An intro to Noetic Field Theory: The quantization of mind, In R. Amoroso & M. Farias et. al. (eds.) Science and the Primacy of Consciousness: Intimation of a 21st Century Revolution, Oakland: Noetic Press.
[159] Wolf, F.A. (1999) The quantum physical communication between the self and the soul, The Noetic Journal, 2:2, 149-158; also (2000) pp. 404-414, R. Amoroso & M. Farias et. al. (eds.) Science and the Primacy of Consciousness: Intimation of a 21st Century Revolution, Oakland: The Noetic Press.
[160] Amoroso, R.L. (2010) The Geometry, Topology and Structural-Phenomenology of the Soul, book in progress.
[161] The Autoimmune Disease Group (Noetic or Spirit-Based Psyconeuroimmunological Medical Etiologies); http://www.mindspring.com/~l.o.v.e.r/autoimmune.pdf
[162] Amoroso, R.L., and Di Biase, F., (eds.) (2005) *A Revolução da Consciência. Novas Descobertas sobre a Mente no Século* XXI, Rio de Janeiro: Editura Vozes.
[163] Di Biase, F. (1981) *Auto-organização nos sistemas biológicos, Ciência e Cult.*, 339: 1155-1159, *Sociedade Brasileira para o Progresso da Ciência*, Brazil.
[164] Di Biase, F. (1995) *O Homem Holístico, a Unidade Mente-Natureza*, Rio de Janeiro: Editora Vozes.
[165] Di Biase, F. & Rocha, M.S. (1998) *Caminhos da Cura, Petrópolis*, Rio de Janeiro: Editora Vozes.
[166] Di Biase, F. & Rocha, M.S. (2004) *Ciência Espiritualidade e Cura, Psicologia Transpessoal e Ciências Holísticas,* Rio de Janeiro: Editora Qualitymark.
[167] Vaz, N. M. & Varela, F.J. (1978) Self and non-sense: an organism-centered approach to immunology, Medical. Hypotheses 4: 231-267.
[168] Jerne, N. K. (1974) Toward a network theory of the immune system, Ann. Immunol. Inst. Pasteur, 125c:373-389.
[169] Jerne, N. K. (1984) Idiotypic networks and other preconceived ideas, Immunological Review, 79: 5-24.
[170] Amoroso, R.L. & Rauscher, E.A. (2009) The Holographic Anthropic Multiiverse: Formalizing the Complex Geometry of Ultimate Reality, Singapore: World Scientific.

Chapter 15

Holographic Wormhole Drive: Philosophical Breakthrough in FTL 'Warp-Drive' Technology

Recent work exploring Faster-Than-Light (FTL) warp drive technologies has proven surprisingly interesting in that science fiction seems on the brink of immanent reality. Most of the new discussion has centered on the Alcubierre spacetime metric, a unique reverse solution derived from Einstein's General Relativity field equations. However these attempts have been troubled by critical problems such as the perceived requirement of near infinite negative or exotic mass-energy to power the FTL drive and creating a free-fall warp bubble containment field to overcome the theoretical limits imposed by Einstein's theory of Special Relativity. In this chapter we offer radical solutions to these conundrums albeit for the purposes of this sophomoric introductory overview primarily in conceptual form; however still bringing a glimpse of the feasibility for FTL warp-drive travel putatively to near term.

15.1 Overview and Current Status

Skeptics say it may take 1,000 years and require an energy source the size of Jupiter to operate superluminal faster than light (FTL) warp-drive technologies. In this work we present ideas on how to change that scenario. We decided to call our FTL model the "Holographic Wormhole Drive" (HWD) for reasons that will become evident as the discussion progresses. Firstly the HWD name is coined after the Holographic Multiverse cosmology paradigm that in our view allows it to operate [1]. The second part *wormhole drive* is misleading because it is suggestive of those models that seek natural or propose creation of large wormholes to pass a spaceship through [2-4]. We define the term wormhole, which shares some general utility with both models, before we proceed further. Wormhole in general:

- A distortion of spacetime that links one location with another through a path or tunnel in the spacetime topology that acts as a shortcut in distance or time compares to the usual path in 3D Euclidean space.

The term wormhole was first coined by Wheeler in 1957 [5], but Weyl had already proposed the theory in 1921. Wormholes are also called Einstein-Rosen bridges or Lorentzian Schwarzschild wormholes after the solution to his field equations discovered by Einstein and Rosen in 1935 [6]. In 1962 Wheeler and Fuller showed that this type of wormhole is unstable, would pinch off instantly after forming so that even light could not pass through. However the existence of this Schwarzschild solution inspired Kip Thorne to consider holding the throat of a wormhole open by exotic matter with negative mass energy. The possibility of traversable wormholes in general relativity was first demonstrated by Thorne and Morris in 1988 [7,8]. This is the origin of the model of traversable wormholes held open by a spherical shell of exotic matter.

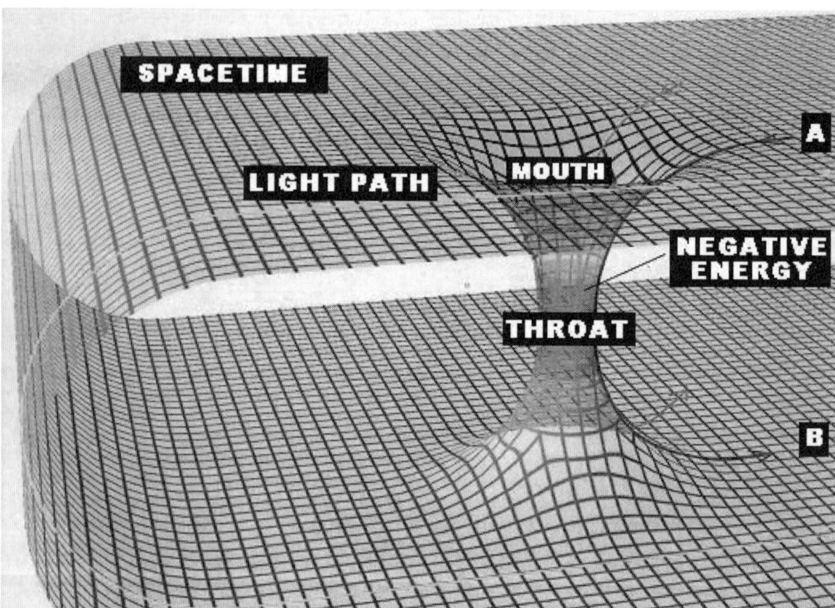

Figure 15.1. Illustration of a spacetime traversable wormhole [9]. It is a 2D conceptualization of what takes 4D or greater dimensionality to describe. Imagine folding a 2D surface into a third dimension that does not exist in the 2D surface space. This is somewhat reminiscent of a Klein bottle where the handle cannot be drawn free in 2D or as a 3D object. Special relativity only applies locally. Wormholes allow superluminal travel by ensuring that the speed of light is not exceeded locally in time because while traveling through a wormhole, subluminal velocities are used.

There are several constructs that make our model unique and to allow FTL warp drive theory to leap to the brink of practicality. Also it should be noted that FTL itself is also a bit of a misnomer; in the EPR sense nonlocal connectivity (as in Newton's law of gravitation) is considered instantaneous. There is still a 'Warp-Factor', however the HWD uses a figure-ground effect in this respect as will be explained in detail below. In the HWD model the local Galilean velocity "Warp-Bubble" is harmonically removed from the background of spacetime that reality is embedded in and dropped back in with a resonant beat frequency. The greater the amplitude and duration of the wave the larger the FTL distance traveled between harmonic beats; this is the warp factor as developed below.

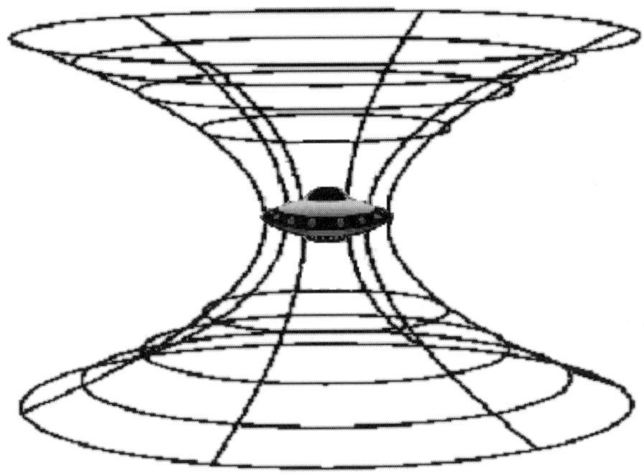

Figure 15.2. One major theoretical FTL method has been the utility of macroscopic wormholes created with a Jupiter size exotic matter (negative energy gravitational mass) or utilizing a natural traversable wormhole in the proximity of a black hole.

This is where the HWD wormhole concept enters in. The warp bubble boundary is covered by a layer of mini-wormholes created by resonant vacuum programming. This layer of mini-wormholes provides the grease so to speak for operation of the figure-ground effect. This figure-ground effect is somewhat reminiscent of the friction table demonstration used in elementary physics education. A puck pushed across the table has the highest coefficient of friction. Hundreds of tiny holes (conceptually our mini-wormholes) are drilled in the table. When air is forced through the holes the pucks then glides across the table with virtually no friction.

The HWD may seem farfetched at this point, but we will develop these new principles as we go along in our dissertation. The key [1] stems from a new 12D understanding of a scale invariant covariant Dirac polarized vacuum concept [1,10] that includes operationally completed models of de Broglie-Bohm-Cramer interpretations of quantum theory [11,12] and dual integrated form of Newton's and Einstein's gravity [1,13]. As in our description for universal bulk quantum computing by surmounting the uncertainty principle [1] most consider quantum computing as merely a fast form of computing utilizing entangled quantum states. This is by no means the case. Our model for universal quantum computing requires a 12D string theoretic background that allows the manipulation of reality itself. Aye there's the rub!

The other primary advance is that the HWD needs no Jupiter size negative energy mass. This feature is the other boon that comes from a greater understanding of the 12D Dirac vacuum [1,11]. Virtually infinite negative and positive energy in equilibrium is inherent in the background of infinite potentia. This apparently is what holds the Multiverse and our individual Hubble sphere, H_R together and allows each H_R in the holographic anthropic multiverse to be fine-tuned.

15.2 The Alcubierre Warp-Drive Metric – Brief Review

In general the Alcubierre solution is used to create an expansion of spacetime behind a spaceship and a spacetime contraction in front of it. The Alcubierre warp-drive metric [14-16] as derived from Einstein's General Relativity field equations,

$$R_{\mu\nu} - (g_{\mu\nu}R)/2 + g_{\mu\nu}\Lambda = (8\pi G/c^4)T_{\mu\nu} \tag{15.1}$$

in general form is:

$$\begin{aligned} ds^2 &= -d\tau^2 = g_{\alpha\beta}dx^\alpha dx^\beta \\ &= -\left(\alpha^2 - \beta_i\beta^i\right)dt^2 + 2\beta_i dx^i dt + \gamma_{ij}dx^i dx^j . \end{aligned} \tag{15.2}$$

The Alcubierre solution is currently considered the most advanced model; and although the solution does not require a wormhole to traverse, it does require exotic matter to generate the spacetime distortion

To quote Alcubierre original paper:

> In this formalism, spacetime is described by a foliation of spacelike hypersurfaces of constant coordinate time, t . The geometry of spacetime is then given in terms of the following quantities: the 3-metric, γ_{ij} of the hypersurfaces, the lapse function, α that gives the interval of proper time between nearby hypersurfaces as measured by the "Eulerian" observers

(those whose four-velocity is normal to the hypersurfaces), and the shift vector, β^i that relates the spatial coordinate systems on different hypersurfaces [14].

Using those quantities, Alcubierre generates the warped spacetime metric written above in Eq. (15.2) and Fig. 15.3 below.

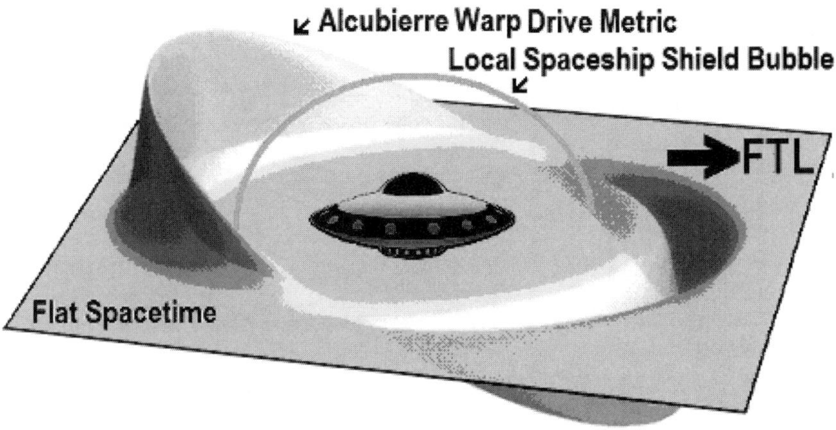

Figure 15.3. The original Alcubierre Warp Drive Metric. Showing how space stretches in a wave. Space ahead of a ship contracts and space behind expands. Inhabitants of the warp-bubble travel along a 'free-fall' geodesic, not moving locally relative to the ships FTL velocity.

15.3 The Philosophy of Potentia in a 12D Stringy Dirac Vacuum

The sub-quantum regime in the Copenhagen Interpretation has been called a stochastic foam, a domain within which time asymmetry is considered more fundamental than quantum theory. Time emerges from a more fundamental domain organizing the structure of and guiding the evolution of events in local reality [17,18]. In our extended interpretations the regime of infinite potentia can be coherently controlled and programmed in a manner essential to practical FTL warp-drive technology. This usage is beyond the usual meaning applied to Heisenberg potentia which only refers to the body of probabilistic states of the wave function before a local measurement is taken. We wish to align with those claiming nothing exists before 'measurement', but in a complex manner in that 4D Minkowski reality itself does not exist and arises from moment to moment as a form of "first collapse" creating the phenomenal basis of the Earth bound observers virtual reality (Fig. 15.4).

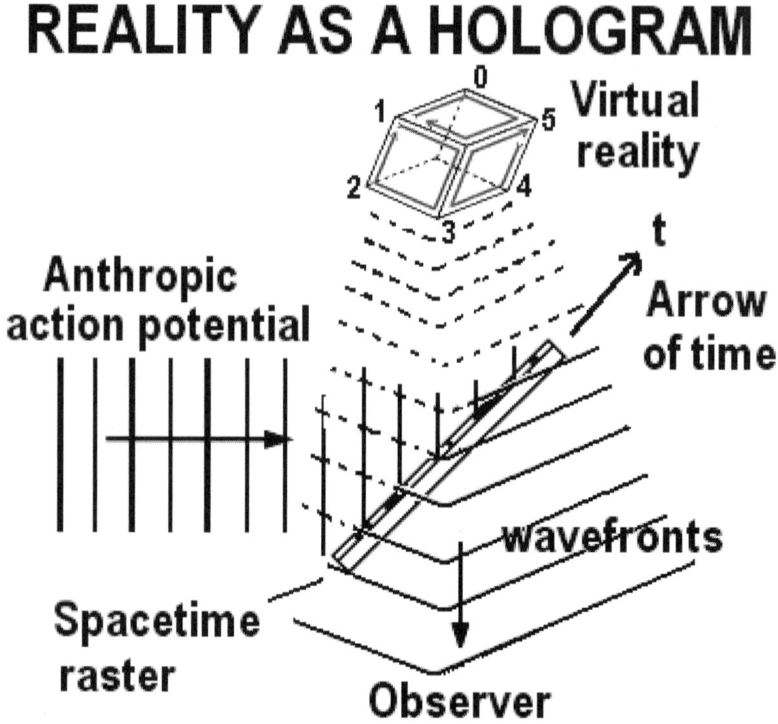

Figure 15.4. Reality is a hyperspherical hologram in an HD background of infinite potentia with the 'laser' being the teleological anthropic unified field action principle 'piloting' its continuous evolution with the unified field as a super-quantum potential.

This multilevel process has properties similar to wave-particle duality except that it applies to the dimensionality and topological boundary conditions of reality not just quanta. A combination of the Heisenberg uncertainty principle and the arrow of time (a subtractive interferometry) mediates this process keeping reality as a virtual subspace of a HD absolute space of infinite potentia. This is a major aspect of the continuous-state process inherent in the new Holographic Anthropic Multiverse (HAM) cosmological paradigm [1]. One could make correspondence to the Hawking-Dewitt model of the wave function of a closed universe only, $H\Psi = 0$ which in a multiverse would apply to our fine-tuned Hubble sphere. Boundary conditions forming reality are continuously created, with the Planck scale, $\hbar + T_0$ quantum stochasticity of matter as its lower bound, ψ_1 for example and the oscillating cosmological constant, $\Lambda + T_0$ as the upper bound, ψ_2.

Observed reality evanesces from a central hysteresis loop [1] of this action which is like an HD standing-wave in the Cramer Transactional Interpretation [7]. This is a radically new idea, that observed reality is an intermediate continuous-state collapse, containing an inherent 'Dirac spinor twist' as an intermediate stage for all the rest of what is considered the microscopic evolution of the quantum wave function to rest on (Fig. 15.15). This is metaphorically as the film in an analog movie projector, a 2D or 3D hologram strip and the bulb in the projector an anthropic laser producing the perceived 3D images on the screen perceived by the observer seated in the theatre.

This is not a popular view because not only does it give prime import to the role of the observer, but it also represents a dualist-interactionist model of awareness [19] unpopular among cognitive theorists who consider mind tantamount to brain because it includes an anthropic teleological action principle giving an inherent importance to the nature and role of the observer We believe this correct and have presented empirical models to support it [20,21].

We look at the zero-point field as interpreted in the Copenhagen Interpretation of quantum theory as 'fog over the ocean'; whereas we require the utility of the full depth of the ocean for our HWD FTL technology. We wish to stick with something that suggests a domain that is truly like a hologram in an HD sense because it seems theoretically the most efficient manner to operate an anthropic multiverse. Especially see our model of the vacuum exiplex which potentially solves numerous open questions in cosmology [1,13].

15.4 Domain Wall Boundaries and Emission-Absorption Advanced-Retarded Wave Loci

We shall consider a static thick domain wall constructed by a scalar field with self-interaction in a Schwarzschild singularity spacetime [22,23].

$$g=-\left(1-\frac{2M}{R}\right)dt^2+\left(1-\frac{2M}{R}\right)^{-1}dR^2+R^2\left(d\vartheta^2+\sin^2\vartheta d\varphi^2\right) \quad (15.3)$$

The metric of the background Schwarzschild black hole is written in terms of the isotropic coordinates, t, r, ϑ, φ, where the new radial coordinate, r is defined by

$$R=r\left(1+\frac{M}{2r}\right)^2. \quad (15.4)$$

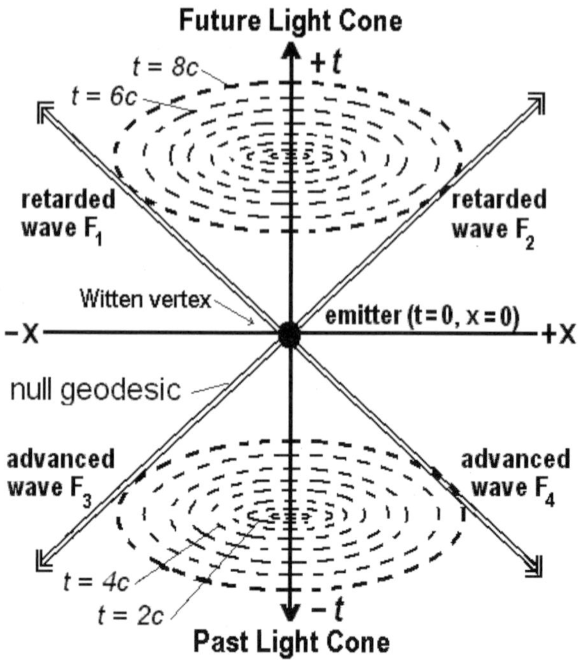

Figure 15.5. Cramer transaction emission locus at x,t = 0,0.We are concerned with the boundary conditions in the region outside the event horizon, where r ≥ M/2 which are of interest even though here applied to a black hole because it might reflect scale invariant principles.

The scalar equation in spherical coordinates of wave motion in spacetime which has spherical symmetry [24,25].

$$\nabla^2\Phi - \frac{1}{c^2}\partial^2\frac{\Phi}{\partial t^2} = 0 \tag{15.5}$$

where Φ is the wave amplitude. The equation has two solutions

$$\begin{aligned} \Phi_{out} &= \frac{1}{r}\Phi_{\max}\exp\left(i\omega t - ikr\right) \\ \Phi_{in} &= \frac{1}{r}\Phi_{\max}\exp\left(i\omega t + ikr\right) \end{aligned} \tag{15.6}$$

which for the programming of spacetime can be applied to the propagation of

Cramer's advanced retarded waves from an emission locus at $x,t = 0,0$ by Eqs. (15.7) and Fig. 15.5.

$$F_{1-Ret} = F_0 e^{-ikx} e^{-2\pi i\omega t}, \quad F_{2-Ret} = F_0 e^{ikx} e^{-2\pi i\omega t}$$
$$F_{3-Adv} = F_0 e^{-ikx} e^{2\pi i\omega t}, \quad F_{4-Adv} = F_0 e^{ikx} e^{2\pi i\omega t} \tag{15.7}$$

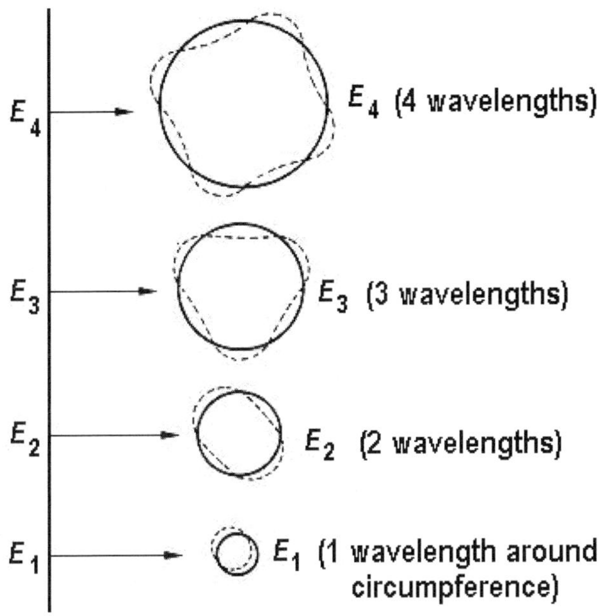

Figure 15.6. A Ring may vibrate with *n* standing wavelengths depending on the relationship of the circumference to the multiple number of whole wavelengths. Simplified here, it is suggested that the topology of spacetime and matter vibrate on and as hyperspherical surfaces.

Traditionally electron standing-waves oscillate about the atomic nucleus. Here we attempt to expand the wave nature of matter itself as static waves centered on the locus of least spacetime units as it is annihilated and recreated in the arrow of time relative to the observer. This requires a conversion of the de Broglie wave equation, $mvr = n(h/2\pi)$ to a static form amenable to the parameters of continuous-state cosmology [26,27]. For Hyperspherical Representation the magnitudes of the radial coordinates of a two-state wavefunction, $\psi(\vec{r}_1, \vec{r}_2)$ in hyperspherical representation are replaced by the hyper-spherical radius, *R* and the hyperspherical angle, α such that

$$R \equiv \left(r_1^2 + r_2^2\right)^{1/2} \quad \text{and} \quad \alpha \equiv \arctan\frac{r_2}{r_1} \tag{15.8}$$

in order that the symmetries may be more clearly shown. The hyperspherical radius, R represents the size of the two-state system and the hyperspherical angle, α is a measure of the radial correlation of the two-state system [28]. It is critical to note that when $\alpha = \pi/4$, $r_1 = r_2$; and when $\alpha = 0$ or $\pi/2$ one of the states is at a greater distance from the least-unit vertex than the other.

15.5 Ising Model Lattice-Gas Energy Increase from Constructive Interference Properties

In terms of the SUSY spacetime lattice represented by close-packed least units functioning as a Riemann 3-sphere Ising model spin lattice, where total energy, $E_T\{s_i\}$ is a function of the spin hysteresis loop

$$E_T\{s_i\} = \sum_i e_i(s_i) = E_0 - \sum_i h_i s_i \tag{15.9}$$

where $e_i(s_i)$ is the energy of an isolated individual least unit, E_0 the ground state and h_i the energy from spin orientation from the external field that allows coherent control of the Ising spin lattice [29]. The external field is the unitary action driving the evolution of the spacetime lattice structure as a putative self-organized complex system.

A surface of constant phase, $\underline{k} \cdot \underline{r} - \omega t = k_x x + k_y y + k_z z - \omega t = \textit{constant}$ is a wavefront [30]. For a surface of constant phase if any wave equation has a time harmonic (sinusoidal) solution of the form $Ae^{i\phi}$ where A is the amplitude and the phase, ϕ a function of position with (x,y,z) constant and phase difference 2π separated by wavelength, $\lambda = 2\pi/k$. The direction cosines of the planes of constant phase are proportional to k and move in the direction of k equal to the phase velocity where

$$\mu = \frac{\omega}{k} = \frac{\omega}{\sqrt{k_x^2 + k_y^2 + k_z^2}}. \tag{15.10}$$

Where $\lambda = 2\pi/k = 2\pi\hbar/p = h/p$ is equivalent to the de Broglie matter wave relations, $E = \hbar\omega$, $\mathbf{p} = \hbar k$ [31].

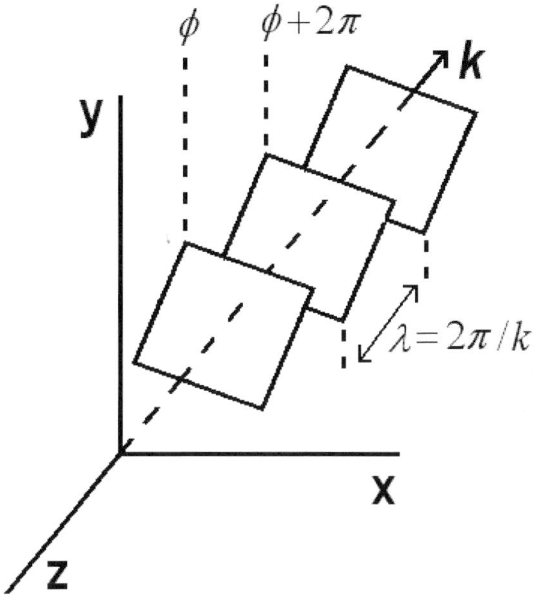

Figure 15.7. Surface of constant phase, in this case to represent orthogonal standing reality waves that can be utilized in figure-ground resonance effects.

15.6 Programmable Vacuum Cellular Automata Topology

Programmable matter is defined as a material that locally adjusts its response to external inputs through programmed control. Amorphous Ising model lattice-gas cellular automata can be used for programming spacetime if designed to mirror the spacetime structure utilized. Each independent computational element in the amorphous or stochastic (accepting all) medium is identically programmed on a topological surface which in this case conforms to the least-unit tori of spacetime. There are too many units to program individually so programming is achieved by neighbor connectedness. Toffoli formed a metaphor to describe this neighbor model [32,33]. Usually a marching band has a leader, this will not work for cellular automata where local self-assembly is internalized for each individual unit which acts as its own agent. This is a fundamental requirement for a massive ballistic response. The nanostructure of the defense shield materials must contain a computing substrate that is composed of fine-grained computing nodes distributed throughout space which communicate using only this nearest neighbor type of interactions [32-36]. According to Drexler [36] the closely packed computational units may be constructed to simulate a fractal system that for us would mean has the required incursive properties.

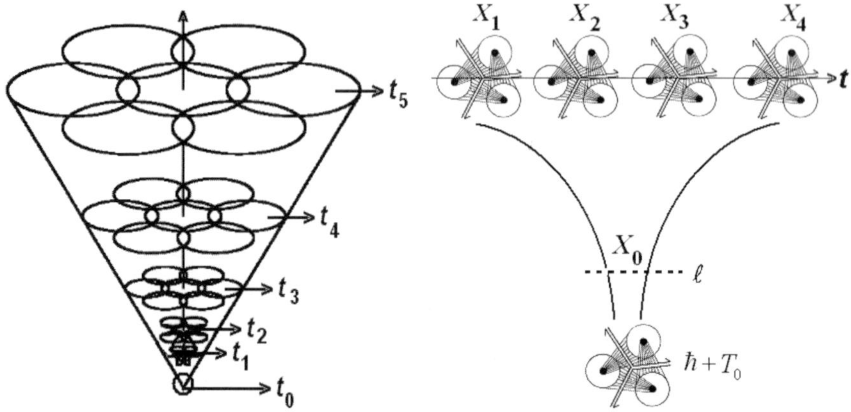

Figure 15.8. A mini-wormhole domain wall is created by coherently controlled conformal scale invariant constructive interference of least-cosmological unit phase relations [1]. The hierarchical nanoscale spacetime cellular automata programmable substrate of modulated cascades is built up into the warp bubble domain wall by static transduction of de Broglie matter-wave resonance configurations into a specified radius of macroscopic Minkowski space.

15.7 Manipulating de Broglie Matter-Waves

De Broglie by considering a material moving object of restmass, m_0 for a stationery observer suggested that a phase wave, or 'pilot' wave, accompanies a particle because the principle of inertia said it should possess an internal energy equal to m_0c^2 [37]. This phase wave arises as an inevitable consequence of de Broglie's assumption of the internal periodic phenomenon of the particle and the Lorentz transformation laws of the special theory of relativity

$$h\nu_0 - m_0c^2, \tag{15.11}$$

with $\nu - \beta c$, $(\beta < 1)$ for total energy $\nu - m_0c^2 / h\sqrt{1-\beta^2}$. De Broglie's result arose from a combination of the principle of Einstein's special relativity and the quantum relationship for the observer which he initially applied to a photon of nonzero restmass, m_γ $(< 10^{-50}\,g)$ which because of its associated internal motion he associated with a piloting phase wave of frequency, v at each point in space.

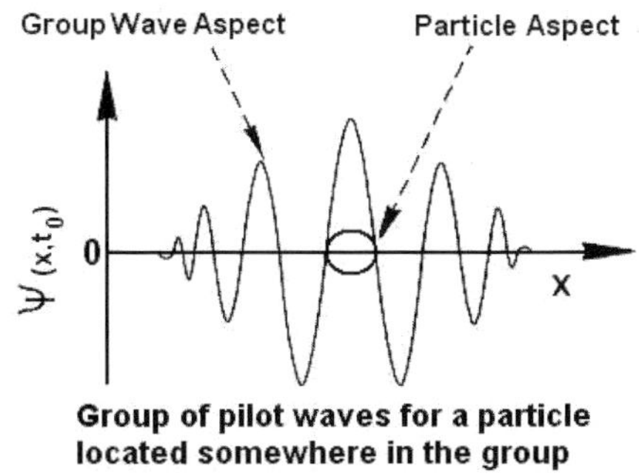

Figure 15.9. The group velocity of de Broglie waves is associated with the velocity of a particle.

MacKinnon [26,27,38] described the de Broglie wave packet for stationery states and nondispersive wave packets of a free particle. He states that the nondispersive wave packet, ψ is a solution of

$$\Box\psi = 0 \tag{15.12}$$

where

$$\Box = \nabla^2 - \frac{1}{c^2}\frac{\partial^2}{\partial t^2}. \tag{15.13}$$

From this MacKinnon shows that the nondispersive wave packet for a particle relative to the observer has the form

$$\psi = \sin(kr / kr)\exp\left[i(\omega t - k_0 x)\right] \tag{15.14}$$

where

$$k = m_0 c / \hbar, \quad r = \left\{ \frac{(x - vt^2)}{1-(v^2/c^2)^{1/2}} + y^2 + z^2 \right\}^{1/2}, \tag{15.15}$$

$$\omega = mc^2 / \hbar \quad k_0 = mv / \hbar.$$

Equation (15.13) is a spherically symmetric solution to Eq. (15.11) after being subjected to the Lorentz transform as initially obtained by de Broglie.

Of critical interest to us is MacKinnon's work to set up a de Broglie wave packet for a stationery state. Although we are interested in relativistic waves, our interest is not for the usual demonstration proving that particles or atoms in general are comprised of de Broglie matter-waves for particles in coordinate motion. Our interest is in de Broglie waves for stationery matter with internal 'continuous-state relativistic annihilation-creation advanced-retarded' effects.

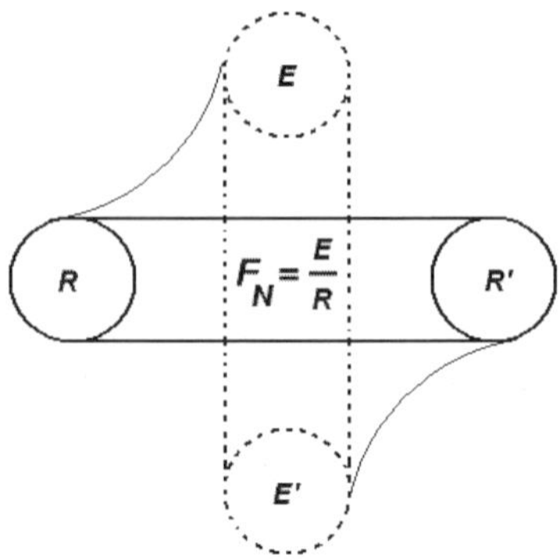

Figure 15.10. Ultimately the control mechanism for standing de Broglie waves depends on applying the noetic field equation, $F_N = E/.R$ to the other programming parameters for the ballistic programming of cellular automata phase modes.

Following MacKinnon consider two identical particles moving in opposite directions relative to an observer at x^* and t^*

$$\psi_1^* = A\cos\left(\omega t^* - kx^*\right), \quad \psi_2^* = A\cos\left(\omega t^* + kx^*\right) \qquad (15.16)$$

which represent standing waves when solved by the Schrödinger equation for a particle in a box and cannot depend on the reference frame [27]. MacKinnon concludes that these stationery states are static and for which Bohm postulated a quantum potential to account for it. MacKinnon carries this point further [26] to suggest that:

> The motion of a particle in spacetime does not depend on the motion relative to it of any observer or any frame of reference [and] if the particle has an internal vibration of the type hypothesized by de Broglie, the phase of that vibration at any point in spacetime must appear to be the same for all observers...Each observer or reference frame will have its own de Broglie wave for the particle. The phase of the particle's vibration must, by definition, be the same as that for all possible de Broglie waves at the point where the particle is. By superimposing all these possible de Broglie waves, a [nondispersive] wave packet is formed centered in space on the particle.

In his original work de Broglie could not properly form a wave packet able to localize a particle; but MacKinnon was able to construct a wave packet from de Broglie's original wave phenomena that is also nondispersive [26].

15.8 Coherent Control of Standing Matter-Waves

If it were possible to conceptually summarize everything required to develop a de Broglie matter-wave FTL Warp-Drive it is illustrated in Fig. 15.11 below which is an exploded conformal scale-invariant view of the continuous-state wave-particle seesaw leapfrog dynamics inherent in the topology of spacetime shown as a template within a brane topological hierarchy amenable to application of programmable resonance modes.

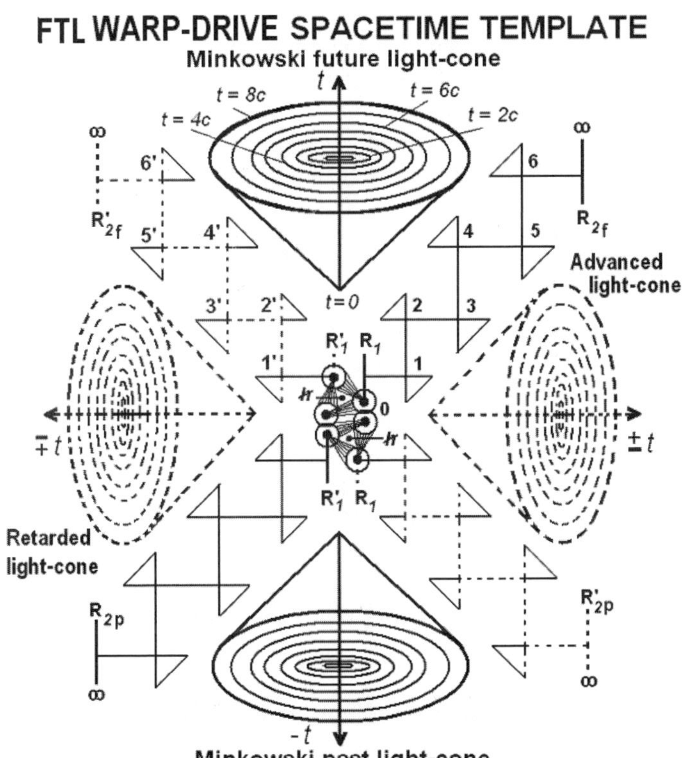

Figure 15.11. Conceptualized schema of the underlying spacetime structure utilized as a template for modulating the matter-wave resonance hierarchy mimicked in the programmable matter of the shield construction materials.

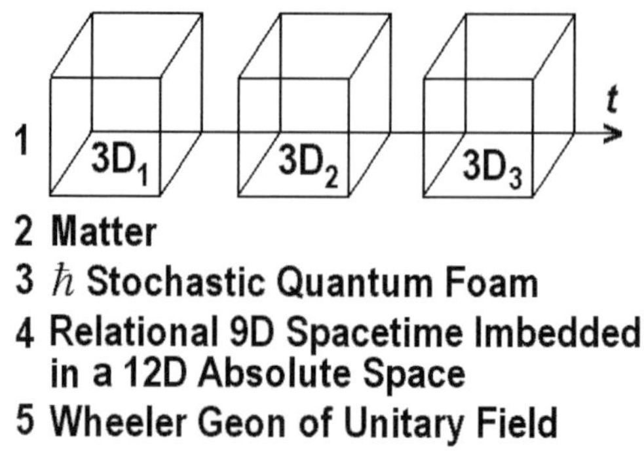

Figure 15.12. Reductionist hierarchic levels of HAM reality from the local standing-wave future-past eternal present to the atemporal geon of unitarity.

We insist up front that this warp-drive technology is impossible to any degree of power without Gödelizing outside the limiting domain of Copenhagen quantum theory into this 5th regime of Fig. 15.12 beyond spacetime to a degree where even the de Broglie-Bohm version is also unsatisfactory and needs further extension to the point of full ontological completion. One must get 'under' or 'beyond' spacetime in order to engineer or program the required full Ising lattice rotations that are able to utilize the 'infinite' power inherent in the vacuum by 'ontologically becoming the vacuum'. The other reason this Gödelization [39] is so important is the requirement not just to summation the phase of stationary de Broglie matter-waves (they are only level 2-3 on Fig. 15.12), but to also coherently control the phases of the topological hierarchy so the mean-free-path will ballistically compute [40,41] in a sufficiently HD regime. The full Gödelization process controls the symmetry of the arrow of time. The ontological foray into level 5 achieved by programming the geometric information of spacetime is before time at the level of the unified field. This is key to controlling the mean-free-path because it is this manipulation that allows the complete control of the Ising model hypersphere spin flips in a manner able to 'reflect the infinity' of the vacuum and be able to manipulate the array of mini-wormholes.

15.9 Reality as an Intermediate Collapse of Potentia

Cramer's transactional model of QT [12] has been ignored by most physicists but is a key element of HAM cosmology and fundamental to the understanding of the HWD. A *Cramer transaction* is based on the Wheeler-Feynman absorber theory of radiation [42] and entails future-past, standing-wave symmetry conditions which when extended to the HD SUSY regime of

Calabi-Yau mirror symmetry readily lend themselves to an HD extension of the Dirac polarized vacuum. Further we suggest this includes additional piloting or super-quantum potential [1] parameters associated with the unified field suggesting two forms, levels or regimes for quantum mechanics – that of the observed 4D phenomenological interaction associated with the uncertainty principle; and a new HD ontological 'piloting' or anthropic guidance regime wherein one is able to program operations that surmount the uncertainty principle. As illustrated in Fig. 15.11 reality itself is a continuous-state transaction phase Because the external world we observe is a limited subspace [1] of a larger contiguous reality some elements are removed from the perception of the observer by subtractive interferometry.

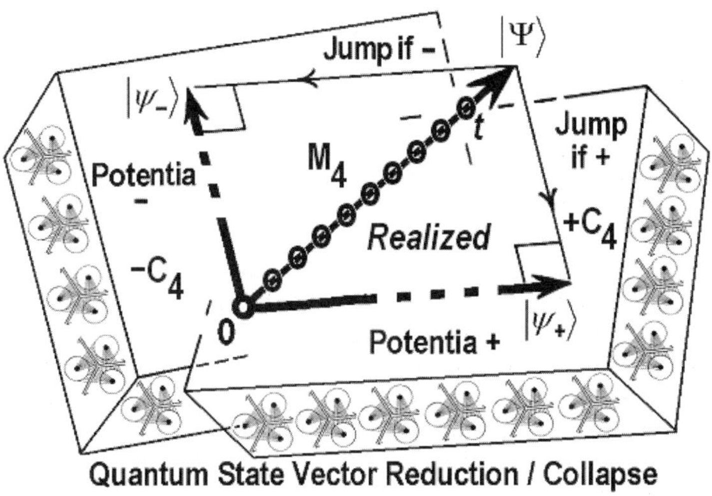

Figure 15.13. A way to look at a Cramer transaction as a collapse, $|\Psi\rangle$ to the 2D Euclidian plane from, in this case, an HD potentia of two possible orthogonal states, $|\psi_+\rangle, |\psi_-\rangle$. But in HAM cosmology this Planck scale action is considered as the microscopic lower bound of a duality whose upper bound creates macroscopic reality and the observed arrow of time.

In the standard Copenhagen Interpretation of QT an event emerges only as a result of measurement and objective reality is considered to be a probabilistic illusion. Cramer considers 'all off diagonal elements of the line element physically real' during the process of the offer-wave-confirmation-wave process preceding a transaction (event) [12]. We may call the final event a resultant of the conditions of Heisenberg Potentia. Here we still wish to consider reality illusory to the Minkowski observer.

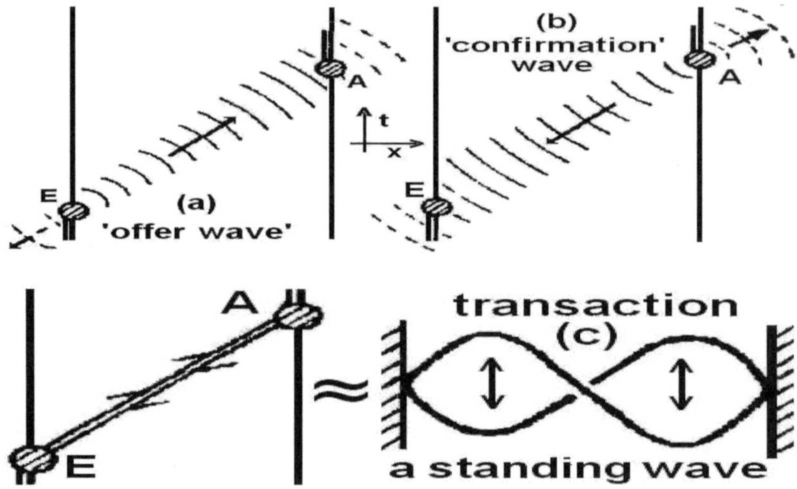

Figure 15.14. Transactional model. a) Offer-wave, b) confirmation-wave combined into the resultant transaction c) which takes the form of an HD future-past advanced-retarded standing or stationary wave. Figs. Adapted from Cramer [12].

Issues of the nature of the fundamental cosmological background continue to be debated with disparate views jockeying for philosophical supremacy; a scenario remaining tenable because experimental avenues for testing physics beyond the standard model have remained elusive. In a companion volume [1] we presented a putative empirical protocol for manipulating the so-called covariant Dirac polarized vacuum (DPV) providing a methodology for both surmounting uncertainty and low energy protocols for testing string theory. The DPV has a sixty year history in the physics literature [43-45] which has for the most part been ignored by the main stream physics community for a number of philosophical conflicts. The problem of surmounting uncertainty is solved by the utility of additional degrees of freedom introduced by utilizing a multiverse cosmology and the associated extended theoretical elements. This understanding allows one to engineer the vacuum.

Everything here relies on new or expanded properties of the covariant polarized Dirac vacuum:

- Programming the Least Cosmological Unit
- Continuous-State
- Dimensional Reduction
- Compactification
- Calabi-Yau Mirror Symmetry
- New Noetic Transformation

15.10 The Holographic Principle

15.10.1 *Usual Formulation of the Holographic Principle*

- The Holographic Principle first proposed by 't Hooft, and given a precise string-theory interpretation by Susskind, is a property of quantum gravity and string theory where a volume of space can be thought of as encoded on a boundary to the region.
- The theory also suggests the entire universe is a 2D information structure "painted" on the cosmological horizon, such that the 3D observed are only an effective description at macroscopic scales and at low energies.
- The holographic principle was inspired by black hole thermodynamics, implying maximal entropy in any region scales with $r2$, not $r4$ as expected. For a black hole, the description of all objects fallen in can be entirely contained in surface fluctuations of the event horizon. The holographic principle resolves the black hole information paradox within string theory.
-

15.10.2 *Extending the Holographic Principle*

- HAM Cosmology introduces a radical extension of the Holographic Principle - our virtual reality is a subspace of a new 12D absolute space.
- The Hubble Sphere is a 12D Hyper-Hologram
- In some interpretations of QT nothing exists before measurement. A duality exists here. Reality itself represents one stage of collapse relative the Earth observer.
- The other aspect of this duality is of course the usual Planck scale quantum effects.
- As seen shortly we will use the extended Holographic Principle as the basis for a 'Figure-Ground Effect' along with the new spacetime transformation to coherently control a 'topological switching' process of the space-like hypersurfaces inherent in the Alcubierre metric…

Newton claimed gravitational effects occurred instantaneously; but Einstein claimed gravitational influences were limited to the speed of light by the tenets of special relativity. Here we introduce a unified Geometrodynamics that is a duality of the two models. This is not farfetched if one considers the quest for a quantum gravity because as one knows by the EPR experiments nonlocal influence has been proven to occur instantaneously.

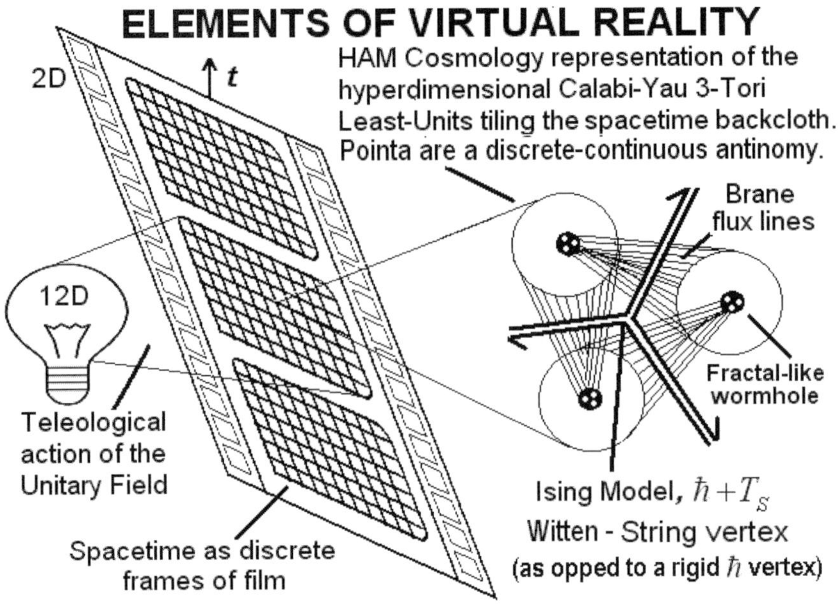

Figure 15.15. Spacetime is virtual in HAM cosmology and the least cosmological units tiling its backcloth are driven by a teleological anthropic action principle. Each 'point' is a continuous-discrete antinomy.

15.11 Essential Properties of Complex Noetic 12-Space

The spacetime programming model relies on a new 12D Absolute Space (AS) (ultimate arena of reality) from which properties of a Wheeler geon [1,5] or 'ocean of light' (unified noetic field) emerge. The noetic AS is an atemporal, highly ordered and symmetric harmonic superspace from which all other space relative to an Earth observer is a composite subspace. The geon domain (9D to 11D) is the first compactification regime; and because of coherence of the unitary field, railroad tracks would not recede but remain parallel.

A set of null lines (complex arrow of time), a loci of eternal points, remains hidden from local observed reality as an eternal present. This is part of the complex, $\pm C_4$ Wheeler-Feynman-Cramer duality of the future-past standing-wave comprising the continuous state present: "a relativistic spin-exchange dimensional reduction compactification process" which represents a new set of transformations beyond Galilean and Lorentz/Poincaré to describe the inherent dynamics of this unitary domain and create the arrow of time [1,13]. This condition results in our $E_3 \subset M_4$ domain being a subspace of eternity; and the essential process for producing the 'synchronization backbone' inherent in the backcloth of HAM cosmology [1].

As in special relativity where *c* remains constant and independent of the velocity of the source; the 12D AS remains static and absolute whether matter is stationery or in relativistic motion. In this context there is a duality in terms of conservation laws, annihilation/creation, advanced/retarded potentials or between space and energy including an asymmetry between the future-past. The new set of transformations makes correspondence with M-Theory and is conceptually considered a higher dimensional extension of Dirac Spherical Rotation [1]. Thus issues of the historical controversy between relational and AS are pushed to the new 12D domain. Within the Classical limit the former 3D Euclidean AS remains relative to the eternal present [1] of the subjective observer. Einstein demonstrated that the application of special relativity to a 3(4)D Minkowski/Riemann manifold makes space relational. The new relational space extends Einstein's view from four to eleven dimensions. In the 12D noetic superspace, S_N the 11D unitary noetic field (and the local 3(4)D $B^{(3)}$ component of the *EM* field) translates longitudinally, but the space (as in water waves) remains fixed because the wave bumps against the close-packed spheres or least units [1,46] (like the water molecules) allowing only transverse displacement while the wave is locally present. This wave cyclically undergoes $m_\gamma = 0$ and $m_\gamma \neq 0$ plus $B^{(3)}$ for certain polarizations.

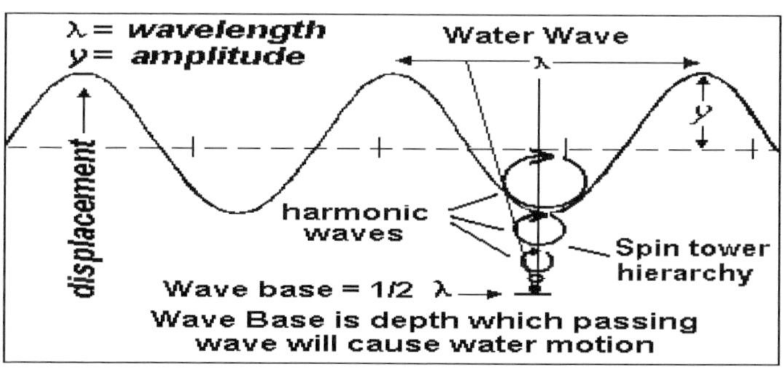

Figure 15.16. Symbolic representation of a 12D Ocean (Dirac sea) of Light (unitary field) modeled after the Wheeler Geon. The "ocean" provides a practical metaphor for 12D space in that polarization of the Dirac sea is believed to have properties similar to water waves.

Current thinking on the topology of space takes three general forms:

- The most commonly accepted 3(4)D Minkowski/ Riemann spacetime manifold; and two putative HD superspace additions,
- Calabi-Yau space preferred by M-Theory and

- Dodecahedral space.

Nature of the true vacuum remains an open question. The 3D absolute space of Newton became the 3(4)D relational spacetime of Einstein. The 12th D of Noetic cosmology represents a new form of absolute space, a periodic superspace where the eternal twelfth dimension has a Wheeler Geon [5] or ocean of 'light' (the unified field) as its $9 \rightleftarrows 11$D subspace. The relational 3(4)D Minkowski/Riemann spacetime manifold is a continuous state standing wave subspace of the 12D noetic superspace; it acts as a topological cover of an eternal present [1] which is not observed and continuously decays into spacetime.

"Space quantization" or the quantization of orientation of atomic systems observed empirically primarily by Stern-Gerlach and secondarily in other phenomena like the Zeeman Effect in an inhomogeneous magnetic field led to the basis for representing spin ½ fermions as a uniform Dirac spherical rotation through a 720° cycle [1] and the commutation relation for angular momentum in quantum theory. We explore extending these properties to 12D, 12D as required for UQC ontological operation.

If the noetic space water wave conception is correct, the continuous-state compactification process contains a tower of spin state Lie groups from spin 0 to spin 4. Spin 4 represents the unified field and makes cyclic correspondence with spin 0 where spacetime lattice Riemann sphere Ising lattice spin flips create dimensional jumps through the helicoids topology. Spin 0, 1/2, 1, & 2 remain in standard form. Spin three is suggested to relate to the orthogonal properties of atomic energy levels and space quantization. Therefore the spin tower hierarchy precesses through 0, 720°, 360°, 180°, 90° & 0 (∞) as powers of *I*, as conceptually illustrated below.

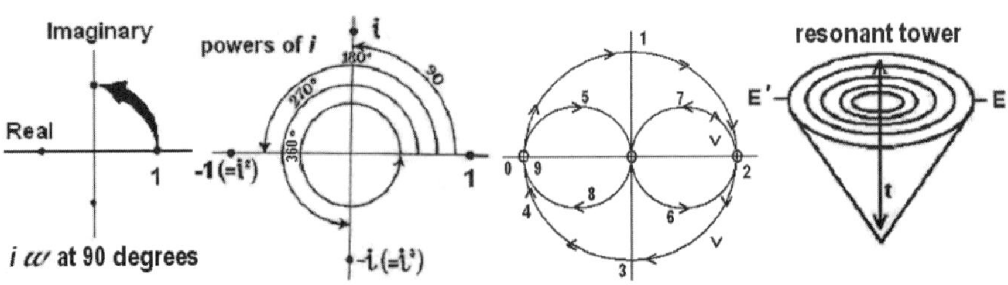

(a) (b) (c) (d)

Figure 15.17. (a) Complex dimension at 90° from the real axis. (b) Powers of i from 90° to 360°. (c) Power of i at 720°. (d) Resonant hierarchy comprised of powers of i in conjunction with the topology of the Genus-1 helicoid "parking-Garage" of the string vacuum with either Ising model, logarithmic spiral or cyclotron resonance hierarchy parameters for applying ladder operators of the resonant modes required to ontologically operate the UQC model.

An instant *t*, for position $r \equiv (x, y, z)$ or for the light cone $r = xdt$, defines a point or event $d = \sqrt{x^2 + y^2 + z^2}$ in ordinary spacetime coordinates, a pseudo-Euclidian metric tensor [48] representing the sixteen points of a 4-sphere (Fig. 11.1c)

$$G_{\mu\nu} = \begin{pmatrix} 1 & 0 & 0 & 0 \\ 0 & -1 & 0 & 0 \\ 0 & 0 & -1 & 0 \\ 0 & 0 & 0 & -1 \end{pmatrix}. \tag{15.17}$$

In summarizing the observers relationship to the *Cosmological Principle* (that the universe is homogeneous and isotropic on average in the large-scale) events are idealized instants in spacetime defined by arbitrary time and position coordinates *t, x, y, z*, written collectively as x^μ where μ runs from 0 to 3. The standard line element is

$$ds^2 = \sum_{\mu\nu} G_{\mu\nu} dx^\mu dx^\nu = G_{\mu\nu} dx^\mu dx^\nu, \tag{15.18}$$

where the metric tensor

$$G_{\mu\nu}(x) = G_{\mu\nu}(x) \tag{15.19}$$

is symmetric. In local Minkowski form all the first derivatives of g_{ij} vanish at the event and equation (15.17) takes the form

$$ds^2 = cdt^2 - dx^2 - dy^2 - dz^2. \tag{15.20}$$

The Cosmological Principle generally suggests that the clocks of all observers are synchronized throughout all space because of the inherent homogeneity and isotropy. Because of this synchronization of clocks for the same world time *t*, for commoving observers the line element in (15.17) becomes

$$ds^2 = dt^2 + G_{\alpha\beta} dx^\alpha dx^\beta = dt^2 - dl^2, \tag{15.21}$$

where dl^2 represents special separation of events at the same world time *t*. This spatial component of the event dl^2 can be represented as an Einstein 3-sphere

$$dl^2 = dx^2 + dy^2 + dz^2 + dw^2 \tag{15.22}$$

which is represented by the set of points (*x, y, z, w*) at a fixed distance *R* from the origin:

$$R^2 = x^2 + y^2 + z^2 + w^2 \tag{15.23}$$

where

$$w^2 = R^2 - r^2 \text{ and } r^2 = x^2 + y^2 + z^2 \tag{15.24}$$

so finally we write the line element of the Einstein 3-sphere as

$$dl^2 = dx^2 + dy^2 + dz^2 + \frac{r^2 dr^2}{R^2 - r^2}. \tag{15.25}$$

By imbedding Einstein's model of the three-sphere in a flat HD space, specifically as a subspace of a new complex 12D superspace, [1] new theoretical interpretations of standard cosmological principles are feasible.

Although the Newton and Coulomb potentials have similar forms the two theories have developed separately. For our purposes, following the Sakharov-Puthoff conjecture [45], that gravity is a product of fluctuation of the zero point field; we unify them with the Amoroso-Vigier methods [1,43] where both fields are represented by 4-vector field densities A_μ. Both phenomena are considered different types of motion within the same real physical field in flat spacetime as two different vacuum types of collective perturbations carried by a single vacuum field (unified).

Maxwell's equations traditionally describe only transverse elements that 'cut-off' at the vacuum. Here for HAM cosmology extended electromagnetic theory is utilized where the Einstein-de Broglie relation, $E = \hbar\nu = mc^2$ allows additional degrees of freedom such as longitudinal components $B^{(3)}$ and polarized vacuum conditions where $m_\gamma \neq 0$ suggests that the photon is piloted. These conditions suggest the need for both the standard *EM* field and extended $\mu\nu$ field coordinates; an understanding of which will be seen to be required for the ontological UQC operations.

In our original integration of *G* and *EM* we chose to fix the $\mu\nu$ field coordinates [1,44]. Here we go a step further. Dirac himself suggested by the rule of coordinate law that the pilot wave and the photon decouples [43]. The two sets of coordinates *EM* or $\mu\nu$ would normally be considered independent of each other. We integrate them in the topology of the Dirac polarized sea and alternate the fixing and decoupling of $\mu\nu$ and *EM* coordinates as an inherent 'leapfrogging' of the nonlocal-supralocal continuous-state standing-wave present [1,12]. Like wave-particle duality of matter, HAM cosmology $EM - \mu\nu$ duality extends to spacetime itself in that the unified field harmonically discretizes into spatial boundary conditions of an Ising model Euclidean point. Two types of computer animation in terms of 'figure' and 'ground' illustrate this. First, the animated figure crosses (arrow of time) the stationery background from left to right, disappears off the screen and reappears cyclically with an inherent frame rate. Each L-R cycle can be considered as one discrete spacetime least-unit quantum to the external observer. However as well known, our so-called quantum is actually comprised of a number of discrete frames that appear continuous to the external observer because of the refresh rate. This could be considered as the properties of quantum phase space and that material Fermi surfaces appear smooth because of the relativistic velocity of the surface electrons.

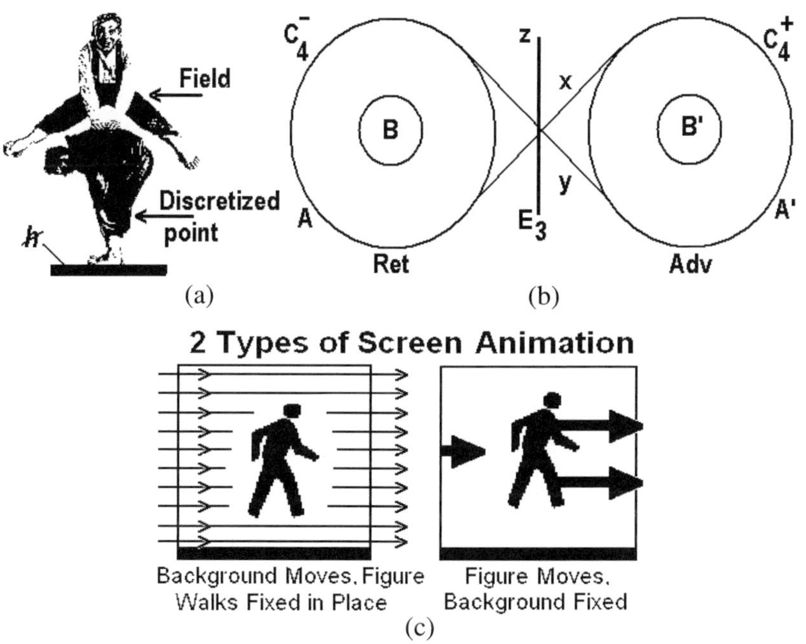

Figure 15.18. (a) Leapfrog metaphor of virtual reality. (b) Adds Ising Model spin-flip properties to the future-past Cramer type transaction Riemann sphere rotation. The central Euclidean point, E_3 is created and annihilated as a standing wave harmonic oscillator within the boundaries (denoted by A:B; A:'B') of two complex 4D tori. (c) The leapfrog duality of the $EM - \mu\nu$ metric also includes two types of spin exchange coupling-decoupling background–foreground interaction topologies. On the left the observer remains coupled, on the right the observer uncouples and reappears; this is part of the wave-boundary duality.

In the second case, the animated figure remains permanently fixed in the center of the screen and the background moves continuously from left to right (Arrow of time again) across the screen. For the sake of the metaphor one can say this latter case is introspective relative to the observer and the first case is objective (quantum) or external to an observer.

Neither of these two views offer a complete description of reality; as noted above, a third case of simultaneity is required. The apparent separateness of the two views; i.e. 'we live in a quantum universe' is the root of the problem because as proposed here we live in a continuous-state universe that is classical, quantum and unitary depending on perspective. The challenge here is to show that by adopting this view a model of vacuum spacetime automata programming can be achieved with coherent control of the continuous-state inherent spacetime synchronization backbone.

The leapfrog mechanism of Fig. 15.18a is essential to understanding Schwinger's concept of static and dynamic forms of the Casimir effect [1]. Metaphorically we like to call these HD topological boundary conditions 'casimirrors'. In our model of Calabi-Yau continuous-state future-past dynamics these conditions imply a cosmological form of the principles of wave-particle duality heretofore only used to describe quanta. But we wish to extend this model to the topological domain walls or boundary conditions of M-Theoretic brane dynamics as scale-invariant properties of the cosmological least-unit tiling the spacetime backcloth of the Dirac polarized vacuum during the continuous-state process. To our model this duality suggests open and closed, field-particle properties where the barrier or domain wall moment is an asymptotic Planck scale lower limit, $\hbar$ that is never reached. In conventional physics $\hbar$ is mathematical artifact of the fact that Gauge theory is an approximation. This continuous-state compactification cycling form large scale to Planck scale occurs with a holophote-like beat frequency in synchrony with the creation, annihilation and recreation of spacetime or space quantization of our virtual reality.

Noetic Space "leapfrogs" from holographic unitarity to discretized reality. This simplifies the boundary conditions and variables needed for UQC operations. The 12D Multiverse surface is considered a new form of Absolute Space (AS) and our observed Euclidian E_3 is a pseudo-AS or subspace of this regime. Because of the leapfrogging which We suppose is a fancy form of Witten's Ising flip [1] of the covariant string vertex. The E_3 pseudo-AS is a periodic discretization or 'frozen moment' of one 4D set of the 12D parameters (when time is included). This gives the least unit of the superspace the geometry of a torus; or in our Wheeler-Feynman future-past model [42] considered as two 4D advanced-retarded tori. This suggests the boundary conditions A:B; A':B' are HD boundary conditions of a harmonic oscillator allowing coherent control of the UQC to be operated with 4D parameters. As well known the usual form of Maxwell's equations in vacuum with $m_\gamma \neq 0$ and $B^{(3)} = 0$ has infinite families of boundary free exact solutions with the Lorentz gauge vector potential $A_\mu = 0$; but in the noetic case with $m_\gamma \neq 0$ where Maxwell's equations do not cut off at the vacuum, there is only one family and one set of boundary conditions, a model justified empirically by existence of the Casimir and Zeeman effects. *EM* theory implies the effects of the *EM* vector four-potential A_μ on the phases, S of quantum mechanical waves

$$\Delta S = \frac{q}{h}\int \phi dt - \frac{q}{hc}\int \vec{A}\cdot d\vec{S}\,. \tag{15.26}$$

For the continuous-state integration the mass term, m_γ is introduced into Maxwell's equations. One may also describe gravity with a four-vector density A_μ^g so that the Newton and Coulomb potentials take the same form but with different coupling constants suggesting both are different aspects of the same fundamental (unified) field with $A_\mu A_\mu \to 0$ where A_μ denotes the total four-potential in a covariant polarized Dirac vacuum.

From the *EM* vector potential $A^\mu(x)$ where $F_{\nu\mu} = A_{\nu,\mu} - A_{\mu,\nu^*}$ the components of E and B form second rank dual antisymetric spacetime field strength tensors $F^{\mu\nu}$(Adv), ${}^*F^{\mu\nu}$(Ret) defined as $F^{\mu\nu} = \partial^\mu A^\nu - \partial^\nu A^\mu$ and ${}^*F^{\mu\nu} = \frac{1}{2}\varepsilon \exp \mu\nu\rho\sigma\ F_{\rho\sigma}$ respectively as matrices

$$F^{\mu\nu} \equiv \begin{pmatrix} 0 & -E^x & -E^y & -E^z \\ E^x & 0 & -B^z & B^y \\ E^y & B^z & 0 & -B^x \\ E^z & -B^y & B^x & 0 \end{pmatrix},$$

$$ {}^*F^{\mu\nu} \equiv \begin{pmatrix} 0 & -B^x & -B^y & -B^z \\ B^x & 0 & E^z & -E^y \\ B^y & -E^z & 0 & E^x \\ B^z & E^y & -E^x & 0 \end{pmatrix}. \tag{15.27}$$

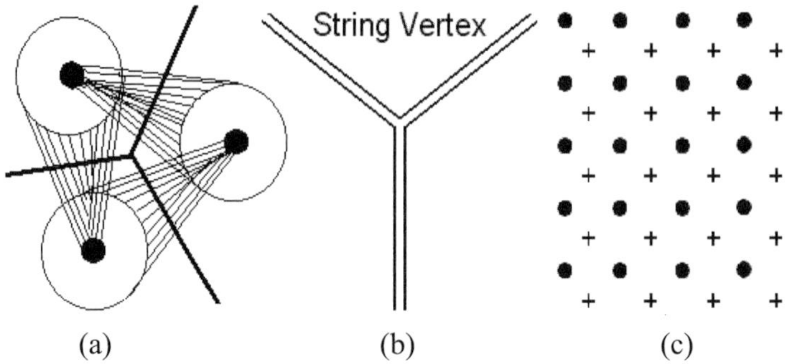

Figure 15.19. a) Least cosmological unit. b) Witten string vertex. c) Ising model background, descriptive elements of symmetry breaking and Riemann rotation.

If properties of the Dirac vacuum are expanded to conform with noetic cosmology Fig. 15.20b graphically represents the top of the Dirac sea where the central point is a space-like radial 4-vector $A_\mu = r_\mu \exp(iS/\hbar)$ with frequency $\nu = m_\gamma c^2/\hbar$. The oppositely rotating dipoles $\pm e$ correspond to gravity and *EM* with each individual subelement 4-momentum $\partial_\mu S$. Figure 15.20a represents one close-packed noetic hypersphere least-unit [1,46] just below this regime which is the vertex at 0 where further unification to the unitary field occurs.

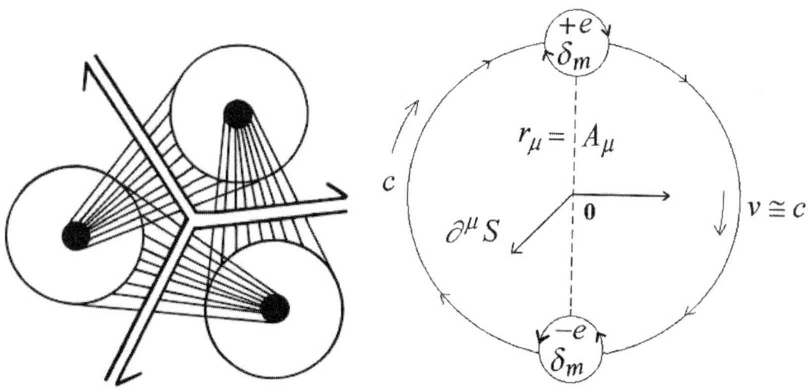

Figure 15.20. Models of least-unit points tiling the Dirac backcloth in the HAM cosmology of 12D noetic superspace. (a) Least cosmological unit with a classical discrete $\hbar$ vertex. Continuous vertex of string theory able to undergo Ising flips as in string theory. Triune nature of Ising least-unit. (b) Conceptualization of two oppositely charged vacuum subelements rotating at $\nu \cong c$ around a central point behaving like a dipole (+ e) EM 'bump' and (-e) G 'hole' on the topological surface of the covariant polarized Dirac vacuum.

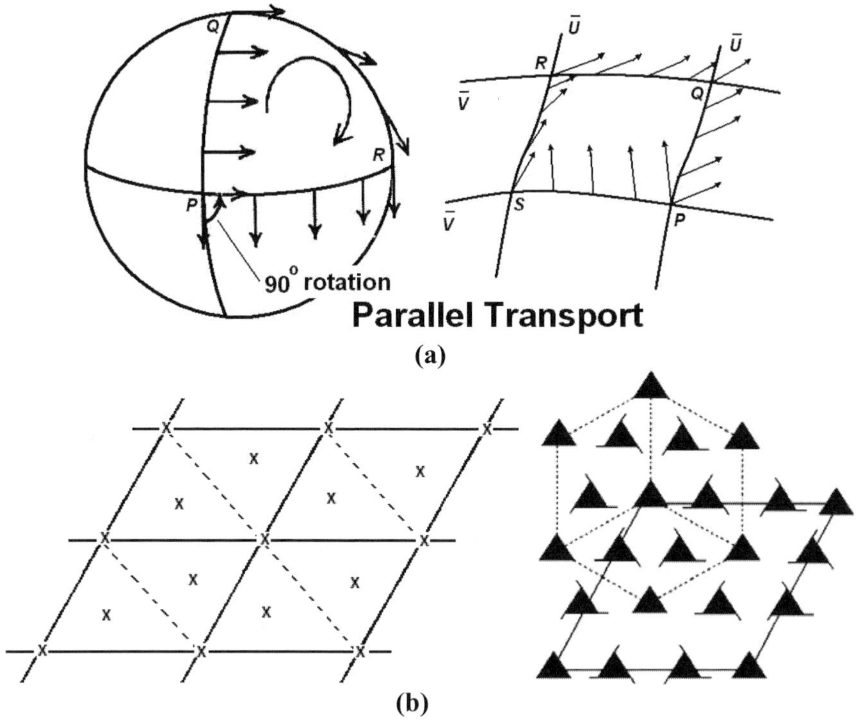

Figure 15,21. a) Parallel transport of a vector or spinor around closed paths generally results in a deficit angle, a mass defect where the vector does not return to the original position *P*. b) Tiling of the spacetime backcloth and projective geometry giving rise to higher dimensionality.

This is only a superficial account of the highly essential relevance of the complementarity of the $G_{\mu\nu} - F_{\mu\nu}$ coordinate systems. Suffice it to simplistically summarize here that the dynamics of the continuous-state SUSY symmetry breaking are key to the ontological properties of this putative model of bulk QC. The G-EM coordinates couple and uncouple fixing one and then the other in a dual seesaw-leapfrogging effect which is like a form of topological wave-particle duality. It is the utilization of this structural-phenomenology as a covariant resonant hierarchy that allows the ontological violation of the Copenhagen regime uncertainty principle. The triune geometry of Fig. 15.20a represents the point 0 in 15.20b shown as an Ising lattice array in Fig. 15.19c. This is similar to the vertex in string theory (Fig. 15.19b) able to topologically undergo spin flips of the Riemann sphere from zero to infinity (Fig. 15.20a). In these continuous-state points the Ising vertices as governed by the super quantum potential (unified field) as described by the noetic field equation [1]. There is a foreground and background duality where the EM and metrics continuously "leapfrog" in the spacetime backcloth. These factors are imposed on spacetime geometry by the symmetry conditions of noetic cosmology. Traditionally parallel transport of a vector or spinor around a closed path P,Q,R (Fig. 15.21a) or P,Q,R,S (Fig. 15.21b) generally results in a deficit angle, a mass deficit that signifies the amount of curvature at that vertex when the Riemann tensor is $\neq 0$ [5].

Tiny loops approximated by a parallelogram of two tangent vectors $\vec{\mu}$ and $\vec{\nu}$ close (no deficit) if $[\vec{\mu},\vec{\nu}]=0$; then the curvature operator is the commutator of covariant derivatives along $\vec{\mu}$ and $\vec{\nu}$, $R(\vec{\mu},\vec{\nu})=[\nabla_{\vec{\mu}},\nabla_{\vec{\nu}}]$ [5]. If $[\vec{\mu},\vec{\nu}]\neq 0$, $[\nabla_{\vec{\mu}},\nabla_{\vec{\nu}}]$ is subtracted from the commutator, the parallelogram doesn't close and the Riemann tensor is $\neq 0$.

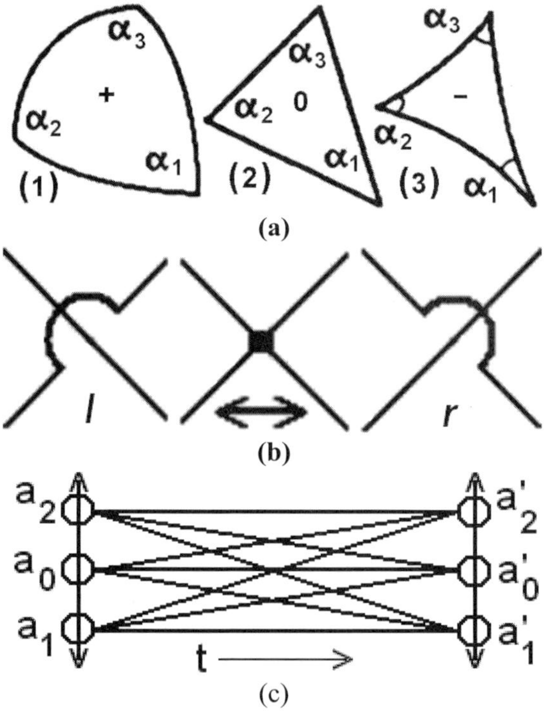

Figure 15.22. (a) Three types of geodesic triangles with Gaussian curvature. 1) Circumsphere with positive curvature, sum of internal angles $> \pi$. 2) Mesosphere, E^3 with zero curvature. 3) Insphere, internal angle sum $< \pi$ so curvature is negative. (b) Chiral properties of a vertex where the coordinate basis topologically switches from fixed to l or r open. (c) Triune elements of an HD transaction in noetic terms where the elements of a least-unit are tertiary.

In Fig. 15.22a, the sum of the three internal angles minus π is the Gaussian curvature integral $(\alpha_1 + \alpha_2 + \alpha_3) - \pi \int KdA$ where K is the Gaussian curvature. Taking Fig. 15.22 triangle (a) for example on a sphere of radius r with $\alpha_1 = \alpha_2 = \alpha_3 = \pi / 2$ the area of the triangle is $(4\pi r^2)/8$ and the Gaussian curvature would be $K = 1/r^2$ which is positive [5].

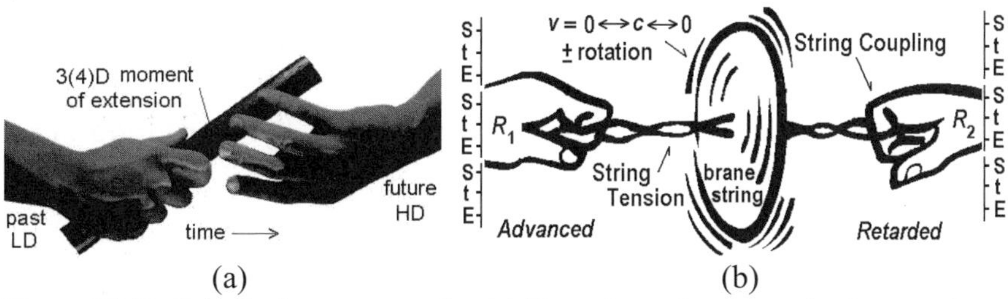

Figure 15.23. Spin Exchange properties (a) The spin exchange mechanism requires a coupling-decoupling moment between the $c \rightarrow q \rightarrow u$ components of the spacetime least-units like the passing of a baton in a relay race. (b) The spinning disk toy illustrates elements of the continuous-state. Imagine an array of disks as in Fig. 15.19c tiling spacetime. When the disk stops momentarily the Ising rotation occurs during the uncoupling, then recouples as spin continues in the opposite direction.

The spin-exchange hierarchy process has many components; more are shown in Fig. 15.24. This detailed plethora of components is what allows the vacuum to be programmed in a controlled manner to engineer the warp-drive parameters. The dominant view among cosmologists regarding extra dimensions is that if they exist they must be microscopic because they are not observed. In noetic cosmology extra dimensions are macroscopic and take part in the creation and recreation of spacetime, the arrow of time and observed macroscopic reality). This scenario arises during the inherent 'continuous-state spin-exchange dimensional reduction compactification process' by parallel transport within the additional context of a dual Dirac spherical rotation of the least-unit topology of subspace elements producing deficit angles during decoupling-coupling allowing relativistic subtraction of supralocal-nonlocal domain components producing the arrow of time. The scaling process begins in the microscopic backcloth without a physical arrow of time and ramps up the helicoid hierarchy to the virtual standing-wave macroscopic present. Because of its relativistic nature the 'baton' passing (coupling-decoupling) between domains appears smooth to the observer is meant to be synonymous with the lightcone rings where the leapfrogging domain frequency provides the context for assigning coupling parameters required for utilizing the synchronization backbone for the UQC.

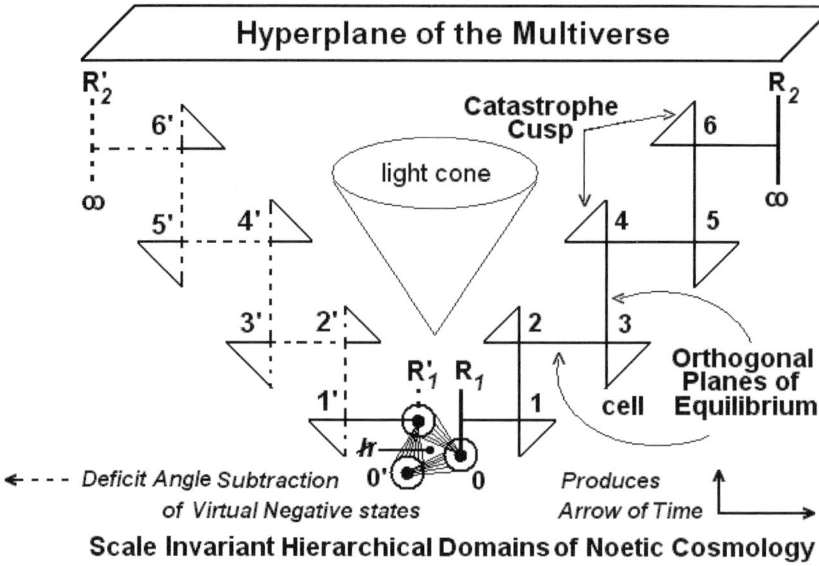

Figure 15.24. Covariant scale invariant hyperplane compactification domains in the hierarchy of continuous-state noetic superspace.

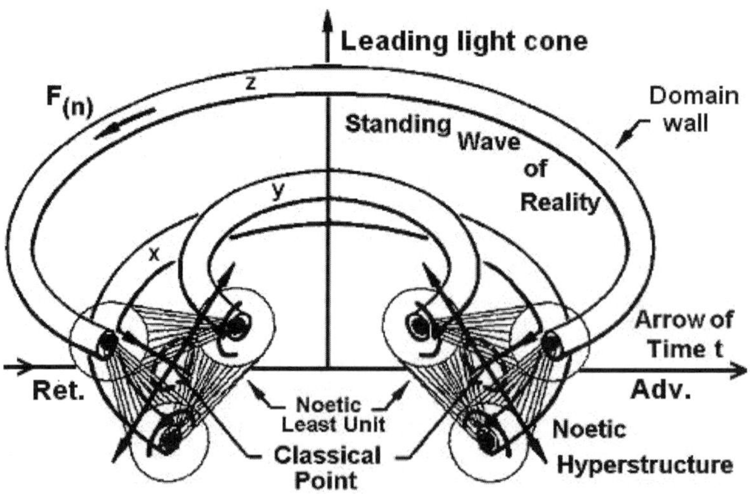

Figure 15.25. Following an extended Cramer transactional model applied to cosmology the Figure represents a single future-past, retarded-advanced Calabi-Yau mirror symmetry domain (brane vibrations) where the properties illustrated in Figs. 15.21,15.23 & 15.24 interplay to produce the observed macroscopic arrow of time by deficit angle subtractive interferommetry.

15.12 The Holographic Wormhole Drive (HWD)

Utilizing advanced principles inherent in a Holographic Anthropic Multiverse (HAM) [1], radical improvements to the Alcubierre warp drive model occur that could lead to FTL travel with existing technological prowess. Based on a radical extension of the Holographic Principle assuming a form of 'Heisenberg Potentia' inherent in an HD absolute space, simplistically the drive mechanism is a holographic figure-ground effect. The local spaceship (figure) is removed (put in free-fall relative to the background) from the infinite potentia. Then through manipulation of a 12D complex form of a covariant Dirac polarized vacuum (string theoretic) the external domain wall or interstice between the ships warp bubble and the Alcubierre metric becomes a coherently controlled stochastic barrier of leapfrogging mini-wormholes key to the drives operation:

- Shield (ship or warp bubble) constructed from inherent infinite vacuum energy by coherently controlled constructive interference of continuous-state 12D brane mirror symmetry resonance modes of the HAM close-packed cosmological least-unit backcloth structure.
- The Alcubierre solution is utilized with minimal curvature because of a new set of unified transformations beyond the Galilean-Lorentz-Poincaré.
- The mini-wormholes arise in a periodic gravitational shock-wavefront interference of the local relativistic (luminal form) and new nonlocal holographic (complex instantaneous form) of the duality of gravitational modes of unified theory. In this interpretation the 'Warp Factor' becomes a beat frequency tier of coupled-uncoupled modes of the spherical shell within the figure-ground leapfrogging oscillation, $R \leftrightarrow R'$.

This warp-drive model is called the Holographic Wormhole Drive, that FAPP does not require negative energy It is Based on a radical extension of the Holographic Principle assuming a form of 'Heisenberg Potentia' inherent in an HD absolute space, simplistically the drive mechanism is a holographic figure-ground effect. The local spaceship (figure) is removed (put in free-fall relative to the background) from the infinite potentia. Then through manipulation of a 12D complex form of a covariant Dirac polarized vacuum (string theoretic) the external domain wall or interstice between the ships warp bubble and the Alcubierre metric becomes a coherently controlled stochastic barrier of leapfrogging mini-wormholes key to the drives operation:

- Shield (ship bubble) constructed from inherent infinite vacuum energy by coherently controlled constructive interference of continuous-state 12D brane mirror symmetry resonance modes of the HAM close-packed cosmological least-unit backcloth structure.
- The Alcubierre solution is utilized with minimal curvature because of a new set of unitary transformations beyond the Galilean-Lorentz-Poincaré.
- The mini-wormholes arise in a periodic gravitational shock-wavefront interference of the local relativistic (luminal form) and new nonlocal holographic (complex instantaneous form) of the duality of gravitational modes of unified theory.

In this interpretation the 'Warp Factor' becomes a beat frequency tier of coupled-uncoupled modes of the spherical shell within the figure-ground leapfrogging oscillation, $R - R'$.

The HWD model provides solutions to the major problems facing the Alcubierre warp-Drive metric based on principles of the Holographic Anthropic Cosmology [1]. The solution relies on a 'Holographic Figure-Ground Effect' where the 'local' free-fall Warp Bubble is separated from the holographic background by covering the domain wall of the free-fall warp-bubble with a system of mini-wormholes created by 'programming certain Calabi-Yau mirror symmetry parameters of the Dirac-type covariant polarized vacuum'. Holophote Manipulation of this figure-ground effect becomes the 'Warp factor'. By utilizing a higher dimensional form of the covariant polarized Dirac Vacuum and an associated operationally completed form of Quantum Theory the need for 'infinite exotic energy' is ameliorated by 'borrowing' the vacuums infinite inherent energy.
New concepts include:

- Operationally completed dual Newton/Einstein model of Geometrodynamics (not a form of quantum gravity)
- Operationally completed 12D form of Quantum Theory that utilizes the unified field as a form of "super quantum potential"
- Extended HD view of the Covariant Dirac Polarized vacuum
- Unique 12D String Theoretic Vacuum Topology
- Alternative utility of the Alcubierre warp-drive metric
- Utility of Mini Wormholes
- Radical utility of the Holographic Principle
- New Set of Noetic Transformations

This essentially entails a New Cosmological Paradigm From the realization that the limits of QT, GR, SR, and the Dirac equation etc. are all inseparable from and rely on understanding cosmology.

15.13 Hyperspherical Holographic Figure-Ground Effect

Utilizing the figure-ground effect requires a new spacetime transformation [1] to cover the domain wall of an Alcubierre warp bubble with an array of mini-wormholes in conjunction with an incursive oscillator for manipulating Alcubierre's α and β functions with minimal external energy input. In considering the utility of wormholes for instantaneous or FTL travel between remote interstellar regions, until now they have been considered only in terms of giant wormholes as might be created near a neutron star or in the vicinity of a black hole in order to sufficiently bend spacetime to form an interconnecting corridor between remote spatial regions that shorten a trip by some light years. Our use of wormholes is very different. Imagine the middle school physics experiment where hundreds of tiny holes are drilled into a smooth table. When air is forced through these holes a heavy puck pushed across the table travels with nearly no friction because it is levitated on the cushion of air. This is not a perfect metaphor except in the sense that the holes represent mini-wormholes to remove the puck from the background, i.e. removing the friction coupling. As a point of interest a similar technique has recently been incorporated into submarine design to reduce drag.

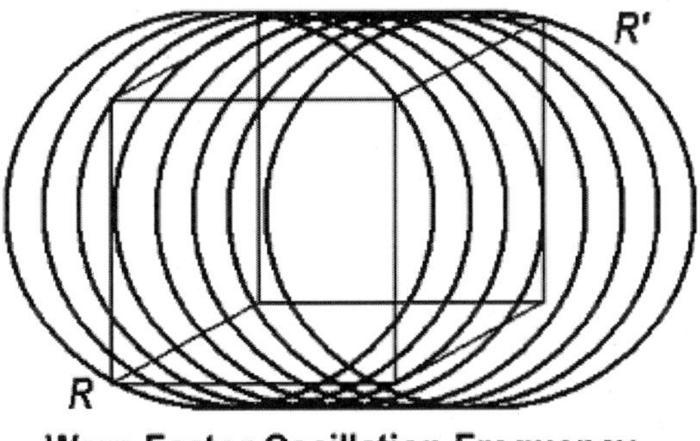

Figure 15.26. Figure-Ground Topological Switching model of warp factor Operation.

Holophote removal of the ship's warp bubble from the background relies on setting up a continuous-state resonant control hierarchy of the 12D Dirac vacuum. A domain wall of mini-wormholes comprised of cosmological least-units that through phase controlled constructive interference produce a topological switching of spacetime topology. The beat frequency of this topological switching is related to both the warp factor and holophote figure-ground oscillation that removes the ship (figure) from the holographic background of the multiverse. A multimode multilevel resonance hierarchy operates the HWD. Some key elements are:

- Alcubierre Metric Orients Direction of Travel
- Creating the Warp Bubble
- Creating the Mini-Wormhole Domain Wall
- Operating the Holographic Figure-Ground effects
- Coherent control of the Warp-Factor Topological Switching Mechanism

15.14 Coherently Controlled Vacuum Drive Mechanism

Genus-1 helicoid 'parking garage' hierarchy representing the advanced-retarded future-past symmetry of a mirror symmetric Calabi-Yau dual 3-form K-K spin tower inherent in the continuous-state fabric of spacetime.

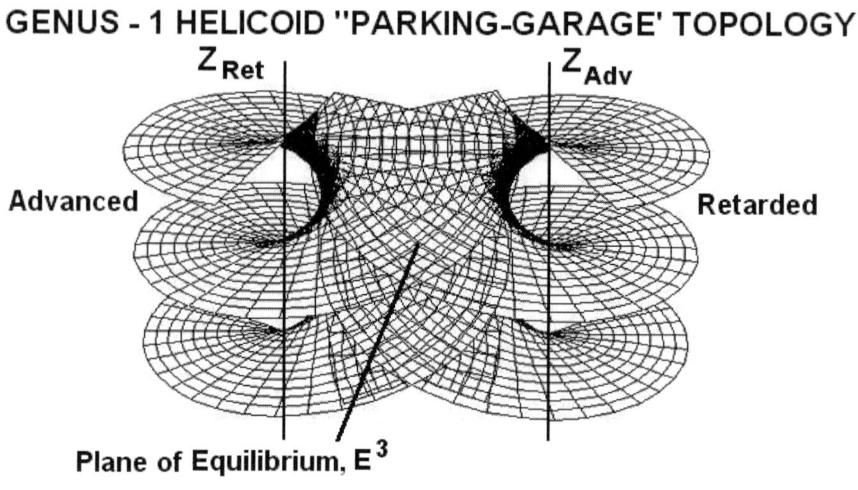

Figure 15.27. The Helicoid, a minimal embedded surface [19], is swept out by a line rotating about and moving down the z axis. Here a double Genus-1 Helicoid is joined into a "parking garage" ramp structure representing the future-past hierarchical topology of noetic space. An ordinary 2D plane can be twisted into a helicoid.

Domain Wall Boundary Conditions & Emission Absorption Loci

for Advanced-Retarded Waves

- We shall consider a static thick domain wall constructed by a scalar field with self-interaction properties as in the Schwarzschild black hole spacetime singularity solution.

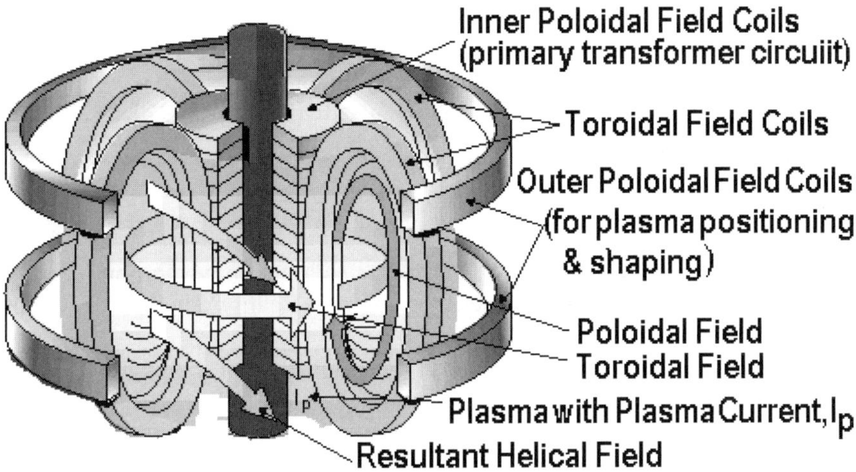

Figure 15.28. Spherical Tokomak reactor for powering the HWD.

Appendix

One of us, Amoroso calls himself a Noeticist, which means essentially that in addition to being a practicing physicist, he utilizes noetic insight or transcendence as a tool in scientific theory formation (see Chap. 14) [47]. This idea originates with the ancient Greek philosopher Plato who taught: 'no matter how vast ones intellect, or how deep ones wisdom, noetic insight is beyond this; it arises as transcendent entelechies from the Anthropic teleology of the cosmos! We risk making such bold claims because we believe now with the discovery of consciousness [47] it is possible to complete the tools of human epistemology: logic, empiricism and now transcendence and we wish to initiate debate on this possibility as a tool in scientific theory formation…

As an example Amoroso offers his path. His noetic side arises from practicing Zen, Hatha Yoga, decades of practicing a personal form of meditation and especially adherence to the strict discipline of being a High Priest in the LDS (Mormon) faith. See end of Chap. 14 for discussion on paths to transcendence. On that note, although obscure, it appears to be LDS doctrine that only agents of the Earth may serve the Earth. This appears to be why stars are so far apart from each other; in order that 'Nazi' marauders like in the Hollywood film, Avatar cannot wreak havoc on other civilizations and interfere with their evolution. Thus putatively perhaps, when a civilization develops FTL technology their millennium begins and soon they are taken off planet. With that wryly said we have done our part in both bringing forward the Judeo-Christian Millennium and presenting a rudimentary design for implementable FTL travel. Of course carrying this reasoning to its penultimate form, once we are off planet we will not need FTL warp-drive technology because as in the Hollywood film, Indiana Jones and the Kingdom of the Crystal Skull, we can travel instantaneously through "inner space"!

References

[1] Amoroso, RL & Rauscher, EA (2009) The Holographic Anthropic Multiverse: Formalizing the Geometry of Ultimate Reality, Singapore: World Scientific.
[2] Visser, M. (1989) Traversable wormholes: Some simple examples. Physical Review D 39, 3182–3184.
[3] Thorne, K. S. (1994). Black Holes and Time Warps, New York: W. W. Norton.
[4] Cramer, J.G., Forward, R.L., Morris, M.S, Visser, M., Benford, G. & Landis, G.A. (1995) Natural wormholes as gravitational lenses, Phys. Rev. D51, 3117.
[5] Misner, CW, Thorne, KS & Wheeler, JA (1973) Gravitation, San Francisco: W.H. Freeman.
[6] Einstein, A. and Rosen, N. (1935) The particle problem in the general theory of relativity, Physical Review 48, 73.
[7] Morris, M. S. & Thorne, K.S. (1988) Wormholes in spacetime and their use for interstellar travel: A tool for teaching general relativity, Am J Physics 56, 395-412.
[8] Morris, M.S., Thorne, K.S. & Yurtsever, U. (1988) Wormholes, time machines, and the weak energy condition, Phys. Rev., 61, 13, pp. 1446 – 1449.
[9] Public domain image, wikicommons.
[10] Petroni, N.C. & Vigier, J-P. (1982) Dirac's aether in relativistic quantum mechanics, Found. Physics, 13: 2; 255-285.
[11] Bohm, D. & Vigier, J-P (1954) Model of the causal interpretation of quantum theory in terms of a fluid with irregular fluctuations, Phys. Rev. 96: 1; 208-217.
[12] Cramer, J.G. (1986) The Transactional interpretation of quantum theory, Reviews of Mod. Physics, 58:3, 647-687.

[13] Amoroso, RL , Kauffman, L.H., Rauscher, EA & Rolands, P. (2011) Search for Fundamental Theory, monograph in preparation.
[14] Alcubierre, M. (1994) The warp drive: hyper-fast travel within general relativity, Classical & Q. Grav., 11:L73-L77; and (2000) arXiv:gr-qc/0009013v1 5.
[15] Cramer, JG (1996) The Alcubierre Warp Drive, November, Analog Science Fiction & Fact.
[16] Obousy, KR & Cleaver, G (2008) Warp Drive: A New Approach, arXiv: 0712.1649v6.
[17] Barbour, J. (1999) *Then end of time - The next revolution in physics*, Oxford Press, Oxford, U.K.
[18] Leibniz, G.W. (1768) *Opera Omnia.* 6 volumes, Louis Dutens, ed. Geneva
[19] Amoroso, RL (ed.) (2010) Complementarity of Mind and Body: Realizing the Dream of Descartes, Einstein and Eccles, New York: Nova Science Publishers.
[20] Chu, M-Y.J. & Amoroso, R.L. (2008) Empirical mediation of the primary mechanism initiating protein conformation in prion propagation, in D. Dubois (ed.) Proceed. CASYS07, IJCAS, Vol. 22, Univ. Liege Belgium.
[21] Amoroso, R.L. (1996) The production of Fröhlich and Bose-Einstein coherent states in in vitro paracrystaline oligomers using phase control laser interferometry, Bioelectrochemistry & Bioenergetics, 41:1, pp.39-42.
[22] Morisawa, Y., Ida, D., Ishibashi, A.& Ken-ichi Nakao, K-I (2002) Thick domain walls around a black hole, arXiv:gr-qc/0209070v2.
[23] Morisawa, Y., Yamazaki, R., Ida, D., Ishibashi, A. & Nakao, K-I (2000) Thick domain walls intersecting a black hole, arXiv:gr-qc/0005022v1.
[24] Wolff, M. (2002) Cosmology, the quantum universe and electron spin, in R.L. Amoroso, G. Hunter, M. Kafatos & J-P Vigier (eds.) Gravitation and Cosmology: From the Hubble Radius to the Planck Scale, pp. 517-524, Dordrecht: Kluwer.
[25] Wolff, M. (2008) Schrödinger's Universe and the Origin of the Natural Laws, Parker: Outskirts Press.
[26] McKinnon, L. (1978) A nondispersive de Broglie wave packet, Foundations of Physics, 8:3-4; 157-176.
[27] McKinnon, L. (1979) The de Broglie wave packet for a simple stationery state, Foundations of Physics, 9:9-10; 787-791.
[28] Starace, A.F. (1988) Hyperspherical description of two-electron systems, in Briggs, J.S., Kleinpoppen, H. & Lutz, H.O. (eds.) Fundamental Processes of Atomic Dynamics, pp. 235-258, NY: Plenum.
[29] Harding, S.L., Miller, J.F. & Rietman, E.A. (2006) Evolution in materio: Exploiting the physics of materials for computation, arXiv:cond-mat/0611462v1.

[30] Peebles, P.J.E. (1992) Quantum Mechanics, Princeton: Princeton Univ. Press.
[31] L. De Broglie, *Théorie générale des Particules à Spin (Méthode de Fusion)*, pp.93-116. Paris, Gauthier-Villars, 1943.
[32] Toffoli, T. Programmable Matter: An introduction, Cambridge: MIT Univ. Press (in preparation).
[33] Toffoli, T. & Margolus, N. (1987) Cellular Automata Machines; A New Environment for Modeling, Cambridge: MIT Univ. Press; Russian translation (1991) *Mashiny Kletochnykh Avtomatov, Izdatelstvo 'Mir'.*
[34] Kodama, T. & Koide, T. (2008) Memory effects and transport coefficients for non-Newtonian fluids, arXiv:0812.4138v1 [hep-ph].
[35] Abelson, H. et al. (2000) Amorphous computing, Communications of the ACM, 43:74-82.
[36] Drexler, K.E. (1992) Nanosystems : Molecular Machinery, Manufacturing and Computation, New York : Wiley & Sons.
[37] de Broglie, L. (1923) Radiation, waves and quanta, *Comptes Rendus*, Vol. 177, pp. 507-510.
[38] MacKinnon, L. (1981) A fundamental equation in quantum mechanics? Let Al Nuovo Cimento, 32:10; 311-316.
[39] Smullyan, R.M. (1992) Gödel's Incompleteness Theorems, Oxford: Oxford University Press.
[40]_Heiblum, M., Nathan, M.I., Thomas, D.C. & Knoedler, C.M. (1985) Observation of Ballistic Transport in GaAs, Phys Rev L, 55:20; 2200-03.
[41] Javey, A., Guo, J., Paulsson, M., Wang, Q., Mann, D., Lundstrom, M. & Dai, H. (2003) High-field, quasi-ballistic transport in short carbon nanotubes, arXiv: 0309/0309242.
[42] Wheeler, J. & Feynman, R. (1945) Interaction with the Absorber as the Mechanism of Radiation, Rev. Mod. Phys. 17, 1578.
[43] Dirac, P.A.M. (1973) New ideas of space and time, *Naturwissenschaften* 32:6; 529-531.
[44] Amoroso, R.L. & Vigier, J-P (2002) The origin of CMBR as intrinsic blackbody cavity-QED resonance inherent in the dynamics of the continuous state topology of the Dirac vacuum, in R.L. Amoroso , G. Hunter, M. Kafatos & J-P Vigier (eds.), Gravitation and Cosmology: From the Hubble Radius to the Planck Scale, Dordrecht: Kluwer.
[45] Puthoff, H.E. 1989, Gravity as a zero-point-fluction force, Physics Review A, 39, 2333-2342.
[46] Stevens, H.H. (1989) Size of a least-unit, in M. Kafatos (ed.) Bell's Theorem, Quantum Theory and Conceptions of the Universe, Dordrecht: Kluwer Academic.

Index

SERIES ON KNOTS AND EVERYTHING

Editor-in-charge: Louis H. Kauffman *(Univ. of Illinois, Chicago)*

The Series on Knots and Everything: is a book series polarized around the theory of knots. Volume 1 in the series is Louis H Kauffman's Knots and Physics.

One purpose of this series is to continue the exploration of many of the themes indicated in Volume 1. These themes reach out beyond knot theory into physics, mathematics, logic, linguistics, philosophy, biology and practical experience. All of these outreaches have relations with knot theory when knot theory is regarded as a pivot or meeting place for apparently separate ideas. Knots act as such a pivotal place. We do not fully understand why this is so. The series represents stages in the exploration of this nexus.

Details of the titles in this series to date give a picture of the enterprise.

Published:*

Vol. 1: Knots and Physics (3rd Edition)
by L. H. Kauffman

Vol. 2: How Surfaces Intersect in Space — An Introduction to Topology (2nd Edition)
by J. S. Carter

Vol. 3: Quantum Topology
edited by L. H. Kauffman & R. A. Baadhio

Vol. 4: Gauge Fields, Knots and Gravity
by J. Baez & J. P. Muniain

Vol. 5: Gems, Computers and Attractors for 3-Manifolds
by S. Lins

Vol. 6: Knots and Applications
edited by L. H. Kauffman

Vol. 7: Random Knotting and Linking
edited by K. C. Millett & D. W. Sumners

Vol. 8: Symmetric Bends: How to Join Two Lengths of Cord
by R. E. Miles

Vol. 9: Combinatorial Physics
by T. Bastin & C. W. Kilmister

Vol. 10: Nonstandard Logics and Nonstandard Metrics in Physics
by W. M. Honig

Vol. 11: History and Science of Knots
edited by J. C. Turner & P. van de Griend

*The complete list of the published volumes in the series, can also be found at
http://www.worldscibooks.com/series/skae_series.shtml

Vol. 12: Relativistic Reality: A Modern View
edited by J. D. Edmonds, Jr.

Vol. 13: Entropic Spacetime Theory
by J. Armel

Vol. 14: Diamond — A Paradox Logic
by N. S. Hellerstein

Vol. 15: Lectures at KNOTS '96
by S. Suzuki

Vol. 16: Delta — A Paradox Logic
by N. S. Hellerstein

Vol. 17: Hypercomplex Iterations — Distance Estimation and Higher Dimensional Fractals
by Y. Dang, L. H. Kauffman & D. Sandin

Vol. 18: The Self-Evolving Cosmos: A Phenomenological Approach to Nature's Unity-in-Diversity
by S. M. Rosen

Vol. 19: Ideal Knots
by A. Stasiak, V. Katritch & L. H. Kauffman

Vol. 20: The Mystery of Knots — Computer Programming for Knot Tabulation
by C. N. Aneziris

Vol. 21: LINKNOT: Knot Theory by Computer
by S. Jablan & R. Sazdanovic

Vol. 22: The Mathematics of Harmony — From Euclid to Contemporary Mathematics and Computer Science
by A. Stakhov (assisted by S. Olsen)

Vol. 23: Diamond: A Paradox Logic (2nd Edition)
by N. S. Hellerstein

Vol. 24: Knots in HELLAS '98 — Proceedings of the International Conference on Knot Theory and Its Ramifications
edited by C. McA Gordon, V. F. R. Jones, L. Kauffman, S. Lambropoulou & J. H. Przytycki

Vol. 25: Connections — The Geometric Bridge between Art and Science (2nd Edition)
by J. Kappraff

Vol. 26: Functorial Knot Theory — Categories of Tangles, Coherence, Categorical Deformations, and Topological Invariants
by David N. Yetter

Vol. 27: Bit-String Physics: A Finite and Discrete Approach to Natural Philosophy
by H. Pierre Noyes; edited by J. C. van den Berg

Vol. 28: Beyond Measure: A Guided Tour Through Nature, Myth, and Number
by J. Kappraff

Vol. 29: Quantum Invariants — A Study of Knots, 3-Manifolds, and Their Sets
by T. Ohtsuki

Vol. 30: Symmetry, Ornament and Modularity
by S. V. Jablan

Vol. 31: Mindsteps to the Cosmos
by G. S. Hawkins

Vol. 32: Algebraic Invariants of Links
by J. A. Hillman

Vol. 33: Energy of Knots and Conformal Geometry
by J. O'Hara

Vol. 34: Woods Hole Mathematics — Perspectives in Mathematics and Physics
edited by N. Tongring & R. C. Penner

Vol. 35: BIOS — A Study of Creation
by H. Sabelli

Vol. 36: Physical and Numerical Models in Knot Theory
edited by J. A. Calvo et al.

Vol. 37: Geometry, Language, and Strategy
by G. H. Thomas

Vol. 38: Current Developments in Mathematical Biology
edited by K. Mahdavi, R. Culshaw & J. Boucher

Vol. 39: Topological Library
Part 1: Cobordisms and Their Applications
edited by S. P. Novikov and I. A. Taimanov

Vol. 40: Intelligence of Low Dimensional Topology 2006
edited by J. Scott Carter et al.

Vol. 41: Zero to Infinity: The Fountations of Physics
by P. Rowlands

Vol. 42: The Origin of Discrete Particles
by T. Bastin and C. Kilmister

Vol. 43: The Holographic Anthropic Multiverse
by R. L. Amoroso and E. A. Ranscher

Vol. 44: Topological Library
Part 2: Characteristic Classes and Smooth Structures on Manifolds
edited by S. P. Novikov and I. A. Taimanov

Vol. 45: Orbiting the Moons of Pluto
Complex Solutions to the Einstein, Maxwell, Schrödinger and Dirac Equations
by E. A. Rauscher and R. L. Amoroso